Synthesis Series in Biomedical Engineering

Volume 1

Sensory Organ Replacement and Repair
Gerald E. Miller
2006

Artificial Organs
Gerald E. Miller
2006

Signal Processing of Random Physiological Signals
Charles S. Lessard
2006

Image and Signal Processing for Networked E-Health Applications
Ilias G. Maglogiannis, Kostas Karpouzis, Manolis Wallace
2006

Volume 2

Fundamentals of Respiratory Sounds and Analysis
Zahra Moussavi
2006

Basic Probability Theory for Biomedical Engineers
John D. Enderle, David C. Farden, Daniel J. Krause
2006

Intermediate Probability Theory for Biomedical Engineers
John D. Enderle, David C. Farden, Daniel J. Krause
2006

Advanced Probability Theory for Biomedical Engineers
John D. Enderle, David C. Farden, Daniel J. Krause
2006

Volume 3

Capstone Design Courses: Producing Industry-Ready Biomedical Engineers
Jay R. Goldberg
2007

BioNanotechnology
Elisabeth S. Papazoglou, Aravind Parthasarathy
2007

Bioinstrumentation
John D. Enderle
2006

Volume 4

Multimodal Imaging in Neurology: Special Focus on MRI Applications and MEG
Hans-Peter Müller, Jan Kassubek
2007

Estimation of Cortical Connectivity in Humans: Advanced Signal Processing Techniques
Laura Astolfi, Fabio Babiloni
2007

Brain–Machine Interface Engineering
Justin C. Sanchez , José C. Principe
2007

Introduction to Statistics for Biomedical Engineers
Kristina M. Ropella
2007

Volume 5

Introduction to Continuum Biomechanics
Kyriacos A. Athanasiou, Roman M. Natoli
2008

The Effects of Hypergravity and Microgravity on Biomedical Experiments
Thais Russomano, Gustavo Dalmarco, Felipe Prehn Falcão
2008

A Biosystems Approach to Industrial Patient Monitoring and Diagnostic Devices
Gail Baura
2008

Volume 6

Neural Interfacing: Forging the Human-Machine Connection
Thomas D. Coates Jr.
2008

Quantitative Neurophysiology
Joseph V. Tranquillo
2008

Tremor: From Pathogenesis to Treatment
Giuliana Grimaldi, Mario Manto
2008

Volume 7

Introductory Medical Imaging
A. A. Bharath
2008

Lung Sounds: An Advanced Signal Processing Perspective
Leontios J. Hadjileontiadis
2008

An Outline of Informational Genetics
Gérard Battail
2008

Volume 8

Basic Feedback Controls in Biomedicine
Charles S. Lessard
2009

Understanding Atrial Fibrillation: The Signal Processing Contribution
Luca Mainardi , Leif Sörnmo , Sergio Cerutti
2008

Volume 9

Landmarking and Segmentation of 3D CT Images
Graham Boag
2009

Phonocardiography Signal Processing
Rasha Bassam
2009

Introduction to Biomedical Engineering: Biomechanics and Bioelectricity Vol 1
Christensen
2009

Introduction to Biomedical Engineering: Biomechanics and Bioelectricity Vol 2
Christensen
2009

Synthesis Series in Biomedical Engineering 1

Sensory Organ Replacement and Repair
Gerald E. Miller

Artificial Organs
Gerald E. Miller

Signal Processing of Random Physiological Signals
Charles S. Lessard

Image and Signal Processing for Networked E-Health Applications
Ilias G. Maglogiannis, Kostas Karpouzis, Manolis Wallace

ISBN: 9781608451845

www.morganclaypool.com

A Publication in the Morgan & Claypool Publishers

SYNTHESIS SERIES IN BIOMEDICAL ENGINEERING

Series Editors: John D. Enderle

Synthesis Series in Biomedical Engineering 1

This volume includes:

Sensory Organ Replacement and Repair

Gerald E. Miller

Artificial Organs

Gerald E. Miller

Signal Processing of Random Physiological Signals

Charles S. Lessard

Image and Signal Processing for Networked E-Health Applications

Ilias G. Maglogiannis, Kostas Karpouzis, Manolis Wallace

Sensory Organ Replacement and Repair

Gerald E. Miller
Virginia Commonwealth University

MORGAN & CLAYPOOL PUBLISHERS

ABSTRACT

The senses of human hearing and sight are often taken for granted by many individuals until they are lost or adversely affected. Millions of individuals suffer from partial or total hearing loss and millions of others have impaired vision. The technologies associated with augmenting these two human senses range from simple hearing aids to complex cochlear implants, and from (now commonplace) intraocular lenses to complex artificial corneas. The areas of human hearing and human sight will be described in detail with the associated array of technologies also described.

KEYWORDS

Human hearing, Audiology, Hearing Aids, Cochlear Implants, Human vision, Intraocular lens, Cataract surgery, Artificial Cornea, Corneal Transplant

Contents

Sensory Organ Replacement and Repair

The senses of human hearing and sight are often taken for granted by many individuals until they are lost or adversely affected. Millions of individuals suffer from partial or total hearing loss and millions of others have impaired vision. The technologies associated with augmenting these two human senses range from simple hearing aids to complex cochlear implants, and from (now commonplace) intraocular lenses (IOLs) to complex artificial corneas. The areas of human hearing and human sight will be described in detail with the associated array of technologies also described.

1 HEARING AIDS

1.1 Anatomy of the Ear and Human Hearing

The human ear is a complex arrangement of mechanical and neurological components that are designed to interpret a multifaceted sound pressure waveform into its individual frequency components and then reconstitute these components together into what we recognize as blended sound. The human ear is shown in Figure 1.

The ear is composed of three sections: the outer ear (also known as the ear canal), the middle ear, and the inner ear. The outer ear is shown in greater detail in Figure 2.

The outer ear is designed to channel sound from a complete hemisphere into a megaphone-shaped channel that is wide at the outside end and funnels down toward the inside end. As was noted above, most sounds are complex waveforms consisting of pressure waves of varying amplitudes and frequencies mixed together into sounds and words that we recognize in our daily lives. A typical sound waveform of spoken words is shown in Figure 3 with the frequency spectrum of those words shown in Figure 4.

The array of pressure waveforms (Figures 3 and 4) results from a series of anatomical elements including the trachea, lungs, vocal cords, tongue, and mouth, as is seen in Figure 5.

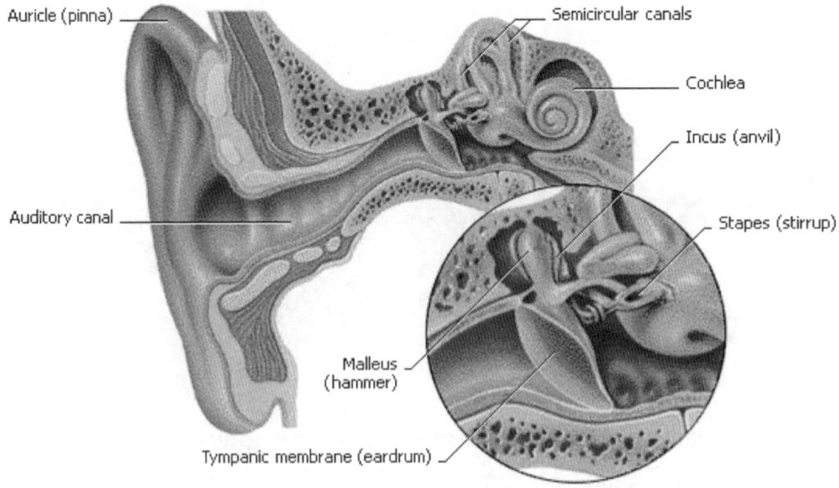

FIGURE 1: Anatomy of the human ear.

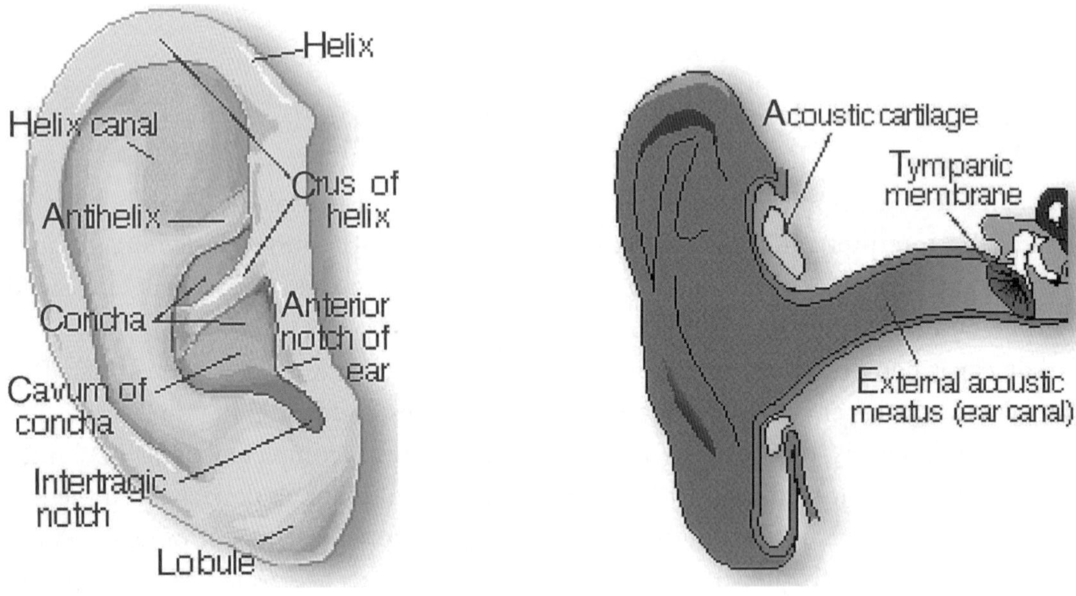

FIGURE 2: The outer ear.

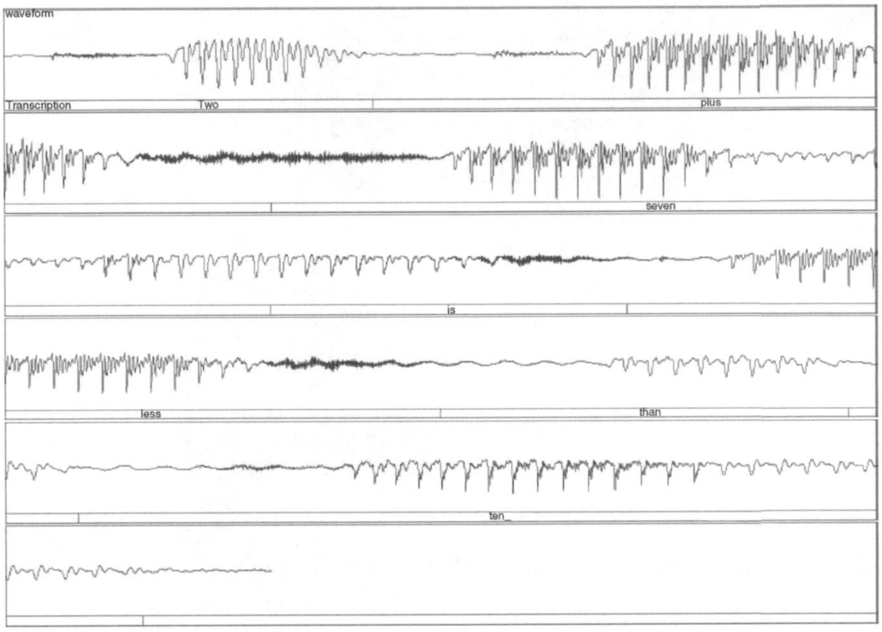

Two plus seven is less than ten

FIGURE 3: Speech waveform of spoken words indicating a breadth of amplitudes and frequencies.

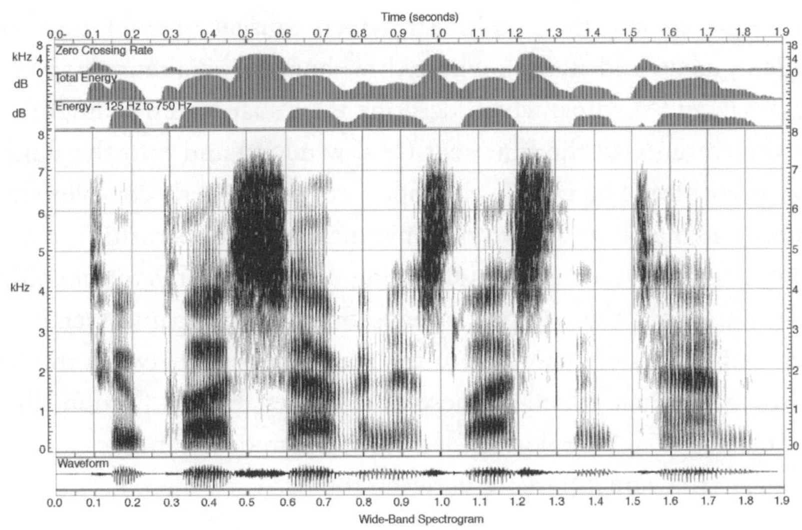

Two plus seven is less than ten

FIGURE 4: Frequency spectrum of spoken words indicating a range of embedded frequencies.

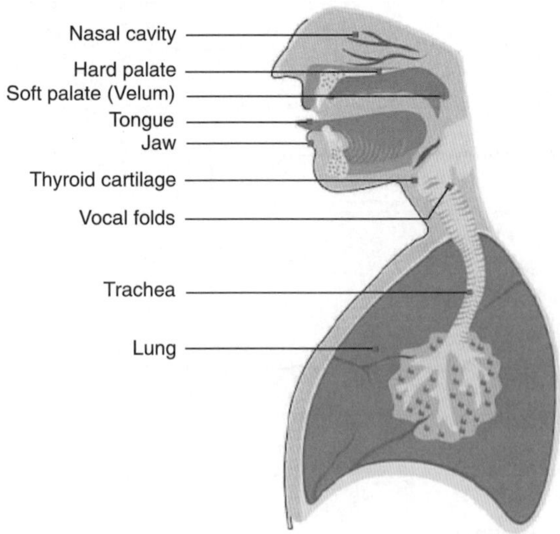

FIGURE 5: Anatomical structures associated with speech production.

The complex waveforms associated with speech and many sounds are channeled into the outer ear and, as noted above, funneled toward the inside, which culminates with the tympanic membrane, also known as the eardrum. This membrane vibrates with the same complex arrangement as the complex sound waveforms that enter the ear canal.

From the tympanic membrane, the vibrations associated with the sound pressure waveform are then transmitted into the middle ear, which consists of three tiny bones called the ossicles (the malleus, incus, and stapes). These three bones are sometimes given the colloquial names hammer, anvil, and stirrups, which describe their shapes. They amplify the sound and send it through the entrance to the inner ear (oval window) and into the fluid-filled hearing organ known as the cochlea. The middle ear appears to be an overly complex structure of three bones that connect one membrane (the tympanic membrane or eardrum) to the oval window leading to the inner ear. However, there are two important elements to the middle ear. Firstly, the bones not only transmit the complex sounds from the tympanic membrane to the oval window, but also amplify the pressure due to the area difference between the two membranes and the size of the bones that attach to these membranes. In this fashion, the middle ear is similar to an impedance-matching device. Secondly, the middle ear is responsible for balance. If there is a significant pressure difference between the two ears, then one's balance is adversely affected.

The sound from the middle ear is transmitted into the inner ear known as the cochlea. It is here that the complex sound waveforms of varying frequencies and amplitudes are separated into individual components. This is accomplished by means of the resonant frequency for a material based upon its geometry. Thicker materials vibrate naturally at lower frequencies, while thinner

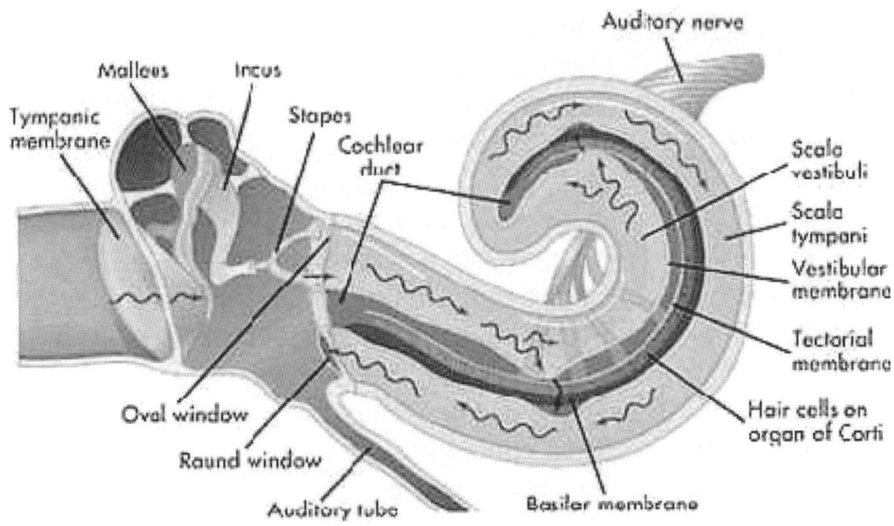

FIGURE 6: Human cochlea with hair cells and auditory nerve.

materials vibrate at higher frequencies. This is why bass speakers are larger than treble speakers in a stereo system. The cochlea is configured as a coiled triangular tissue with a wide thicker end and a narrow thinner end with a continual taper along its length. As complex sound enters into the cochlea, each relevant section of the cochlea would vibrate at its resonant frequency based upon its thickness. Thus, the complex sound is then separated into its individual frequency elements. The cochlea is thus a reverse auditory sound mixer. Along the cochlea, there are thousands of hair cells that are connected to tiny nerves. As the various sections of the cochlea vibrate at each resonant frequency at an amplitude dependent on the incoming sound amplitude (at that frequency), so do the attached hair cells. These hair cells excite their associated nerves. These nerves are then connected *en masse* to the auditory nerve or eighth cranial nerve, which sends the recombined electrical signal to the brain. The brain translates these impulses into what we experience as sound. It is unusual that the cochlea separates sounds into their individual components, converts them from vibrations into electrical signals, and then reconstitutes them into a blended sound, similar to its original construct. Figure 6 indicates the cochlea with hair cells leading to the auditory nerve.

The wide variety of sounds that we routinely hear vary from a whisper to a roar with pressure variations covering several orders of magnitude. To provide some uniform measurement system, a logarithmic sound scale has been established in units known as decibels. A decibel scale for many common sounds is shown in Figure 7.

The frequency range of human hearing is 20–20 000 Hz with the lowest end being more of a vibratory feeling than true auditory hearing. The range of human hearing as compared to other species is shown in Figure 8.

Decibel	Sound
0–9	One hand moving in air
10–19	Softest sound
20–29	Softest whisper
40–49	Quiet office (no typing)
50–59	Average home with A/C
60–69	Normal conversations
70–79	Living room music, radio, busy traffic
80–89	Food blender, noisy restaurant
90–99	Motorcycle at 25 ft, subway
100–109	Airplanes 1000 ft away
110–119	Rock concerts
>120	Jets, rockets

FIGURE 7: Decibel scale representing common sounds at various amplitudes.

Hearing is a very powerful sense—far more powerful than most people realize. Hearing allows individuals to determine where a sound is coming from, even if it is behind them or they cannot see the source of the sound. In addition, in a noisy environment, individuals can selectively filter out "extraneous" sound and concentrate on one voice or sound source. Try this

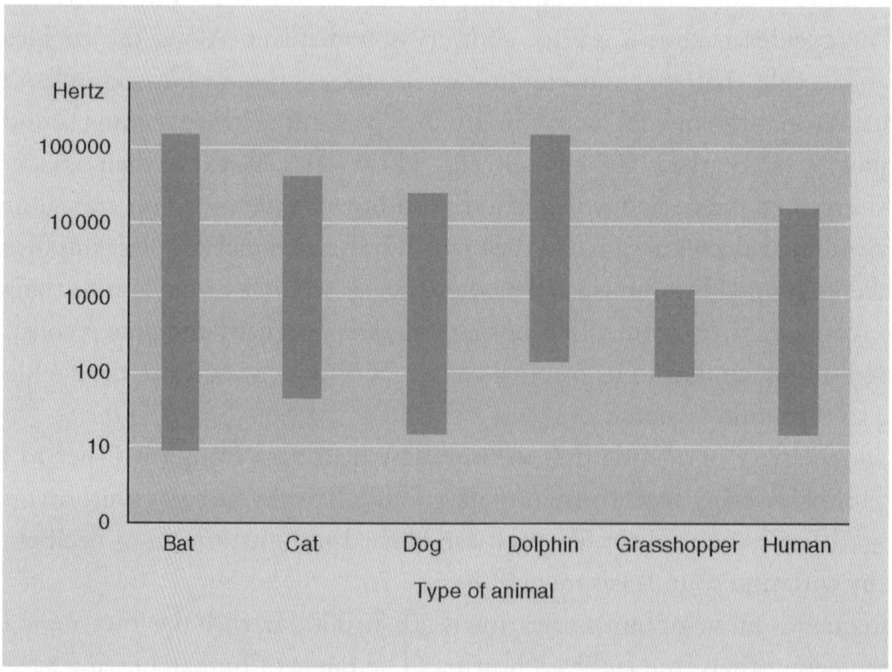

FIGURE 8: Frequency range of human hearing compared to other species.

experiment: In a crowded room, close your eyes. Then try to concentrate on one person's voice and determine where that person is merely from the sounds reaching your ear. As such, hearing is truly three dimensional and, of more importance, can be selective among several blended sounds. This is important when one has a hearing impairment that requires augmentation with a hearing aid or cochlear implant. One may regain sound amplitude and frequency processing, but loses the spatial resolution feature that is such a powerful aspect of normal hearing.

1.2 Hearing Loss

When the problem is in the inner ear, a *sensorineural* hearing loss occurs. Sensorineural hearing loss is the most common type of hearing loss. More than 90% of all hearing aid wearers have sensorineural hearing loss. The most common causes of sensorineural hearing loss are age-related changes, noise exposure, inner ear blood circulation, inner ear fluid disturbances, and problems with the hearing nerve. *Conductive* hearing loss occurs when sound is not conducted efficiently through the ear canal, eardrum or the tiny bones of the middle ear, resulting in a reduction of loudness of sound. Conductive loss may result from earwax blocking the ear canal, fluid in the middle ear, middle ear infection, obstructions in the ear canal, perforations (hole) in the eardrum, or disease of any of the three middle ear bones. People with conductive hearing loss may notice that their ears seem to be full or plugged. They may speak softly because they hear their own voice loudly. Crunchy foods, such as celery or carrots, seem very loud to the person with a conductive hearing loss and this person may have to stop chewing to hear what is being said. All conductive hearing losses should be evaluated by an audiologist and a physician to explore medical and surgical options.

Audiometry is the testing of a person's ability to hear various sound frequencies. The test is performed with the use of electronic equipment called an *audiometer*. This testing is usually administered by a trained technician called an *audiologist*. Audiometry testing is used to identify and diagnose hearing loss. The equipment is used in health screening programs, for example, in grade schools, to detect hearing problems in children. It is also used in the doctor's office or in hospital's audiology department to diagnose hearing problems in children, adults, and the elderly. With correct diagnosis of a person's specific pattern of hearing impairment, the right type of therapy, which might include hearing aids, corrective surgery, or speech therapy, can be prescribed.

The person being tested wears a set of headphones that blocks out other distracting sounds and delivers a test tone to one ear at a time with the amplitude slowly increasing. When the patient hears the sound of a tone, he holds up a hand or finger to indicate that the sound is detected. The audiologist lowers the volume and repeats the sound until the patient can no longer detect it. This process is repeated over a wide range of tones or frequencies from very

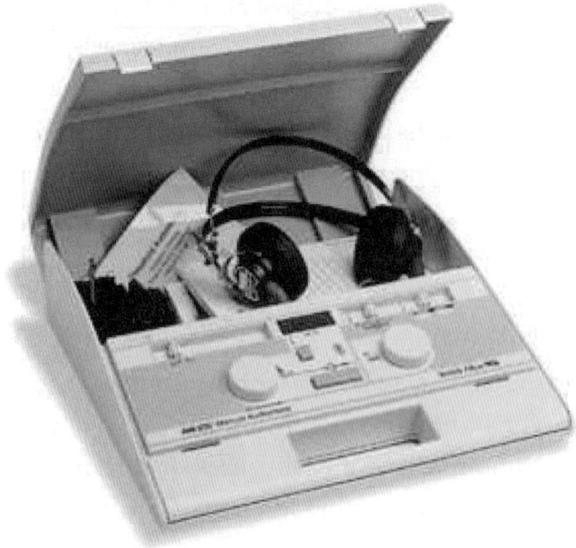

FIGURE 9: Small desktop audiometer.

deep, low sounds, like the lowest note played on a tuba, to very high sounds, like the pinging of a triangle. Each ear is tested separately. It is not unusual for levels of sensitivity to sound to differ from one ear to the other. The results of the audiometry test may be recorded on a grid or graph called an audiogram. This graph is generally set up with low frequencies or tones at one end and high ones at the other end, much like a piano keyboard. Low notes are graphed on the left and high notes on the right. The graph also charts the volume of the tones used, from soft, quiet sounds at the top of the chart to loud sounds at the bottom. Hearing is measured in a unit called decibels. Most of the sounds associated with normal speech patterns are generally spoken in the range of 20–50 dB. An adult with normal hearing can detect tones between 0 and 20 dB.

A typical desktop audiometer is shown in Figure 9 with a computer-driven system shown in Figure 10.

Some hospitals and schools utilize an acoustic chamber type audiometer that further isolates interfering sound. This chamber is sometimes called an anechoic chamber and is shown in Figure 11.

The audiograms noted above can be computer generated as seen in Figure 12 or hand drawn as with a desktop system as seen in Figure 13.

As can be seen from the audiograms given in Figures 12 and 13, there are data points at various frequencies for both ears. Normal hearing is designated by a fairly parallel set of points for each ear as is shown in Figure 14.

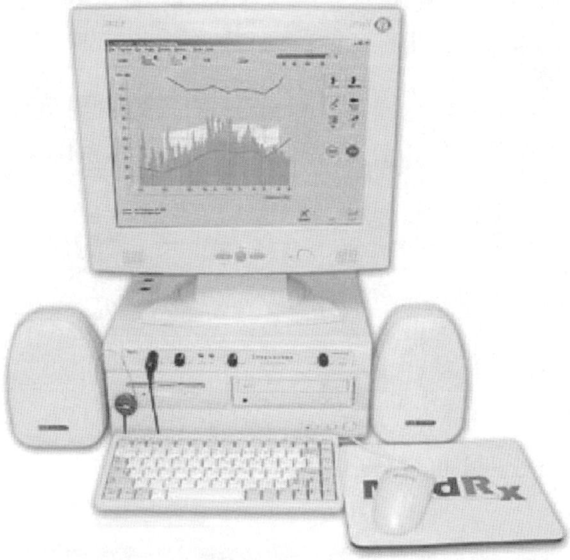

FIGURE 10: Computer-controlled audiometer.

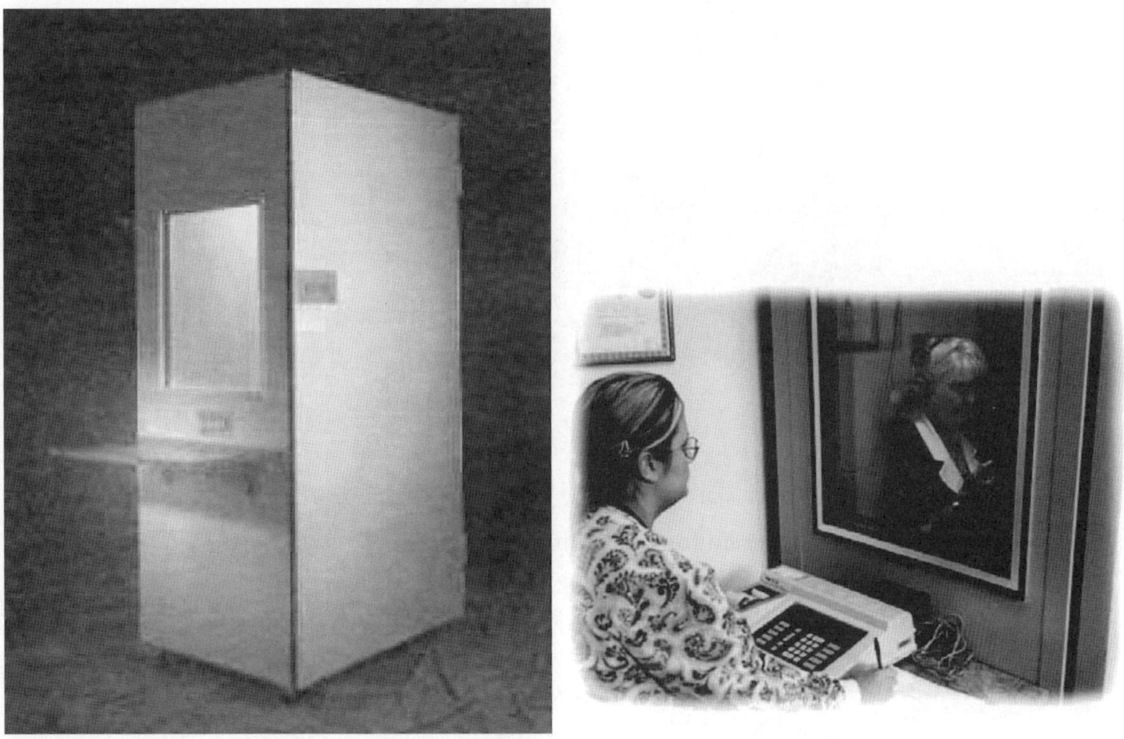

FIGURE 11: Acoustic chamber type audiometer.

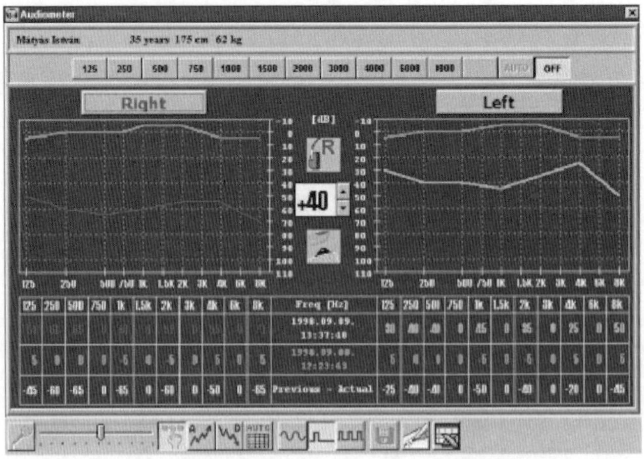

FIGURE 12: Computer-generated audiogram.

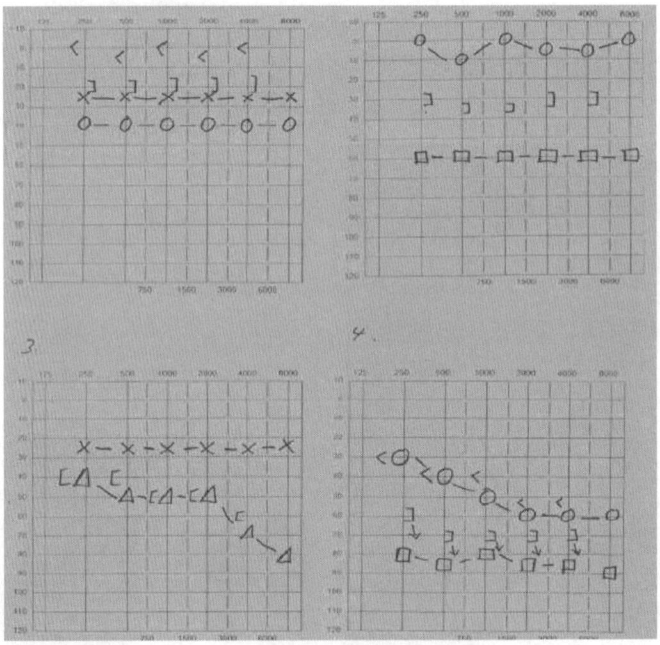

FIGURE 13: Hand-drawn audiogram from a desktop audiometer.

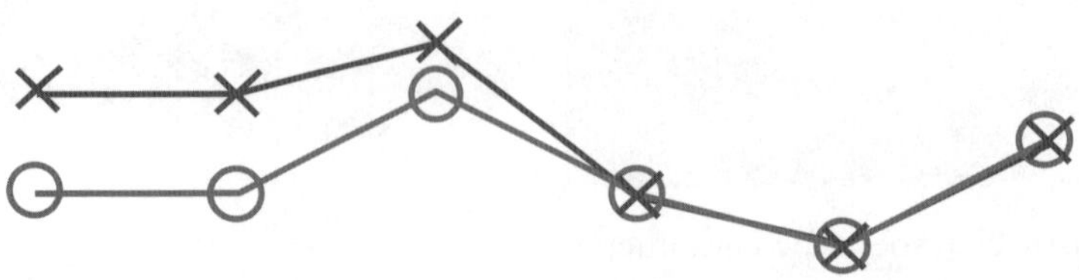

FIGURE 14: Normal bilateral hearing with fairly parallel data points at all frequencies.

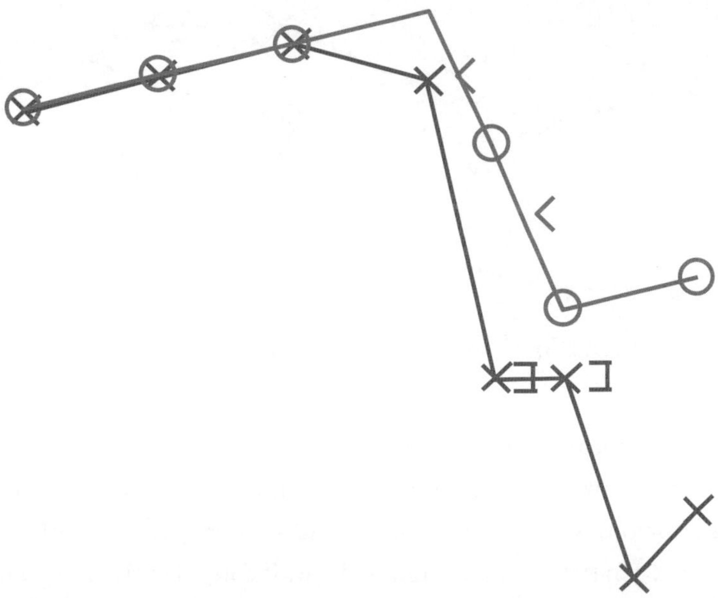

FIGURE 15: Upper frequency hearing loss in both ears.

Elderly individuals or those who have been repeatedly exposed to loud sounds often lose hearing at the upper frequencies as shown in Figure 15. Still other patients may have hearing loss in mid-frequency ranges due to childhood illnesses or recurring tympanic membrane problems, as is seen in Figure 16.

There are sometimes hearing losses in only one ear, typically due to recurring infection in one ear or damage to the outer or middle ear in one ear. When both ears are affected, the patient may utilize hearing aids in both ears, although that is not required. Obviously, when only one ear is affected, a single hearing aid may be needed. Figure 17 depicts one-sided hearing loss with a patient who has a recurring right ear infection.

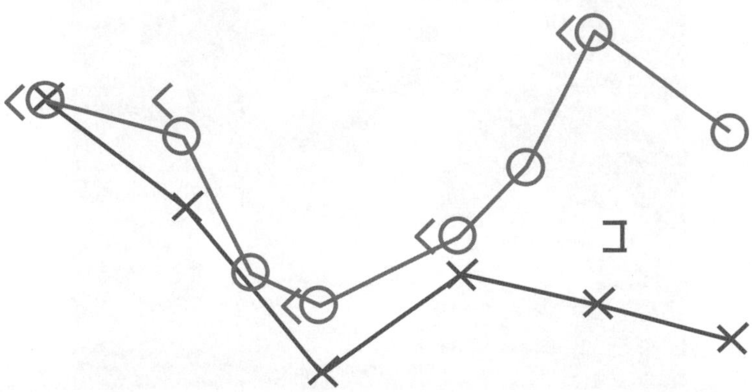

FIGURE 16: Mid-range hearing loss in both ears.

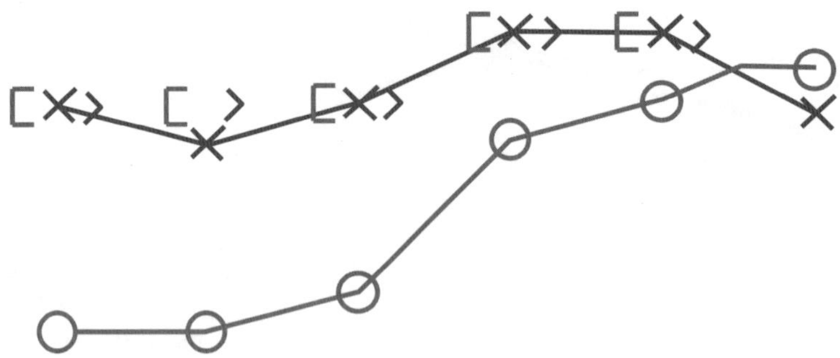

FIGURE 17: One-sided hearing loss at most frequencies due to a recurring right ear infection.

1.3 Hearing Aid Technologies

Hearing aids are actually miniature sound systems with a microphone, speaker, audio amplifier, PA and associated electronics. As with many sound systems, the amplitude of the amplified sound can be adjusted as can the frequency range. In addition, as with many stereo systems, issues such as total harmonic distortion, signal to noise ratio, overall gain, and other signal processing factors are also elements in hearing aids. Since any individual patient has an individualized hearing loss, which can be frequency and amplitude dependent, there are settings for gain at varying frequencies that are preset for each patient, and then are fixed. The patient can have control over the overall gain and whether the unit is turned on or off, but cannot normally set the gain at individual frequencies, as these are internally set within the casing of the hearing aid within the electronics. A small dime-sized battery, which is normally on a pivot holder to allow for ease of replacement, powers the device. A typical hearing aid is shown in Figure 18 that depicts an ear mold (not used in all types of hearing aids), volume control, and on–off switch.

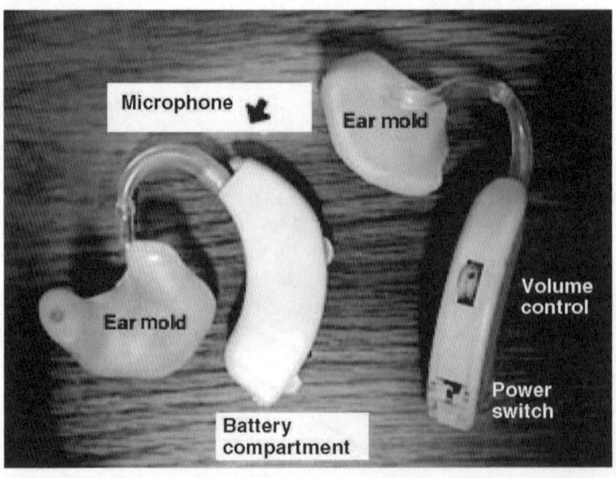

FIGURE 18: Typical hearing aid configuration with volume control and on–off switch.

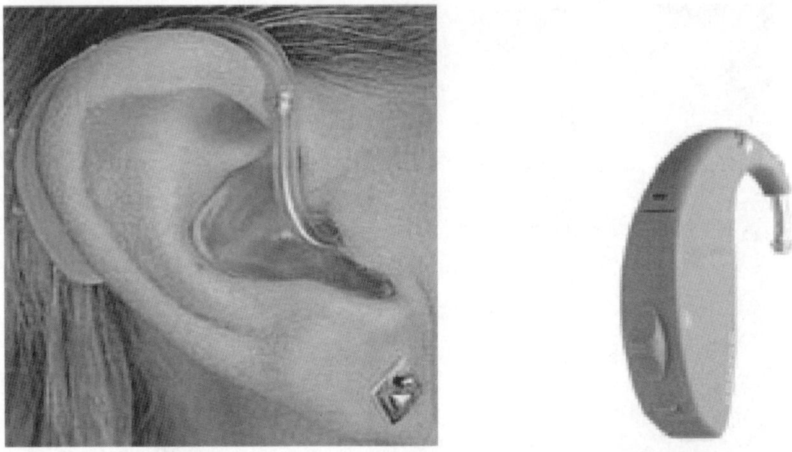

FIGURE 19: BTE hearing aid with clear tube fitting into the ear canal.

There are four styles or configurations of hearing aids. These include the "behind-the-ear" (BTE) style, the "in-the-canal" (ITC) style, and the "completely-in-the-canal" (CIC) style. The BTE hearing instruments are extremely flexible for all types of hearing loss.

The hearing device is housed within a curved shell that sits behind each ear and delivers sound through a clear tube, as is seen in Figure 19. The clear tube fits into a mold that has been customized to comfortably fit inside each ear.

The in-the-ear (ITE) hearing instruments are very easy to operate, even if the user has poor dexterity. The hearing device is housed within a custom-made shell that fits comfortably inside each ear and delivers sound directly to the ear. This style is shown in Figure 20 and is

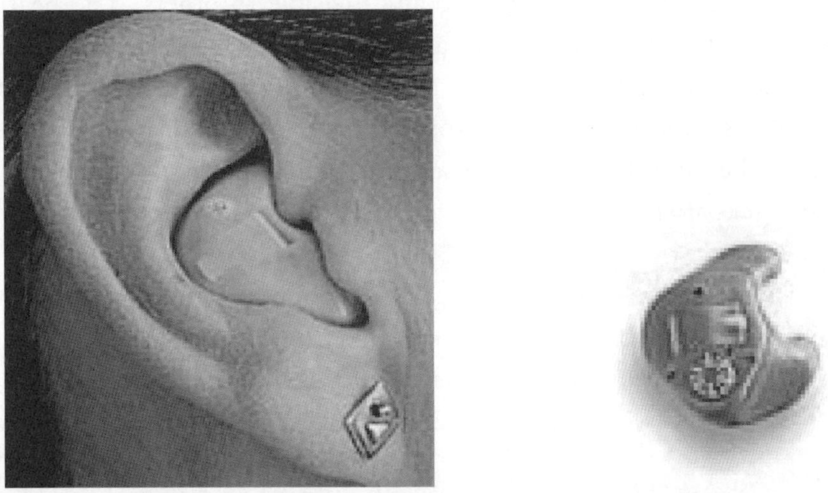

FIGURE 20: ITE style of hearing aid with the device easily accessible and removable.

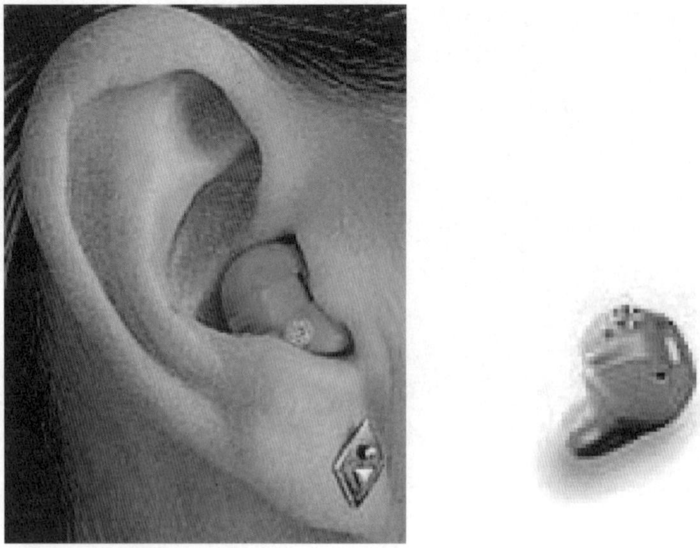

FIGURE 21: ITC style hearing aid with the device positioned further into the canal.

one of the most common styles of hearing aids, particularly for the elderly or for individuals wearing glasses who cannot easily utilize the BTE style.

The ITC hearing instruments can barely be seen and are very easy to operate, even if the user has poor dexterity. The hearing device is housed within a custom-made shell that fits comfortably inside each ear canal and delivers sound directly to the ear. However, this style is not as accessible as the ITE style. This version is seen in Figure 21 and is often utilized by younger patients as opposed to elderly patients.

The CIC hearing instruments are virtually invisible to others. The hearing device is housed in a tiny shell that fits comfortably and completely into each ear canal. The device is removed from the ear canal by pulling a tiny cord. Where these miniature instruments are both powerful and cosmetically appealing, some features such as manual volume control are not available simply because the devices are so small. These types of devices are worn almost exclusively by younger patients who wish to avoid the appearance of a hearing aid. This style is shown in Figure 22.

Hearing aid batteries are small dime-sized devices housed within a pivoting holder within the shell of the device. As is the case with many electronic devices, there are many types of hearing aids. However, in order to avoid confusion for a health care device, hearing aid batteries are color coded, so that it is easy to purchase the correct replacement battery, as is seen in Figure 23.

Typically, batteries last 7–14 days based on 16 h per day use cycle. Batteries are inexpensive, costing less than a dollar each. Generally, the smaller the battery size, the shorter the battery

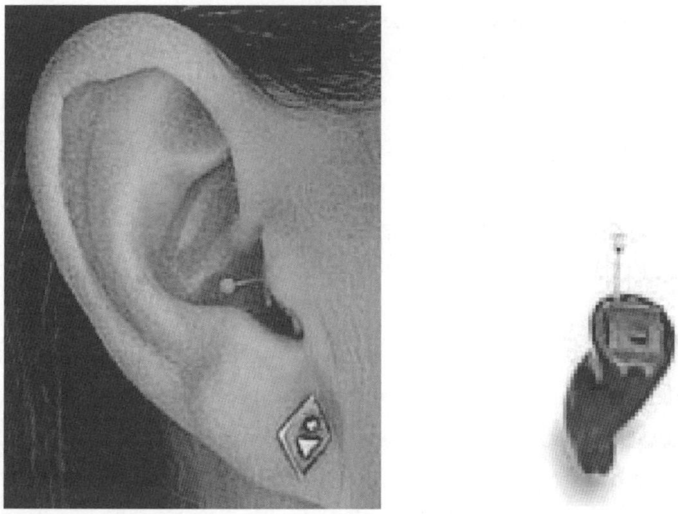

FIGURE 22: CIC style with a protruding cord to remove the device.

life. The sizes of hearing aid batteries are listed below along with their standard numbers and color codes:

Size 5: Red
Size 10 (or 230): Yellow
Size 13: Orange
Size 312: Brown
Size 675: Blue

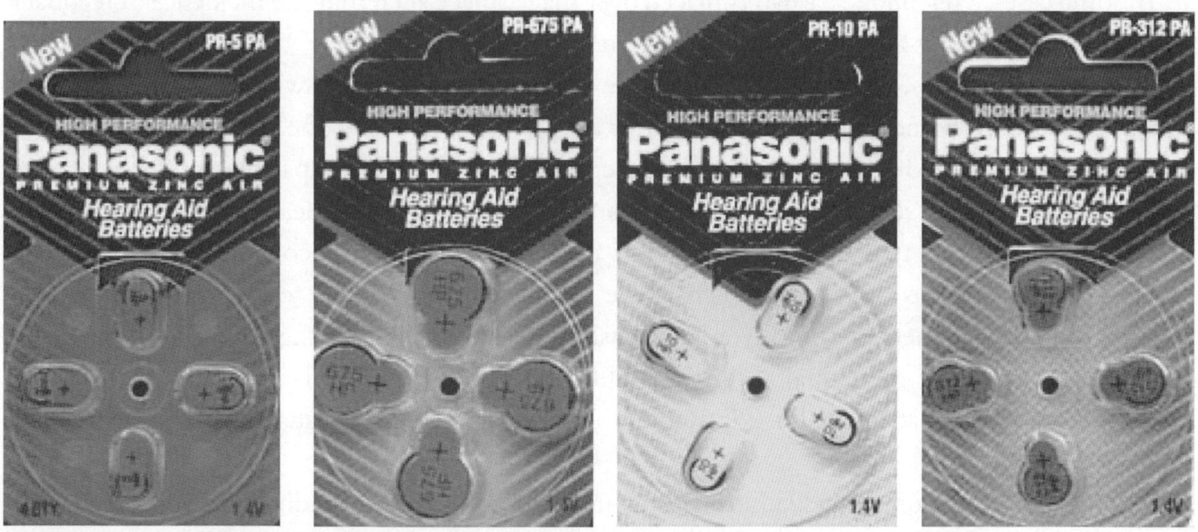

FIGURE 23: Color-coded hearing aid batteries with easy punch out sections and easily seen plus/minus poles.

Today's hearing aid batteries are "zinc-air." Because the batteries are air activated, a factory-sealed sticker keeps them "inactive" until you remove the sticker or push the battery from within its sealed package. Once the sticker is removed from the back of the battery, oxygen in the air contacts the zinc within the battery, and the battery is activated. Zinc-air batteries have a "shelf life" of up to 3 years when stored in a cool, dry environment. Storing zinc-air hearing aids in the refrigerator has no beneficial effect on their shelf life. In fact, quite the opposite may happen. The cold air may actually form little water particles under the sticker.

Hearing aid electronics may be analog or digital. The analog versions may be programmable, but the digital versions are all programmable. Most people recognize that digital hearing aids provide better quality to the user; however, some people still use analog hearing aids for the simple fact that digital models are considerably more expensive. Analog hearing aids usually cost anywhere from $200 to $1000 and digital hearing aids usually cost between $800 and $2000. This means that digital hearing aid models can potentially cost more than twice as much as their analog counterpart. In general, digital hearing aids provide a higher quality, more "realistic" sound with the ability to filter out ambient, nonspeech sounds. In addition, they are more easily programmable in terms of frequency ranges to be augmented and/or depressed and are also smaller in size than their analog counterparts.

There are various auxiliary devices used by individuals with hearing impairments who also use hearing aids. These include modified telephones, alarm clocks, television audio controls and interfaces, and radio/stereo interfaces. One vital component to hearing aids is the use of a telecoil (T-coil), which provides an interface to a telephone while avoiding feedback to the hearing aid. In some cases, a telephone earpad is placed over the earpiece such that the incidence of feedback to the hearing aid is also reduced, as is shown in Figure 24.

In most cases, an amplified telephone is used by individuals with hearing aids to allow for selectively increased amplification of the spoken voice over the phone. The telephone bandwidth is 300–3000 Hz. Amplified telephones are T-coil compatible, often have tone controls, may have audio jacks for headsets or attachment to cochlear implants (to be discussed below), and they amplify the speech up to 40 dB, which is considerable. The ringer is also adjustable up to 95 dB, which is the sound level equivalent to a motorcycle or subway. In addition, there is typically a large flashing visual display for the ringer to further alert a hearing impaired individual to an incoming call. There are third-party add-ons to standard telephones that allow for adjustable volume control and a visual display as shown in Figure 25.

In order to avoid outside interfering sounds and also to provide adjustable volume control, those with impaired hearing or who wear hearing aids sometimes employ an infrared wireless system that can attach to a standard television as shown in Figure 26.

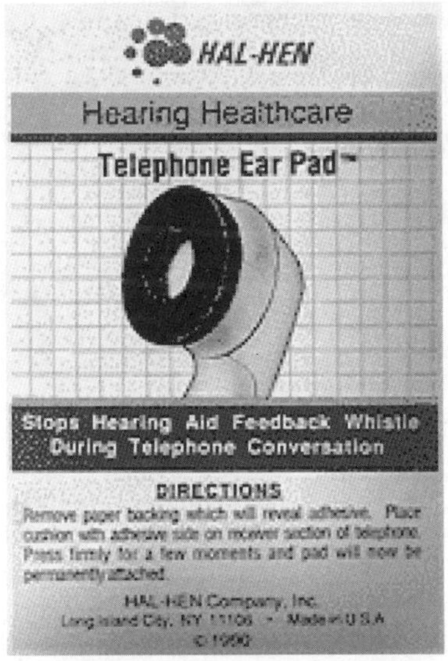

FIGURE 24: Telephone earpad to allow the hearing aid to be more distant to the telephone speaker and reduce interference and feedback.

As was noted above, special alarm clocks can be purchased that allow for adjustable volume increase up to 40 dB and have a large (sometimes projecting) visual display when the alarm sounds.

The Federal Trade Commission (FTC) is responsible for monitoring the business practices of hearing aid dispensers and vendors. The FTC can take action against companies that mislead or deceive consumers. Such companies may use misleading sales and advertising

FIGURE 25: Adjustable volume control as an attachment to a standard telephone.

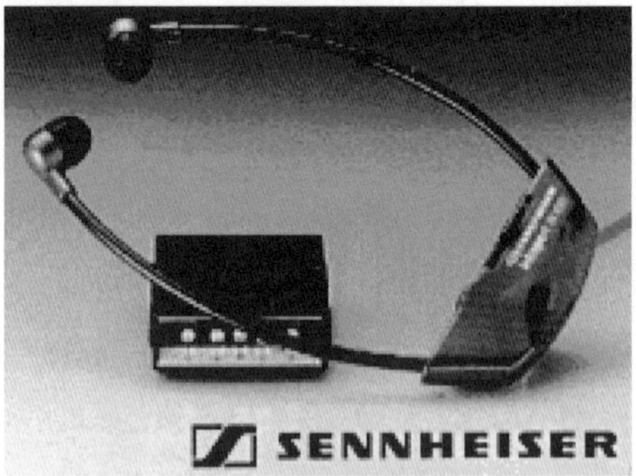

FIGURE 26: Infrared wireless headset for television listening and interface.

practices, giving inaccurate information about hearing loss, hearing aid performance, refund policies, or warranty coverage. The law further requires companies offering warranties to fully disclose all terms and conditions of their warranties.

The Food and Drug Administration (FDA) enforces regulations that deal specifically with the manufacture and sale of hearing aids. According to the FDA, the following conditions must be met by all dispensers before selling a hearing aid:

1. Dispensers must obtain a written statement from the patient, signed by a licensed physician. The statement must be dated within the previous 6 months, state that the patient's ears have been medically evaluated, and that the patient is cleared for fitting with a hearing aid.
2. A patient of age 18 years or older can sign a waiver for a medical examination, but dispensers must avoid encouraging the patient to waive the medical evaluation requirement. Dispensers also must advise the patient that waiving the examination is not in his best health interest.
3. Dispensers must advise patients who appear to have a hearing problem to consult promptly with a physician.
4. The FDA regulations also require that an instruction brochure be provided with the hearing aid that illustrates and describes its operation, use, and care. The brochure must list sources for repair and maintenance, and include a statement that the use of a hearing aid may be only a part of a rehabilitative program.

The FDA Web site that notes standards for hearing aids is at http://www.accessdata.fda.gov/scripts/cdrh/cfdocs/cfStandards/Detail.CFM?STANDARD_IDENTIFICATION_

NO=14730. Hearing aids are regulated by the FDA within the Center for Devices and Radiological Health, which can be accessed at http://www.fda.gov/cdrh/.

Numerous studies have been performed regarding hearing aid use, design, and evaluation. Studies involving the use of hearing aids by the elderly have been published by Cohen-Mansfield and Infeld (2005), and van Hooren *et al.* (2005), among others. Studies regarding the evaluation of hearing aid electronic design include those by Bentler (2005), Lewis *et al.* (2005), Moore *et al.* (2005), and Ricketts and Hornsby (2005), among others. Still other studies have examined the methods of fitting hearing aids in patients or evaluation of first-time users, which include Aarts and Caffee (2005), Gustav Mueller (2005), Gustav Mueller and Bentler (2005), Killion and Gudmundsen (2005), Reber and Kompis (2005), Reese and Hnath-Chisolm (2005), and Uriarte *et al.* (2005), among others. General reviews regarding the use of hearing aids include those by Vuorialho *et al.* (2005).

2 MIDDLE EAR REPLACEMENT

2.1 Introduction

Otosclerosis is a condition that affects hearing as a result of hardening of a bone or bones in the middle ear. The hearing loss associated with this disease is called a conductive hearing loss because the extra bone growth around the middle ear bones prevents sound from being conducted into the inner ear in a normal way. Otosclerosis is inherited and it tends to run in families although you may not know who in your family passed it on to you. About 80% of people with otosclerosis will have the disease in both ears. If you have otosclerosis, it is estimated that there is less than a one in four chance of passing it on to your children.

Sound vibrations that reach the eardrum are usually relayed to the inner ear by way of three small bones in the middle ear. These tiny bones, called the hammer (malleus), anvil (incus), and stirrup (stapes), act as a kind of transformer to change sound waves into liquid waves in the inner ear. The stapes bone is the final link in the hearing chain of bones and is the bone most often affected by otosclerosis. When the stapes hardens, a conductive hearing loss occurs. As the hardening continues over time, your hearing worsens. When the hardening spreads to the inner ear, a sensorineural hearing loss occurs. A sensorineural hearing loss is a nerve hearing loss that is usually permanent and can only be helped with a hearing aid.

Hearing loss is the most common symptom of otosclerosis. Some people also have tinnitus, a ringing noise in the head or ear, and almost half of all people with this disease have dizziness. A woman with otosclerosis that becomes pregnant might find that her hearing loss becomes worse.

Unfortunately, there is no medicine that will help or stabilize the hearing loss in people who have otosclerosis. For many people, however, surgery can help or even overcome the hearing

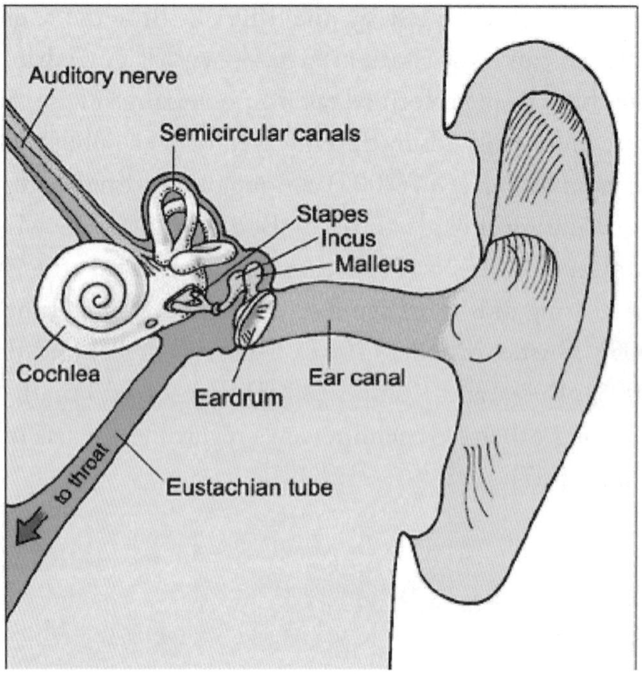

FIGURE 27: Bones of the middle ear.

loss they have. There are several surgery techniques used to correct the hearing loss associated with otosclerosis. A stapedectomy or stapedotomy is usually recommended. These operations are usually done in a hospital or surgery center with a local anesthetic. After your ear has been numbed, it is cleaned and then a cut is made down inside the ear canal and the eardrum is lifted up to uncover the middle ear. The diseased stapes bone is removed, the inner ear is sealed with tissue, and a new stapes is inserted. Sometimes a laser is used to open the bottom of the stapes and a pistonlike stapes replacement is used. It is also possible that a replacement malleus and/or incus is required. Figure 27 depicts the middle ear among the various anatomical features of the human ear.

2.2 Technology and Replacement Components

When the incus is eroded, broken, or absent, the ossicular chain is reconstructed with an incus replacement prosthesis. The one depicted in Figure 28 is a cylinder with a notch that fits under the handle of the malleus and a circular groove that sits on the head of the stapes.

When both the incus and the malleus are eroded or absent, the ossicular chain is reconstructed with a partial ossicular replacement prosthesis (PORP). The one shown in the Figure 29 is a cylinder with a circular flange that fits under the drum. A piece of cartilage is

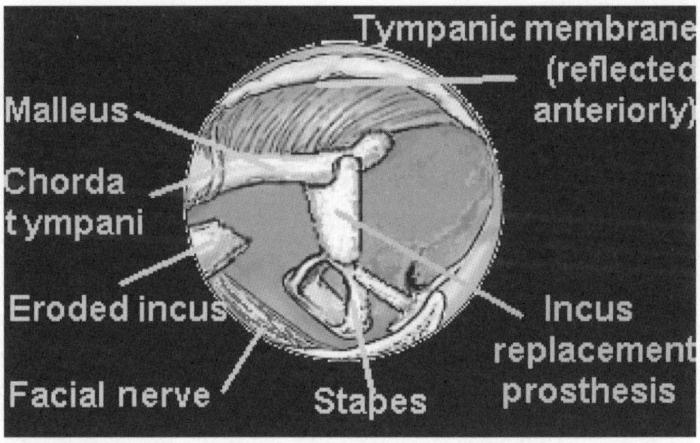

FIGURE 28: Incus replacement with artificial incus.

removed from the tragus and inserted between the prosthesis and the inner surface of the drum to minimize rejection.

When the incus and arch of the stapes are eroded, or when the malleus, incus, and arch of the stapes are absent, the ossicular chain is reconstructed with a total ossicular replacement prosthesis (TORP). The one depicted in Figure 30 has a circular flange that fits under the drum and a slender shaft that is placed over the footplate of the stapes. Here again, a piece of cartilage is inserted between the drum and the flange.

Surgical replacement of the malleus is shown in Figure 31 and surgical replacement of the incus in Figure 32.

The actual size of an artificial stapes, malleus, or incus is shown in Figure 33 as compared to a common dime.

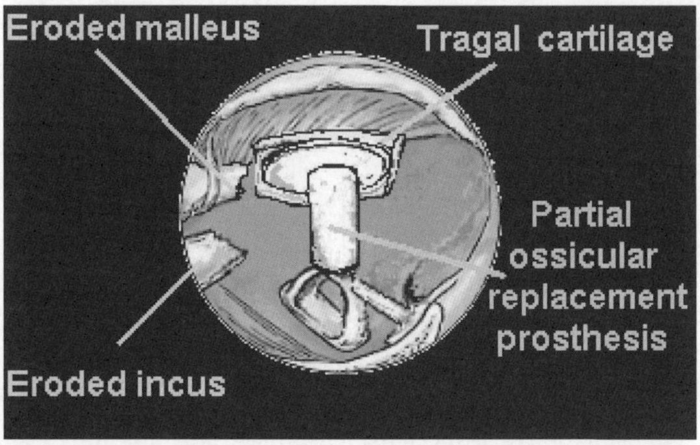

FIGURE 29: PORP replacement of both the incus and the malleus.

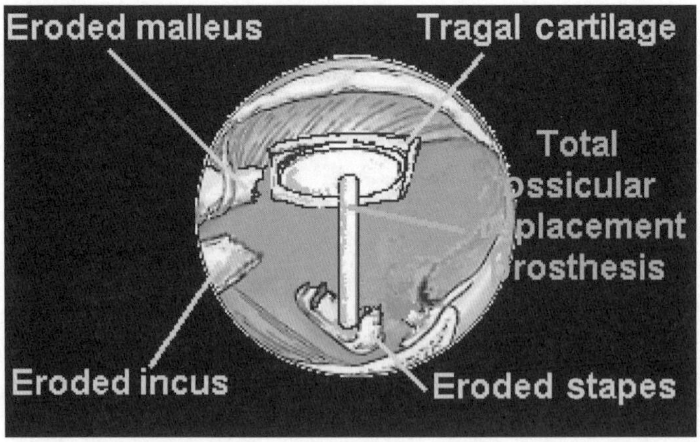

FIGURE 30: TORP replacement of both the stapes and the incus.

2.3 History of Ossicle Surgery and Replacement

Otosclerosis surgery has developed through three eras. The mobilization era began in the late 1800s when Kessel attempted stapes mobilization without ossicular chain reconstruction in cases where it was noted to be fixed (Heermann, 1969). Later, Jack removed the stapes, leaving the oval window open (Schuknecht, 1968). Both techniques allowed increased transmission of sound through the oval window but did not use middle ear amplification structures. Furthermore, fatal cases of meningitis from intraoperative exposure of perilymph to bacteria occurred, and

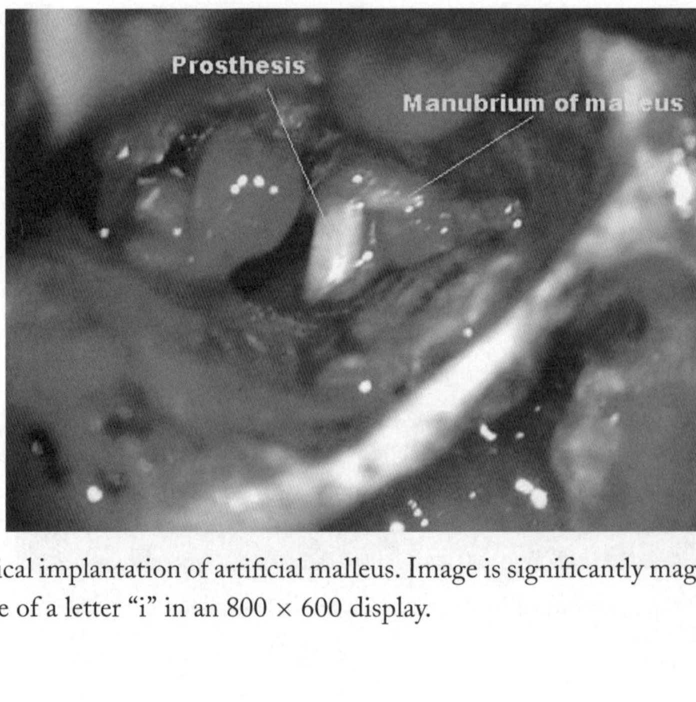

FIGURE 31: Surgical implantation of artificial malleus. Image is significantly magnified with the actual prosthesis of the size of a letter "i" in an 800 × 600 display.

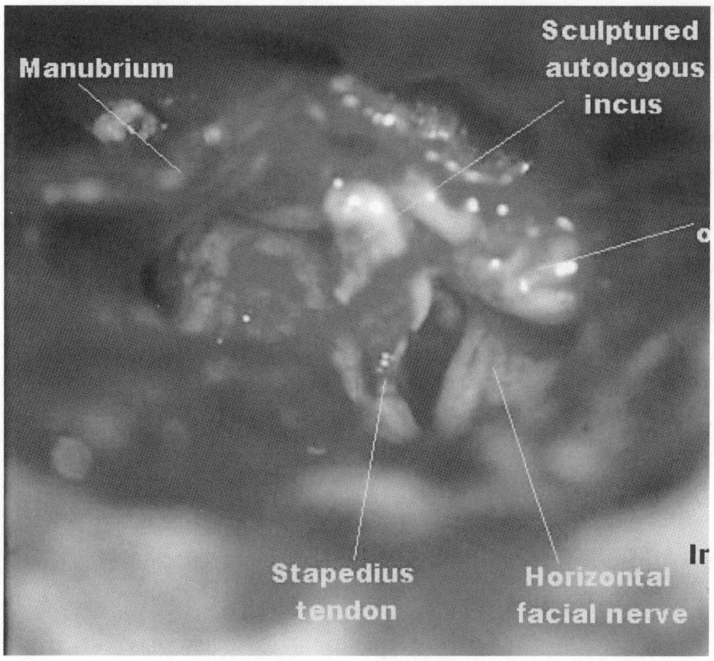

FIGURE 32: Surgical implantation of an artificial incus. Image is significantly magnified.

any gains in hearing frequently were temporary because any remaining stapes footplate often refixed.

The fenestration era began in 1923, when Holmgren created a fistula in the horizontal semicircular canal and sealed it immediately with periosteum (Shea, 1998). This procedure allowed sound conduction preferentially through the fistula, rather than the ossicular chain. Sourdille popularized the procedure when his three-stage technique was widely published during the 1930s (Shea, 1998). Lempert developed a one-stage technique for horizontal semicircular

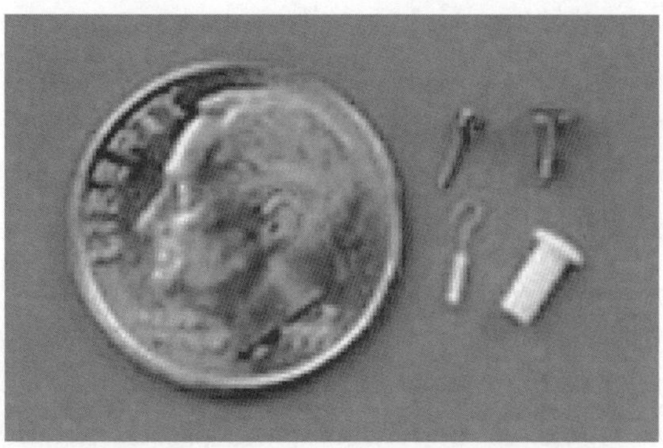

FIGURE 33: Actual size of artificial ossicle (middle bone) components.

fenestration, which went on to gain worldwide acceptance after it proved to enhance hearing (Lempert *et al.*, 1956). Results, however, were short-lived because the fenestra often resealed with bone.

The stapedectomy era began before the fenestration era closed. Rosen revisited stapes mobilization in 1952 (Shea, 1998). Later, Shea removed the stapes, sealed the oval window with an autograft vein wall, and then reconstructed the sound-conducting mechanism with an artificial prosthesis (House *et al.*, 1960; Shea, 1976).

This technique gained wide acceptance and has been improved since inception. In the 1970s, Myers conducted stapedotomy using a piston prosthesis (Myers *et al.*, 1970; Myers and Myers, 1968). In the early 1980s, Perkins began using the laser for stapedotomy in a procedure in which a small hole is made in the footplate, as opposed to complete or subtotal removal (Perkins, 1980). Several techniques and approaches are commonly used today, with largely excellent results. A few challenges remain, such as those patients enduring sensorineural hearing loss and unsteadiness, but many think that surgical treatment for otosclerosis has reached perfection.

Evidence has recently mounted that the measles virus plays an important role in gene activation of otosclerosis (Ferlito *et al.*, 2003; Karosi *et al.*, 2004; Karosi *et al.*, 2005; Niedermeyer *et al.*, 2001). This hypothesis is supported by a declining incidence of otosclerosis since measles vaccinations became widespread.

3 COCHLEAR IMPLANT

3.1 Introduction

Individuals who have extreme sensorineural hearing loss, are born completely deaf, acquire deafness through illness or injury, or who cannot be adequately treated with the use of a hearing aid or artificial ossicle are candidates for a cochlear implant. The cochlear implant is a prosthetic replacement for the inner ear (cochlea) and is appropriate only for people who receive minimal or no benefit from a conventional hearing aid. The cochlear implant bypasses damaged parts of the inner ear and electronically stimulates the hair cells and adjacent nerves within the cochlea. Part of the device is surgically implanted in the skull behind the ear and tiny wires are inserted into the cochlea at set intervals depending on the number of channels or number of frequency bands to excite. The other part of the device is external and has a microphone, a speech processor (that converts sound into electrical impulses), and connecting cables. It is battery powered, adjustable, and expensive.

A surgical procedure places the implant on the outside of the skull and under the skin just above and behind the ear. An electrode array attached to the implant is inserted into the cochlea. Within a few weeks of the surgery, the user is fitted with external devices—a microphone, processor, and transmitter coil. Sound waves are received by the microphone that

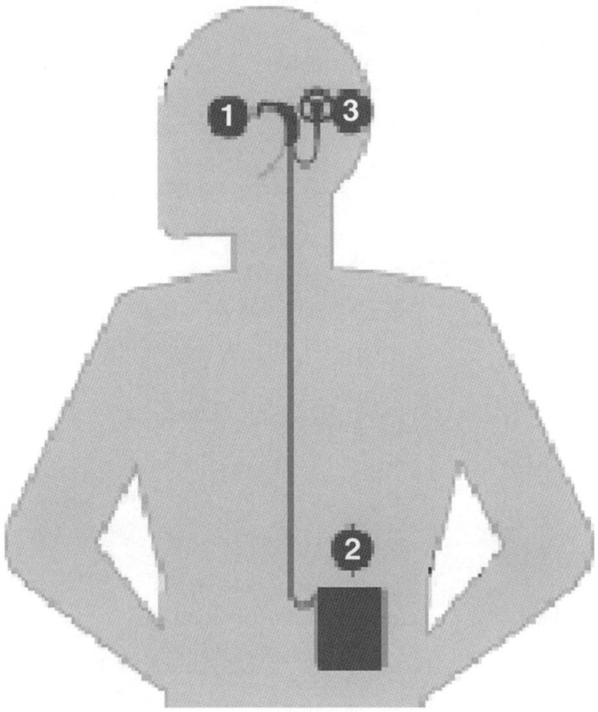

FIGURE 34: Schematic view of a cochlear implant with (1) BTE speech processor or (2) on-the-body speech processor. Either is connected to (3) the connector to the electrode array.

is located either above the ear (Figure 34, point 1) or on the transmitter coil (Figure 34, point 3). The signal from the microphone is sent to the speech processor, which comes in two designs. It may be either a BTE model, which looks like a hearing aid (Figure 34, point 1), or a body-worn device (BWD) that rests on the belt (Figure 34, point 2). In either case, the speech processor deciphers the sound and decides how it will be presented to the ear by means of which electrodes (representing frequencies/regions of the cochlea) will be excited. There are many ways that a processor can translate sound into an electronic code, each of which is called a speech strategy. New speech strategies are being constantly developed to help cochlear implant recipients hear more accurately. A schematic of a cochlear implant depicting both of the potential sites for the speech processor is shown in Figure 34.

3.2 Cochlear Implant Components and Surgery

As can be seen in Figure 35, the electronic code is sent to the transmitter coil (Figure 35, point 3) that is held above the implant (Figure 35, point 4) by a magnet. The transmitter coil uses an FM radio signal to transmit the signal through the intact skin into the implant. The implant package decodes the signal and sends a pattern of very rapid, small electrical pulses to

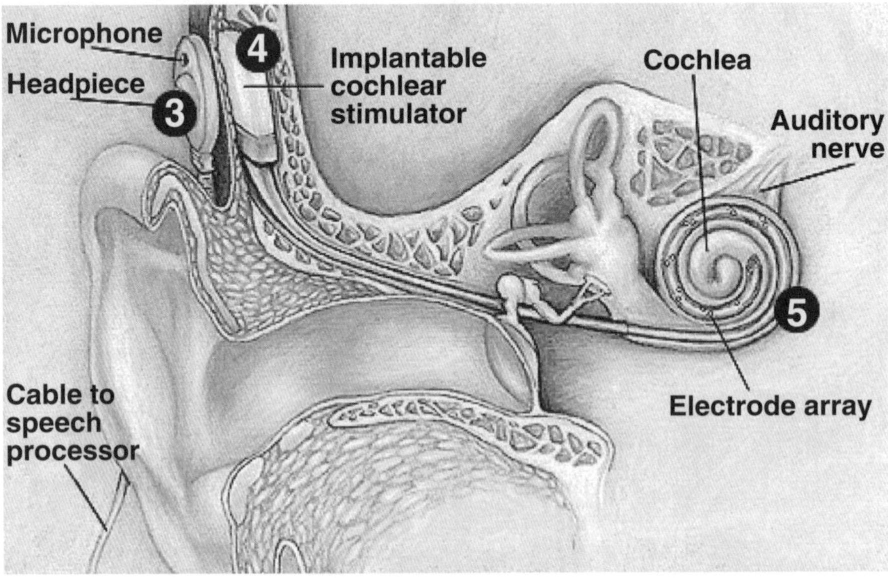

FIGURE 35: Cochlear implant components (see text for details).

the electrodes (Figure 35, point 5) in the cochlea, which stimulate the auditory nerves directly. Different parts of the nerve are stimulated according to the pitch of the sound waves. In response, the auditory nerve carries out its natural function and conducts nerve impulses to the brain. The brain receives the nerve impulses and interprets them as sound. This whole process takes place within a few milliseconds, corresponding to the time delay in the normally functioning ear.

The circuit is attached to a bundle of tiny wires that are inserted into the cochlea. At the end of the wires are as many as 24 electrodes that cover a distance of 25 mm along the length of the cochlea. Stimulation of each electrode usually causes a different pitch perception. A close-up view of the headpiece is shown in Figure 36, which indicates that the headpiece is often covered by the patient's hair and thus it is not normally evident. This is done on purpose to avoid the stigma associated with this relatively new technology for a profoundly deaf individual. The reasons for such a stigma will be discussed below.

Sounds are picked up by the small, directional microphone located in the headset of the ear. The microphone picks up all sound from the environment whether it is speech, environmental sounds, or music. Unlike a hearing aid, a cochlear implant is routinely used for the profoundly deaf individual. Although processing of speech is important, the transmission of other sounds is often as important, unlike a hearing aid where speech amplification is often the primary goal. Many profoundly deaf individuals may have never heard music or environmental sounds if deaf from birth. Thus the overall processing of sound with a cochlear implant has a greater impact

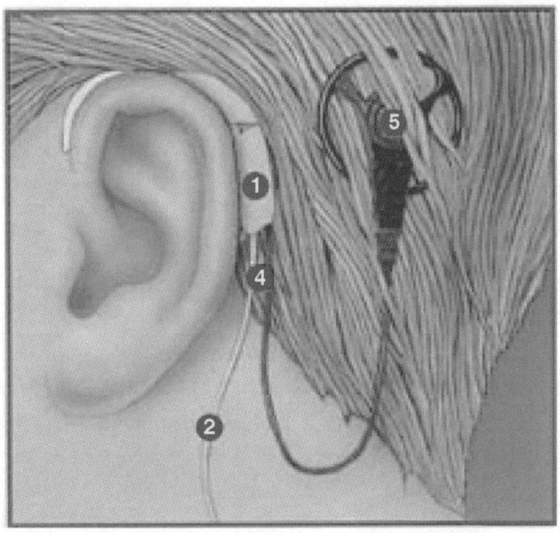

FIGURE 36: Close-up view of the headpiece connected near the mastoid bone.

and a broader goal than that of a hearing aid. An excellent review of the basics of cochlear implants has been published by Loizou (1998).

Sound received by the microphone must be processed to determine how the electrodes should be activated. The simplest way of sound processing would be to divide the auditory signal by the number of electrodes in the implanted portion of the device and apply appropriate voltage(s) to the appropriate electrode(s). In actuality, sophisticated processing algorithms are used, since applying voltage to each of the electrodes at the same time would cause currents to flow between the electrodes, which would stimulate multiple nerves and thus produce multiple (unwanted) frequencies in the "heard" sound. Modern sound waveform processing strategies often utilize bandpass filters to divide the signal into different frequency bands. Algorithms are employed to select a number of the strongest outputs from the filters, which often emphasize transmission of the time-domain aspects of uttered speech. Specialized feature extraction strategies use vowels and formants to evaluate each spoken utterance, the latter being an expanded set of vowel-like sounds. Fricative sounds are an expanded set of consonants and are utilized to fine tune the vowel and formant coding to allow proper microprocessor recognition of a spoken word or utterance. Normally, it is the time-varying aspects of the spectral content of speech that are employed in such coding. An example of this time-varying spectral content is shown in Figure 4. Figure 37 depicts speech coding strategies and elements of a cochlear implant including bandpass filters, envelope detectors, pulse generators, and culminating in the stimulating electrodes for a four-channel system (Loizou, 1998). The waveforms at each stage of this speech coding process are also displayed in Figure 37. The sound is processed through a set of four bandpass filters that divide the acoustic waveform into the four channels. Current

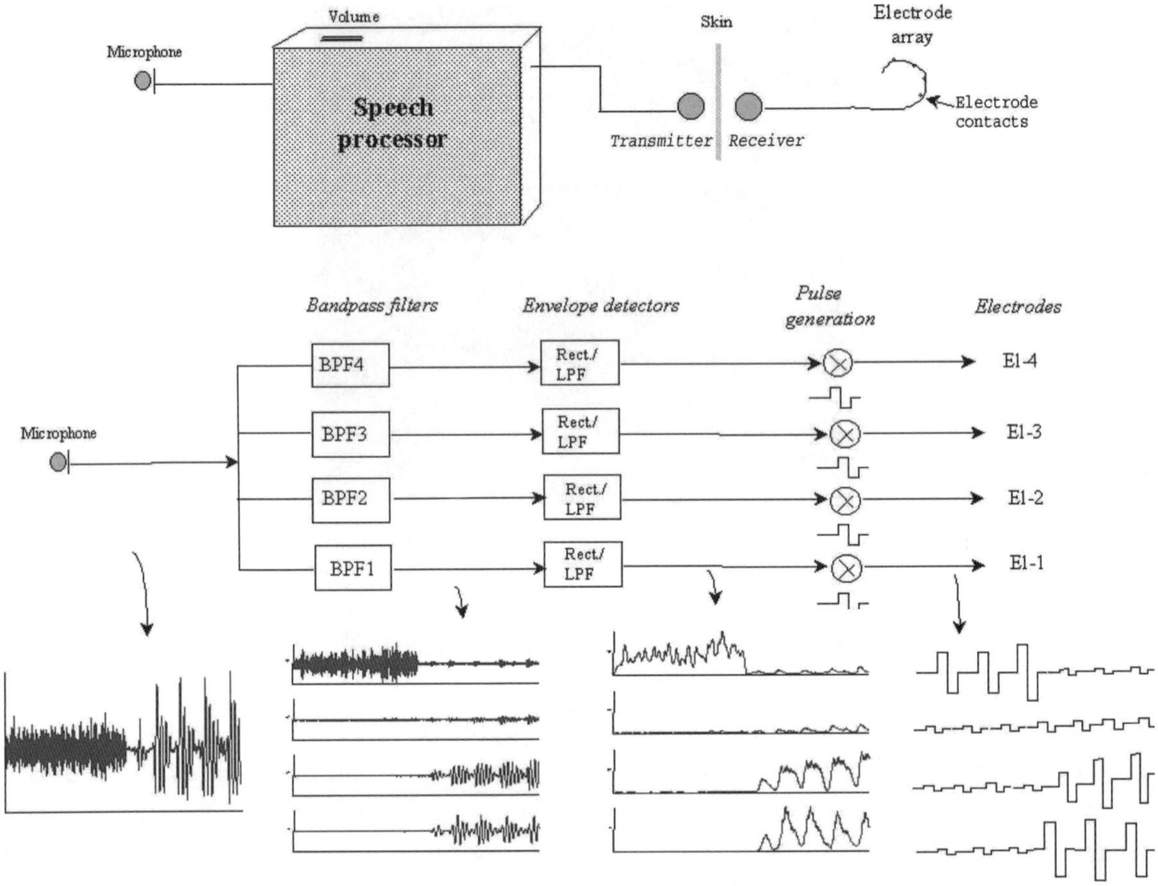

FIGURE 37: Stages of the speech processor in a typical cochlear implant and the associated processing waveforms at each stage for a four-channel device (Loizou, 1998).

pulses are generated with amplitudes proportional to the energy of each channel, and transmitted to the four electrodes through a radio frequency link. Channel 1 represents the lowest of the frequencies and channel 4 the highest. The amplitudes of each of the current pulses delivered to each of the electrodes reflect the spectral content of the input signal, with each appropriate electrode receiving excitation relative to the other electrodes. As an example, if the speech signal mostly contains high-frequency information, then the pulse amplitude of the fourth channel will be larger relative to the pulse amplitudes of the other channels. Similarly, if the speech signal mostly contains low-frequency information, then the pulse amplitude of the first channel will be larger relative to the amplitudes of the higher channels. The electrodes are therefore stimulated according to the energy level of each frequency channel, in a manner similar to that one in the natural cochlea.

Fundamental elements of a cochlear implant include electrode design (number of electrodes and electrode configuration), method of stimulation (analog or digital), as well as the method of feature extraction signal processing, as noted above. The design of electrodes for a cochlear implant has been the focus of research for over many years since the inception of the device (Clark *et al.*, 1983; Hochmair-Desoyer *et al.*, 1983). Early designs incorporated single electrodes in order to test the concept of a cochlear implant, such as that by Tyler (1998), with the number of electrodes employed rising with advances in the science and technology associated with cochlear implants. Modern versions employ 16, 24, or 32 electrodes. Issues associated with electrodes include electrode placement, number of electrodes, spacing of contacts, and orientation of electrodes with respect to the excitable tissue. A larger number of electrodes would provide better reproduction of sound to the brain, while a smaller number of electrodes would provide a more artificially sounding reproduction. However, more electrodes might introduce cross talk within the cochlea as closely spaced electrodes might produce excitation in neighboring tissue (below another electrode), even if that neighboring electrode did not fire. One issue is the sliding of the electrode set along the cochlea to rest at the appropriate place, with, as an example, the electrode representing the band 800–900 Hz being at the proper place in the cochlea where that would represent the same frequencies. This is both a surgery issue and a signal processing issue. It is vital that the electrodes are placed in close proximity with corresponding auditory neurons that lie along the length of the cochlea, or the entire sound spectrum will be shifted in frequency as "heard" by the brain. In most cases, the electrode arrays can be inserted in the scala tympani to depths of 22–30 mm within the cochlea.

The method of frequency encoding and electrode stimulation is constrained by the number of surviving auditory neurons that can be stimulated in the cochlea. It would be ideal to have surviving auditory neurons lying along the length of the cochlea, which would support a good frequency representation through the use of multiple electrodes, with each stimulating a different site in the cochlea as is done in a normal human ear. If the number of surviving auditory neurons is restricted to a small area in the cochlea, then the number of electrodes and frequency pattern is limited by pathophysiology. In that situation, only a few electrodes are needed near the surviving area of the cochlea. As such, it is important to have electrode design and numbers along with firing pattern and stimulation linked to the pathophysiology of the cochlea. It is pointless to use electrodes where there is no chance of providing neural response. Therefore, using a large number of electrodes will not necessarily result in better performance, because frequency coding and electrode stimulation are constrained by the number of surviving auditory neurons that can be stimulated.

In addition, electrode stimulation is constrained by the spread of excitation caused by standard electrical stimulation of tissue. When electric current is injected into tissue such as the cochlea, it tends to spread out symmetrically from the source. This is the basis for the

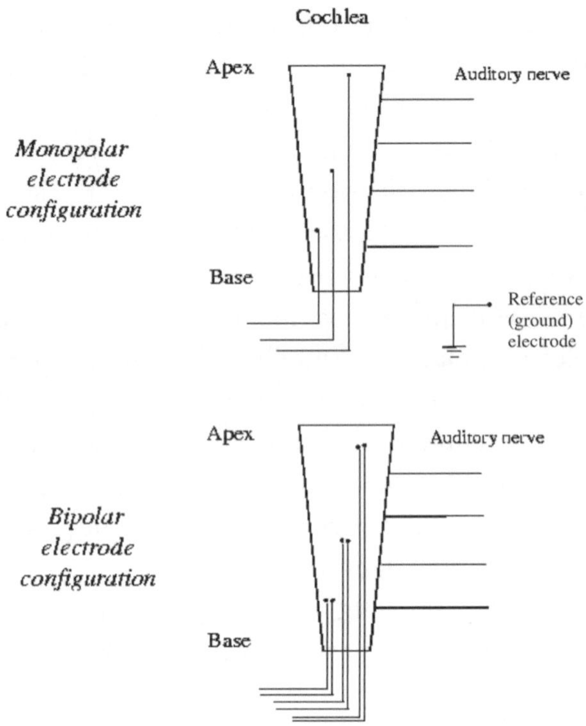

FIGURE 38: Monopolar and bipolar configurations of cochlear implant electrodes (Loizou, 1998).

electrocardiogram, as an example. As a result, the current stimulus in the cochlea does not stimulate just a single auditory neuron, but several. Such a spread in excitation is more prominent in a monopolar type of electrode configuration. In such a configuration, the active electrode is located far from a reference electrode, which acts as a ground for all electrodes in a multiple electrode configuration. The spread of excitation can be controlled to a degree by using a bipolar electrode configuration. In the bipolar configuration, the active and the reference (ground) electrodes are placed in close proximity to each other. Various studies have shown that bipolar electrodes produce a more localized stimulation than monopolar electrodes, such as Merzenich and White (1977) and van den Honert and Stypulkowski (1987), among others. Figure 38 depicts both types of electrodes.

There are generally two types of stimulation depending on how information is presented to the electrodes: analog or digital. In analog stimulation, an electrical analog of the acoustic waveform itself is presented to the electrode. As an example, in multichannel implants, the acoustic waveform is bandpass filtered, and the filtered waveforms are presented to the appropriate electrodes simultaneously in analog form. An advantage of analog signal processing is that the electronics are usually simpler and cheaper. A disadvantage of analog signal processing is that the simultaneous stimulation may result in cross talk. In digital signal processing, the information is delivered to the electrodes using a set of pulses. In some approaches, the timing and amplitude of these pulses are obtained from the envelopes of the filtered waveforms, as is shown in Figure 37. The advantage of this type of signal processing is that the pulses can be delivered in a nonoverlapping approach, thereby minimizing cross talk. The rate at which these pulses are delivered to the electrodes has been found to affect speech recognition performance (Wilson *et al.*, 1995). High pulse rates tend to yield better performance than low pulse rates. Pulse rates as low as 100 pulses per second and as high as 2500 pulses per second have been used. Wilson *et al.* reported that some patients obtained a maximum performance with a pulse rate of 833 pulses per second and a pulse duration of 33 ms, while other patients obtained significant increases in performance as the pulse rate increased from 833 to 1365 pulses per second, and from 1365 to 2525 pulses per second, again using 33-ms pulses (Wilson *et al.*, 1995). It is apparent that an optimal pulse rate depends on a particular patient, but that the pulse width is dependent on the tissue sustaining the electrical excitation. A digital representation of the speech signal often results in a more accurate representation, but the technique may be more expensive. Modern cochlear implants employ digital signal processing techniques for multiple electrode configurations. Another benefit of the digital technique is that the digital pulse stimulation order (to the various electrodes) can be varied to minimize possible cross talk between channels. Electrodes can be sequentially excited (in a rapid fashion), randomly excited, or excited in groups with the spacing far enough apart to minimize cross talk. As with the pulse rate, the optimal electrode firing pattern may vary from patient to patient.

Currently, some implant devices employ monopolar electrodes, while some devices employ bipolar electrodes, and there are other devices that employ both types of electrodes. Loizou reported that the Symbion cochlear implant used six electrodes spaced 4 mm apart in their bipolar configuration with only the four most apical electrodes used in the monopolar configuration (Loizou, 1998). The Nucleus cochlear implant used 22 electrodes spaced 0.75 mm apart for the monopolar configuration with electrodes that were 1.5 mm apart used as bipolar pairs. The Clarion cochlear implant used both monopolar and bipolar configurations. Eight electrodes spaced 2 mm apart were used. The Med-El cochlear implant used eight electrodes spaced 2.8 mm apart in a monopolar configuration. Further details regarding the electronics and signal processing elements in various cochlear implants have been reported by Loizou (1998).

3.3 Current Devices and Cochlear Implant Companies

In 2005, the top three cochlear implant devices were manufactured by Cochlear Corporation (Australia), Advanced Bionics (United States), and MED-EL (Austria). These are similar devices. Each manufacturer has adapted some of the successful innovations of the other companies to their own devices. There is no clear-cut consensus that any one of these implants is superior to the others. Users of all three devices display a wide range of performance after implantation.

Of the three companies, Cochlear Corporation has the longest history in cochlear implantation and the high reliability of their devices has been demonstrated over three decades, as the cochlear implant is generally known to have originated in Australia. Advanced Bionics has been noted to have more advanced electronics, resulting in the more advanced speech processing capabilities. MED-EL has the longest electrode array of the three companies, enabling insertion into the deepest part of the cochlea.

Since the devices have a similar range of outcomes, other criteria are often considered when choosing a cochlear implant: usability of external components, cosmetic factors, battery life, reliability of the internal and external components, customer service from the manufacturer, the familiarity of the user's surgeon and audiologist with the particular device, and anatomical concerns.

There is a great variability in the speech recognition performance of cochlear implants as reported by patients, technicians, and physicians. For a given type of implant, auditory performance (in terms of word or utterance recognition) may vary from zero to nearly 100% correct. The factors responsible for successful word/utterance recognition have been the focus of research for many years. Some of the factors that have been found to affect auditory performance are listed below:

1. *Outside factors*: These include duration of deafness, age at implantation, and residual hearing, as well as issues that are beyond our control. There is a huge variance in how well people do with implants, largely due to the wide degree of differences of these factors. Those with residual hearing respond better to cochlear implants and require less training. Those patients who are deaf from birth require more training and have a checkered success rate. Younger patients do better than older patients who have been deaf for longer periods. Very young patients, who have not yet formulated language, respond better. A discussion of cochlear implants in the very young patients is noted below as are factors associated with the deaf community response to cochlear implant usage.
2. *Attitude and determination*: The amount of work and practice that the patient puts into making it work is very important to making the most of the implant. Cochlear implant recipients normally receive speech and/or auditory therapy to make the most of their implants. The implant is not a miracle cure for deafness, since the number of channels is far less than the number of frequency excitations in the natural ear. Many profoundly deaf (or deaf from birth) patients find it frustrating and discouraging to undergo the long and difficult process of learning to hear with the implant.
3. *Skill of audiologist and surgeon*: A good surgeon and audiologist can help ensure that recipients get the most that the implant has to offer. One should determine how much

experience the audiologist and the surgeon have with implants, and get a sense for how dedicated they may be to help you in the years to come with any challenges that arise.

4. *Type of device*: Newer generations of cochlear implants have improved the chances that people will hear well with them, with digital signal processing techniques, improved electrode design, improved support electronics, and improved surgical techniques.

3.4 Issues Associated with Implant Use

As was noted above, there have been issues raised regarding the use of cochlear implants, which has limited its implantation in a large number of patients. For the elderly patient, the risk of surgery in the older patient must be weighed against the improvement in quality of life. As the devices improve, particularly the sound processor hardware and software, the benefit is judged to be worth the surgical risk particularly for the newly deaf elderly patient (Waltzman *et al.*, 1993).

The use of cochlear implants is objected to by some, particularly in the deaf signing community (people who use sign language to communicate). Many members do not view deafness as a disability that has to be "fixed" by cochlear implants, but rather just a different way of living. Cochlear implants for children work best when implanted at a young age, when the brain is still learning to interpret sound, and hence are implanted before the recipients can decide for themselves. There is a debate about the age that would be the best or safest for children to receive cochlear implants. However, as cochlear implants have improved and proven themselves to work well, the objections are getting less tenable. Proponents of cochlear implants believe that, since mammals are meant to have a hearing sense, deafness is a disability to be corrected. To them, objecting to a cochlear implant is akin to objecting to a heart transplant, prosthetic arm, or glasses.

Opponents of cochlear implants often compare implantation to cultural genocide, as the common ground for the signing community is deafness, either through personal experience or by personal relationships with others who have hearing impairments. Opponents of implantation tend to object not to the medical benefits of implantation, but to the perceived loss of the sense of community that has defined deaf history. Statistics show that the majority of deaf infants are born into hearing families. For such children becoming a member of the deaf community is not automatic, nor is it initially a decision for them to make; rather it is a choice made by their parents, in what they perceive to be the children's best interests, as they understand them. In 2000, an academy-award-nominated film *Sound and Fury* depicted this cultural divide. In addition, the implantation of cochlear implants in very young children has become a topic of much consternation in the deaf community and has spawned Web sites such as www.cochlearwar.com and www.ragged-edge-mag.com, both have numerous citations and editorials opposing the implantation of cochlear implants in small children. Of all artificial organs and all sensory assist or replacement devices, the use of cochlear implants has created the largest amount of controversy and vitriol. This is primarily due to the strongly cohesive nature of the deaf community that does not feel that deafness is a disability like paraplegia or blindness, and thus does not necessitate the use of a cochlear implant. Although there is less controversy with the implantation of cochlear implants in adults, there is still not widespread support for such implantation, which has reduced the number of individuals seeking an alternative to profound deafness. Opponents of cochlear implants cite the prevalence of acquired meningitis as a result of implantation surgery. However, data from the FDA and the Center for Disease Control (CDC) indicates that the prevalence of meningitis for the general population varies from 1 in 10 000–15 000 and for those with cochlear implants, it is 1 in 9000–13 000. There is thus statistically no significant difference. However, the data for the general population utilize a far greater sample size than that for the cochlear implant patient data.

3.5 History of the Cochlear Implant

Professor Graeme Clark of the University of Melbourne is generally credited as the creator and developer of the world's first multichannel implant and is considered by many to be the father of the cochlear implant. In 1967, Professor Clark began researching the possibilities of an electronic implantable hearing device. His endeavor would eventually lead to the creation of the implantable "bionic ear." Inspired by his close relationship with his father, who had been deaf throughout his life, Professor Clark's goal was to find a way to improve hearing, and the quality of life for people who are deaf. Professor Clark has been the Research Leader at the Department of Otolaryngology at the University of Melbourne since 1970. His first cochlear implant surgery took place in 1978. In 1979, Nucleus, a group of medical equipment manufacturers, became interested in the commercial potential of Professor Clark's work at the University of Melbourne. Two years later, the University of Melbourne, the Australian Government, and Nucleus set out to develop a commercially viable cochlear implant, and to bring it to market through a worldwide clinical trial. Cochlear Ltd. was established in 1982, as a corporate entity in order to continue commercial operations. Professor Clark is the Director of The Bionic Ear Institute in Melbourne (Australia), which was established in 1984 to support and undertake research to better understand deafness and improve the "bionic ear."

The basis for the cochlear implant is predicated on the spatial frequency representation of the cochlea, which was first elucidated by the pioneering work of Georg von Bekesy in the 1950s, which showed that the basilar membrane in the inner ear is responsible for analyzing the input signal into different frequencies (Hachmeister, 2003; Shampo and Kyle, 1993). Different frequencies cause maximum vibration amplitude at different points along the basilar membrane. A famous diagram of this spatial frequency representation is shown in Figure 39.

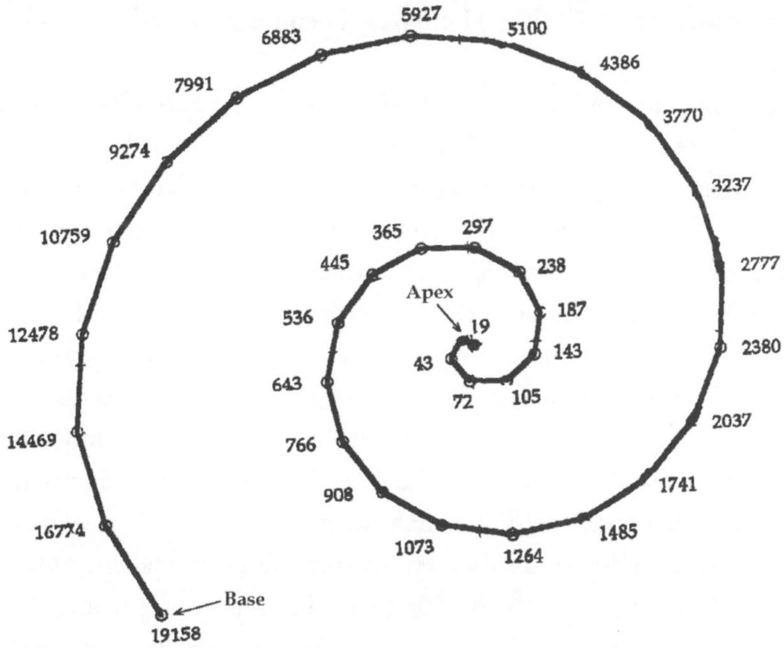

FIGURE 39: Spatial variation by frequency in the human cochlea, first represented by von Bekesley.

3.6 Regulatory Issues for Cochlear Implants and Recent Research

For a relatively new technology such as a cochlear implant with few implants compared to the entire deaf population, the issues regarding regulatory controls for cochlear implants are of paramount importance. This is very evident when one looks at the FDA Web site at www.fda.gov and then to its medical devices section, where cochlear implants are one of the key topics on the right (as of January 2006). The FDA has a primer on the topic complete with movies. From 1984 through 2001, there were seven different applications approved for cochlear implants including the following:

Cochlear Corp	1984	3M Brand
Cochlear Americas	1985	Nucleus Mutichannel
Cochlear Americas	1990	Nucleus 22 Channel
Advanced Bionics	1996	Clarion Multistrategy
Advanced Bionics	1997	Clarion Multistrategy
Cochlear Americas	1998	Nucleus 24
Med-El Corp	2001	Combi 40+

The Web site listing these approved applications is at http://www.accessdata.fda.gov/scripts/cdrh/devicesatfda/index.cfm, which also notes any warning letters, MedWatch reports, or enforcement reports for each device. There have been no approved devices by the FDA since 2001.

In 2004, Advanced Bionics issued a voluntary recall of its Clarion cochlear implants because there were noted moisture problems prior to implantation for some of their devices. This report is reported at http://www.fda.gov/bbs/topics/news/2004/NEW01119.html. The FDA also issued a report on the risk of bacterial meningitis in small children receiving cochlear implants. This report is at http://www.fda.gov/cdrh/safety/cochlear.html.

As with any technology that has received few approved applications over 20 years including none over the last 5 years, there are still significant questions to answer. Is the use of the device hampered by the nonsupport of the deaf community or due to the checkered success rate for profoundly deaf patients (from birth) who have no basis to understand spoken language as processed by a cochlear implant? The FDA Web site that analyzes risks and benefits ishttp://www.fda.gov/cdrh/cochlear/RiskBenefit.html.

There have been numerous published reports regarding clinical and empirical research on cochlear implants. These can generally be categorized as reports regarding the use of cochlear implants in children, the use of cochlear implants in the elderly, a comparison of cochlear implants versus hearing aids, and studies regarding design issues, such as electrode configuration and signal processing techniques. Recent studies regarding the use of cochlear implants in small children include those by Beadle *et al.* (2005), Berg *et al.* (2005), Biernath *et al.* (2006), Flipsen and Colvard (2005), Gordon *et al.* (2005), Henkin *et al.* (2005), Higgins *et al.* (2005), Hyde and Power (2005), and Tomblin *et al.* (2005), among others.

Studies regarding the use of cochlear implants in the elderly include those by Haensel *et al.* (2005) and Hay-McCutcheon *et al.* (2005), among others, while the studies regarding design considerations include Fabry (2005), Fu *et al.* (2005), Hong and Rubinstein (2006), Laneau *et al.* (2005), Munson and Nelson (2005), Stickney *et al.* (2005a), Stickney *et al.* (2005b), Vermeire *et al.* (2006), and Xu *et al.* (2005), among others.

Despite the relatively small numbers of FDA approved devices, the number of worldwide implants is growing, and the number of clinical studies and published reports number in the thousands.

4 INTRAOCULAR LENS

4.1 Anatomy of the Eye

An IOL is used to replace a damaged lens in the eye, most normally due to a cataract, which results in cloudy vision. The human eye, shown in Figure 40, consists of the multilayer outer covering (the cornea), the iris and associated pupil opening, the lens inside a lens sac, and the retina. There are fluid layers between the cornea and the lens and again between the lens and the

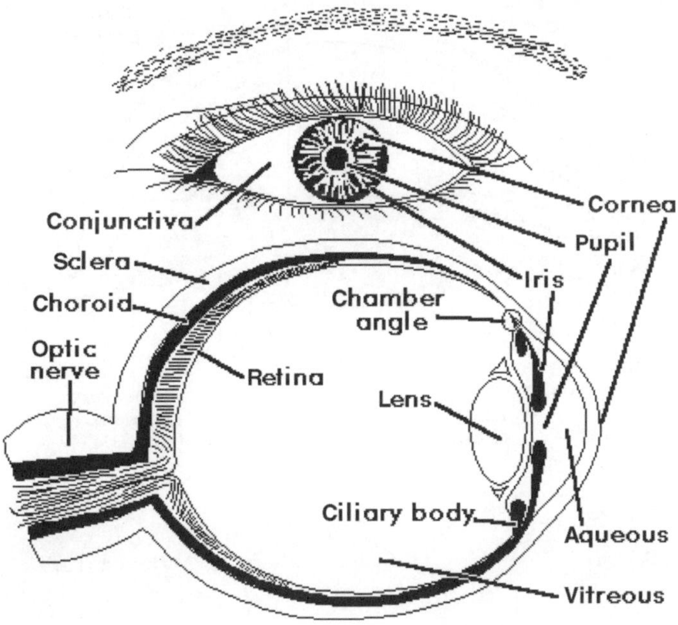

FIGURE 40: Anatomy of the human eye.

retina. The retina ends with the optic nerve. The cornea provides 75% of the overall focusing power of the eye with the lens proving the other 25%.

However, it is the ability of the natural lens to *accommodate*, to change focal length, and the fine tuning of accurate sight that is key to visual field acuity. The human lens is a gelatinous substance, which is flexible. This is important as the sac in which it resides is connected on either end by ciliary muscles. These muscles can rapidly move in and out. As your visual attention moves closer or farther away, the ciliary muscles (sometimes called the ciliary body) attached to your lens sac move in and out, which changes the shape of the lens to make it more or less concave and provide variable focus. This process is called accommodation. When one looks directly at one's feet and then suddenly attempts to focus 50 yards away, the focus remains true and one's visual field remains perfectly clear. This could not happen without the lens being able to change its shape as the lens sac is pulled by the attached muscles. The continuing change in the shape of the lens allows the focal point of vision to fall on a section of the retina known as the *fovea*. A more detailed view of the human eye showing the fovea on the retina is shown in Figure 41.

4.2 Cataracts and Their Determination

A cataract is a clouding of the human lens with the resulting vision becoming opaque in that eye. Often, if only one eye is affected, it may be possible that an individual does not realize that a

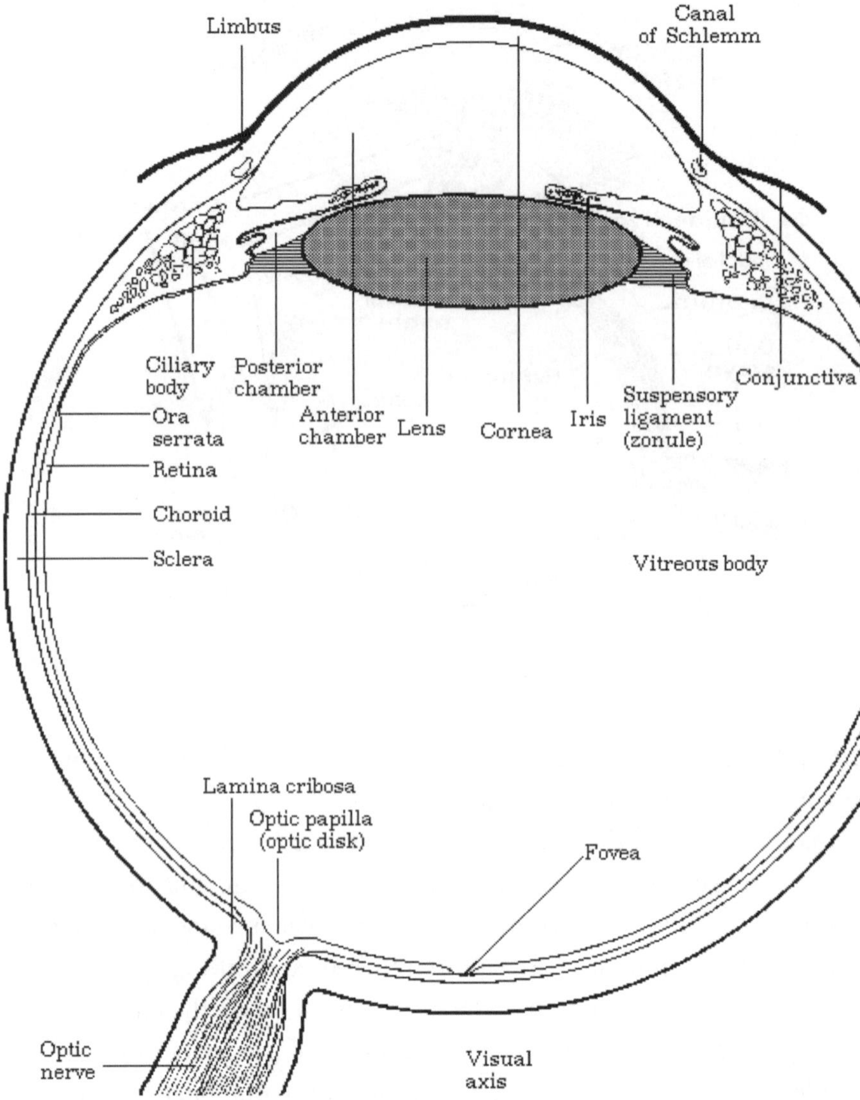

FIGURE 41: Anatomy of the human eye showing the fovea portion of the retina.

cataract is evident. Over 40 million people in the United States suffer from cataracts. A cataract can result from many sources, including age, diabetes, glaucoma, blunt injury, excessive infrared light exposure, excessive steroids, and genetic effects. The natural lens, which is crystalline in nature, becomes opaque and less flexible with an advanced cataract. This results in the lens becoming less able to accommodate by changing shape. However, it is the opaque vision effect of a cataract that is often more telling in both the diagnosis and the patient complaint. A routine eye examination with a slit lamp would indicate a cataract, since the eye would show whiteness in the lens indicating cloudy vision. A typical slit lamp examination is shown in Figure 42.

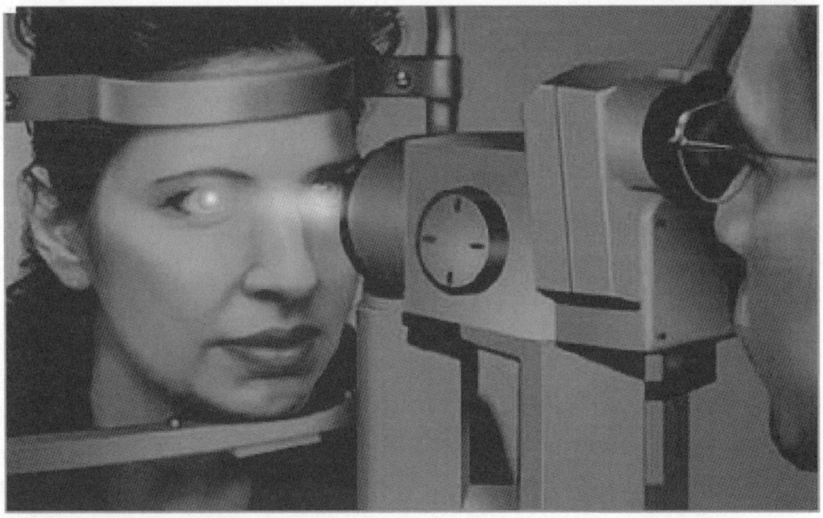

FIGURE 42: Typical slit lamp examination to examine the interior of the eye.

Figure 43 shows both a normal eye (front view) and an eye with a cloudy lens. It is obviously easy to see this in a routine eye examination. The resulting visual field is different for a normal lens and a cloudy lens, as is shown in Figure 44. The opaque nature of the visual field is easier to notice by the patient in sunlight and is less easy to notice in artificial (indoor) light.

Once a cataract has been determined to exist, it is necessary to evaluate the location of the lens, its distance from the cornea and from the retina, the thickness of the lens, and the size of the sac. This will eventually allow the clinician to determine the correct power and size of an artificial lens, the *intraocular lens*. This geometric and anatomical information of the eye requires detailed measurements within the fluid-filled eye with a resolution that is on the order of 0.1 mm. The use of ultrasound at a high frequency is perfect for this determination for several reasons: 1) ultrasound transmits well within a liquid; 2) at a 10–15-MHz emission frequency, the resolution is on the order of the required 0.1 mm; and 3) the attenuation factor for ultrasound

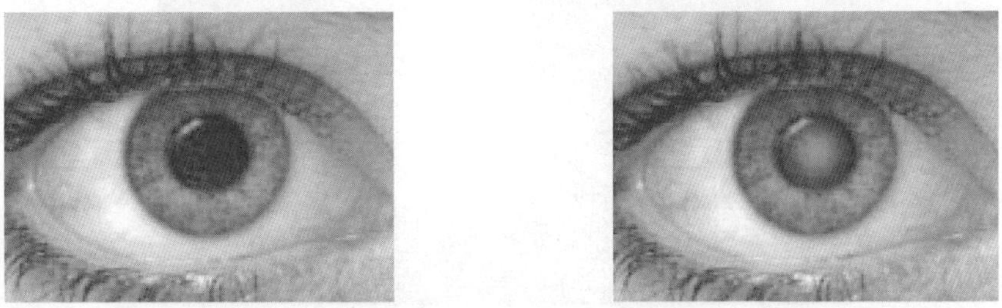

FIGURE 43: Views of human lens: normal lens on the left and cloudy lens on the right.

FIGURE 44: Visual field for a normal lens on the left and a cloudy lens on the right.

is lessened due to the small depth of the eye. A typical A-scan ultrasound examination of the eye is shown in Figure 45.

The ultrasound probe is a small 3-mm rod attached to a spring-loaded micrometer device, sometimes called a Goldman tonometer holder. The clinician slowly rotates a knob and the probe slowly nears the patient's eye. When the probe contacts the cornea and meets resistance, it automatically springs backward away from the patient. This actually gives the sufficient time to make the appropriate measurement, since the 10-MHz emission frequency is actually pulsed 1000 times per second. As a result, each 1-ms pulse width contains 10 000 emitted pulses at the 10-MHz frequency (each emitted wave being 0.1 μs). Thus, in as little as 1 ms of contact

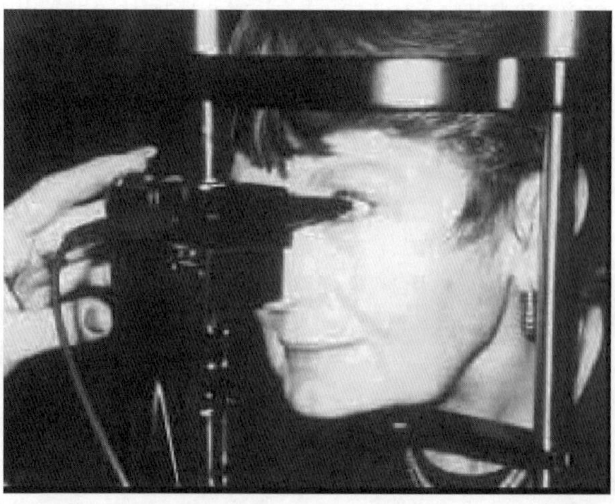

FIGURE 45: A scanner for the eye.

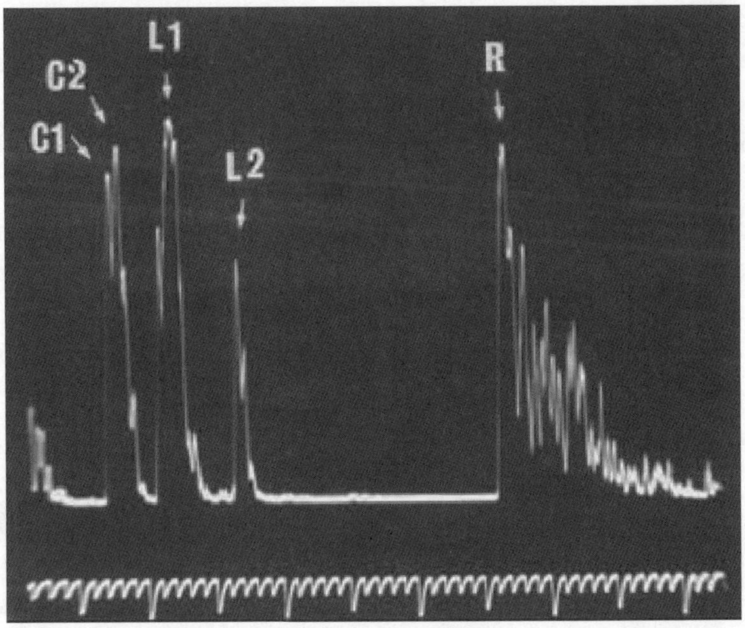

FIGURE 46: A-scan ultrasound pulse-echo pattern in the eye.

with the cornea, the A-scan probe would have emitted 10 000 pulses with the returning echoes processed for turnaround time. This turnaround time duration is equivalent to the distance from the probe (and thus the cornea outer edge), since a constant speed of sound is assumed. A typical A-scan ultrasound pulse-echo pattern is shown in Figure 46.

The C1 and C2 echoes in Figure 46 refer to the anterior and posterior surfaces of the cornea, while the L1 and L2 echoes refer to the anterior and posterior surfaces of the lens, and the R echo refers to the anterior surface of the retina. The various distances from the cornea to the lens and from the lens to the retina dictate the power of whatever replacement lens is required. Any variation in measurements of these distances can affect the final power of the implanted lens, and this depends on the axial length (depth) of the eye. In an average 23.5-mm eye, a 0.1-mm difference in axial length measurement (of the eye) affects the final postoperative refraction by 0.25 D (D = diopter, the standard optical power for lenses). In a long 26.0-mm eye, a 0.1-mm difference in the AL measurement affects the final postoperative refraction by only 0.20 D. In the short 21.0-mm eye, a 0.1-mm difference in the axial length measurement affects the final postoperative refraction by 0.31 D.

4.3 The IOL and Implantation Surgery

An IOL is a replacement device for a cloudy natural lens. Because of the small size of the natural lens and the lens sac, along with a need to have very small incisions for removal of the old lens and insertion of the replacement lens, there is a size restriction on the artificial lens and any device to hold this lens. Furthermore, as this is a device that will be in close contact with liquids and tissues inside the body, this device must be biocompatible and nonreactive. Obviously, this device must have the correct optical power as determined by the A-scan ultrasonic examination of the eye. As will be described below, the standard IOL is pliable and can be folded in half during insertion, so as to limit the size of the incision.

However, first things first. The cloudy natural lens must somehow be removed in order to have space for the replacement IOL. As the natural lens is a crystalline, gelatinous material, we can employ various technologies to remove this lens from the eye. The most commonly employed technique is to use high-frequency, therapeutic level ultrasound, often called *phacoemulsification*. Unlike the A-scan ultrasound used to measure eye distances, the ultrasound used to emulsify the natural lens is not a pulse-echo device, but rather a continuous wave device. The pulse-echo device is actually "on" for 1% of the time with the other 99% in the echo mode to detect the time course of returning echoes. In the continuous wave mode, particularly with higher amplitudes than the pulse-echo mode, the tissues or fluids can vibrate and even heat up. Therapeutic ultrasound is sometimes employed to ease aching muscles by heating the tissue and increasing blood flow to the region. In phacoemulsification, the gelatinous lens is liquefied (emulsified) and then aspirated out, leaving an empty sac behind.

The cataract surgical procedure consists of the following steps:

1. An outer incision to the edge of the cornea
2. An inner incision in the lens sac
3. Phacoemulsification of the natural lens
4. Aspiration of the liquefied natural lens
5. Insertion of the folded IOL
6. Suture the lens sac with dissolvable sutures
7. Suture the outer incision with nondissolvable sutures

Both incisions are approximately 3 mm long. Figure 47 depicts the miniature scalpel used to create the microincisions. The surgeon creates an opening in the lens capsule, which is a microthin membrane surrounding the cataract. This procedure, called capsulorhexus, requires extraordinary precision since the capsule is only about four-thousandths of a millimeter thick.

A typical phacoemulsification probe is shown in Figure 48 with the ultrasound waves seen emanating within the lens.

Figure 49 shows the aspiration portion of the process once the lens has been fully emulsified (liquefied). The aspiration probe is moved around the sac to completely aspirate all loose materials within the sac, as is shown in Figure 49.

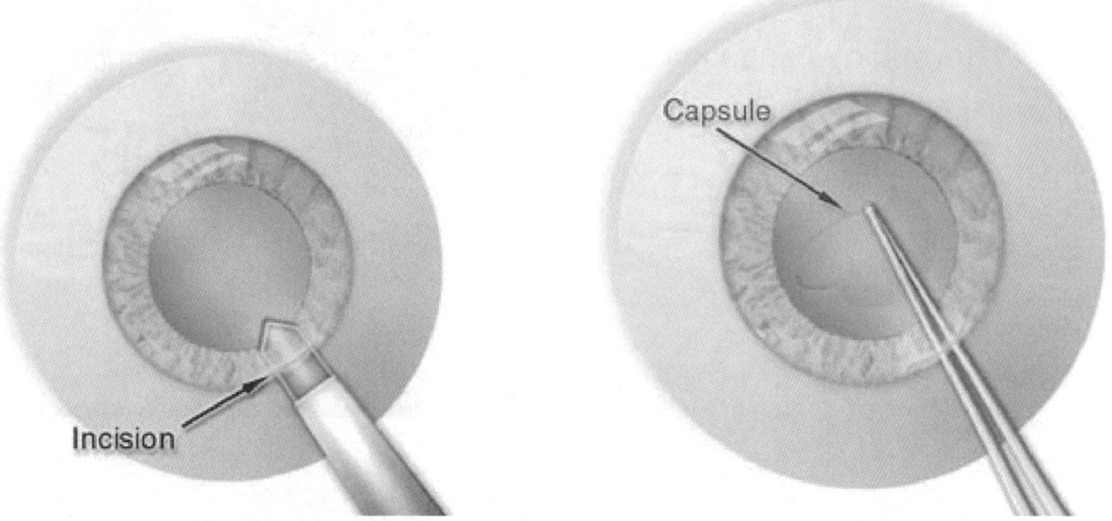

FIGURE 47: Incisions in cataract surgery: left—on the side of the cornea with a miniature scalpel and right—on the side of the lens sac.

Once the original cloudy lens has been removed, it is then possible to insert the IOL. A typical IOL is shown in Figure 50 and consists of a biconvex, clear lens with "wings" to hold the lens in place within the lens sac/capsule.

The springlike wings pop out when the IOL is unfolded during insertion. A diagram with the IOL in place within the eye as compared to the natural lens is shown in Figure 51.

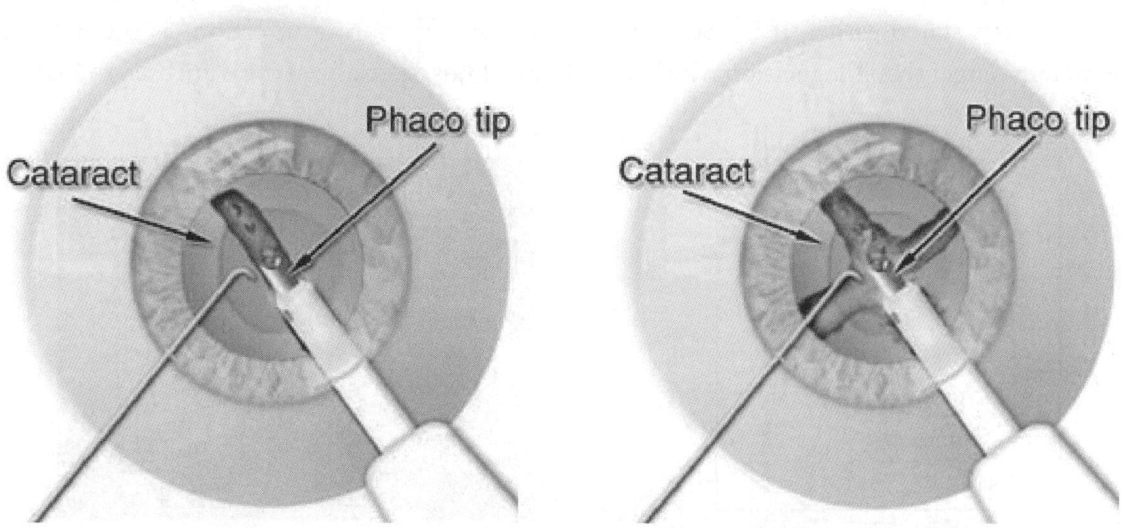

FIGURE 48: Phacoemulsification showing the sound waves with breakup of lens to left.

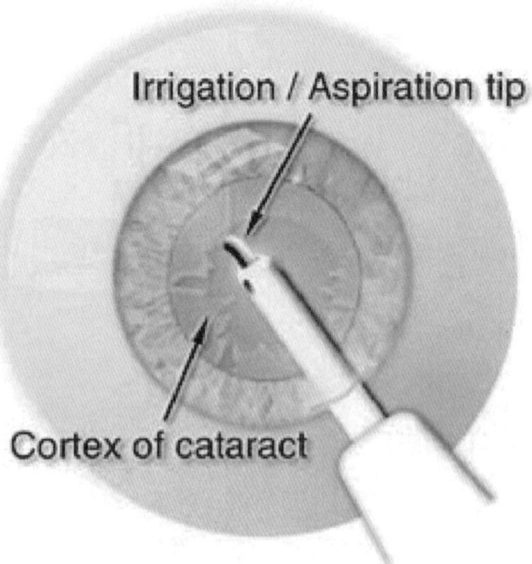

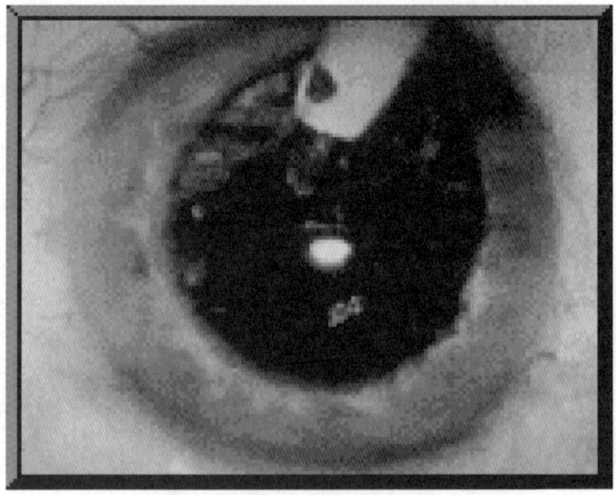

FIGURE 49: Irrigation and aspiration of the emulsified lens. Aspiration probe is moved around the lens sac to aspirate all particles and lens pieces.

A typical IOL is a clear lens made of an acrylic polymer with 25–30% water content, so that it will not absorb water when inserted into the fluid-filled environment of the human eye (aqueous and vitreous regions as noted in Figures 40 and 41). An IOL has various optical powers based upon patient needs. It is typically 6–8 mm in diameter and is available in a power range between 10 and 30 D in 0.5-D increments. Other IOL's are made of silicone, poly(methyl methacrolate) (PMMA), or a clear polymer known as perspex. Many IOL designs have ultraviolet light blockers incorporated into the lens material.

As was stated earlier, it is necessary to fold the IOL during insertion, as the 6–8-mm diameter is larger than the incision width of 3–4 mm. Thus, specialized tools are used to fold and

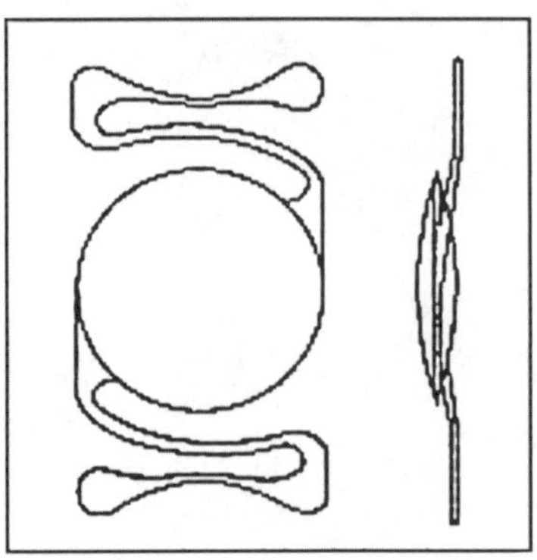

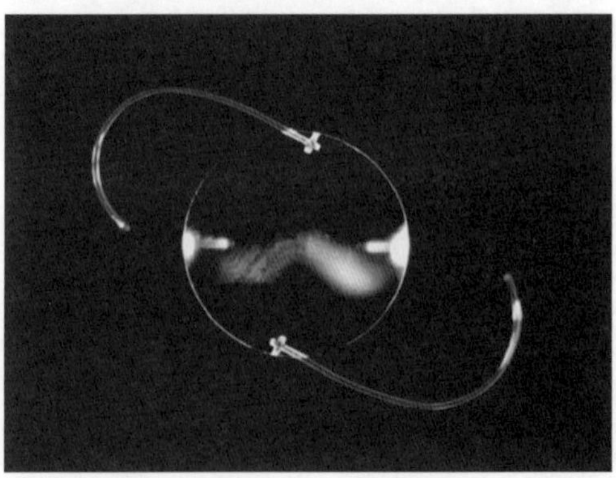

FIGURE 50: Typical IOL with wings.

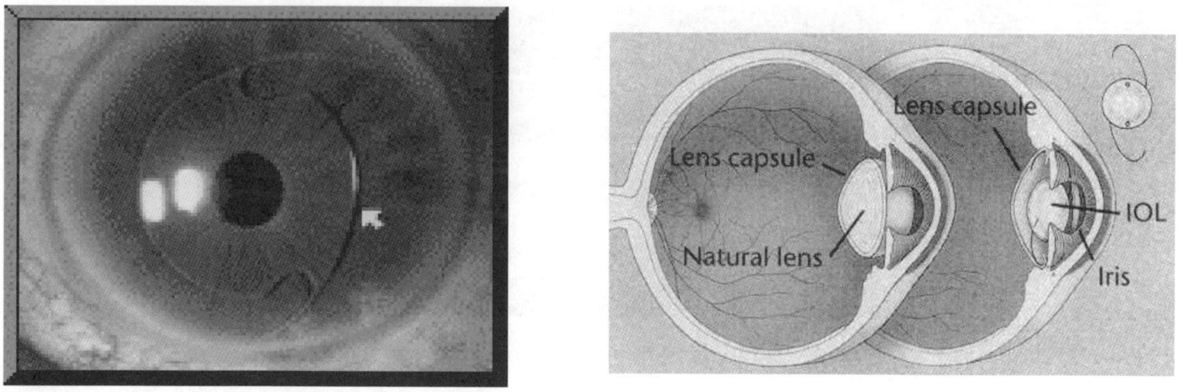

FIGURE 51: IOL in place within the eye and lens sac/capsule.

hold the IOL as well as release it once it is placed within the lens sac/capsule. A forceps design is offered by various IOL manufacturers which folds and holds the lens while it is inserted into the capsule. Typical forceps designs are shown in Figure 52.

Once the IOL is placed within the lens capsule, the IOL is then released and it pops open within the capsule with the spring-loaded wings extending outward. Other forms of IOL holders include injection-type devices with push button or screw-turned hold and release mechanisms, as is shown in Figure 53.

Following implantation, the slit at the edge of the lens sac/capsule is often sutured with dissolvable sutures. The outer slit at the edge of the cornea is closed with nondissolvable sutures. The patient is given strong wraparound sunglasses, as there will be a strong light sensitivity for

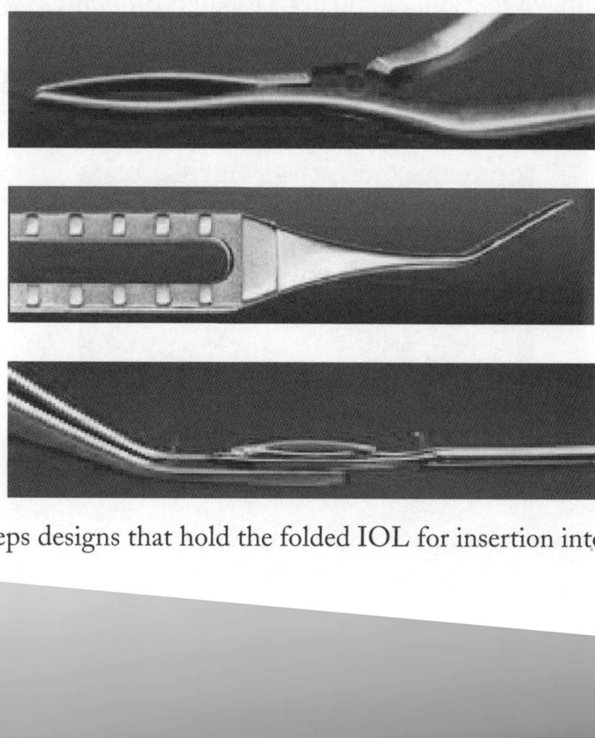

FIGURE 52: IOL forceps designs that hold the folded IOL for insertion into the lens sac.

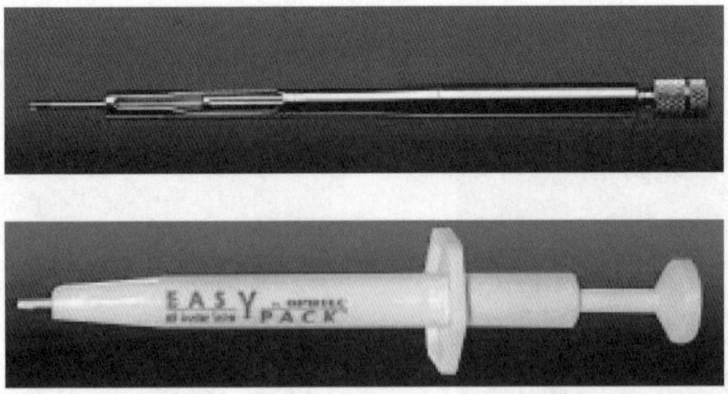

FIGURE 53: IOL injection style holders with push button or screw release.

a few days. In addition, the patient is given a plastic disk that must be taped over the affected eye at night. This is to protect the eye from accidental rubbing during the night. Normally, the patient is seen by the ophthalmic surgeon the day after surgery, in order to check the incision, the healing process, and to determine if there is any swelling or infection. Approximately 1 week after surgery, the patient returns to have the outer sutures removed. The ophthalmic surgeon carefully snips the ends of the sutures and uses a small tweezers to pull the sutures out. The patient can feel the light pulling of the sutures, but rarely feels any pain.

On some occasions, the lens sac/capsule may become cloudy at some point following the surgery as is shown in Figure 54. This is normally remedied by means of an out-patient visit to the surgeon's office. An excimer laser is used to burn small holes in the sac at various points. The patient is placed in a head mount device and the affected eye is held open. The patient sees red flashes and hears "pops," but does not feel anything during the process. The holes in the sac supply sufficient light such that the clear vision is restored. The sac is otherwise still intact.

At times, a patient may experience a "floater." Floaters, or *muscae volitantes* (Latin—"flying flies"), are characterized by shadowlike shapes that appear singly or together with several others in one's field of vision. They can take the form of spots, threads, or fragments of cobweblike

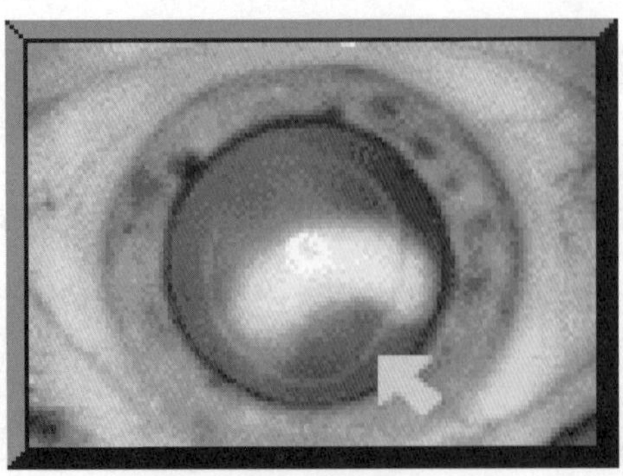

FIGURE 54: Lens sac becomes cloudy some time after surgery.

shapes that float slowly before one's eyes. Floaters are suspended in the vitreous humor, the thick fluid or gel that fills the eye. Thus, they generally follow the rapid motions of the eye while drifting slowly within the fluid. When they are first noticed, the natural reaction is to attempt to look directly at them. However, attempts to shift the gaze toward them are frustrating, because the floaters follow the motion of the eye, and remain to the side of the direction of gaze. Although the blood vessels of the eye also obstruct light, they are invisible under normal circumstances (and thus not annoying) because they are fixed in the location relative to the retina, and the brain "tunes out" stabilized images. This does not occur with floaters and they remain visible, and, in some cases, when large and numerous, are very irritating. Despite the name "floaters," many of these specks have a tendency to sink toward the bottom of the eyeball, in whichever way the eyeball is oriented. Floaters are not uncommon, although they rarely cause problems for those who have them. Floaters can be a nuisance and a distraction to those who suffer from severe cases, as the spots seem to drift through the field of vision. Normally, there is no treatment indicated, particularly for mild cases, as the floater will eventually sink toward the bottom of the vitreous body and out of the field of vision. Surgical or laser treatments have been employed for severe cases with mixed results.

4.4 History of the IOL and Current Research

Nicholas Harold Lloyd Ridley, most commonly known as Harold Ridley, was a British ophthalmologist who pioneered IOL surgery for cataract patients. He was famous as a Surgeon at Moorfields Eye Hospital and St. Thomas' Hospital in London, specializing in ophthalmology. While working with Royal Air Force casualties during World War II, he noticed that when splinters of perspex from aircraft cockpit canopies became lodged in the eyes of wounded pilots, they did not trigger rejection, leading him to propose the use of artificial lenses in the eye to correct cases of cataracts. He had a lens manufactured using an identical plastic, and on November 29, 1949, at St. Thomas' Hospital, he achieved the first implant of an IOL, although it was not until 1950 that he implanted an artificial lens permanently in an eye. He went on to develop widespread programs for cataract surgery with intraocular implants and pioneered this treatment in the face of initially strong opposition from the medical community. He continually refined the technique; until by the late 1960s, with his pupil Peter Choyce, he eventually achieved widespread support for the technique. The IOL was finally approved for use in the United States by the FDA in 1981. IOL implantation and cataract surgery is now a common type of eye surgery.

Although the most common form of an IOL is for the treatment of a cataract, there is now another form of IOL used for the treatment of myopia. The alternative IOL is placed in the anterior chamber (with respect to the lens capsule) and is primarily used to treat nearsightedness, as opposed to a cataract. In this fashion, it is similar to that of glasses or contact lenses, but is permanently placed within the eye, as is done for more standard IOLs. This new IOL is the latest type to receive FDA approval as noted at http://www.fda.gov/bbs/topics/ANSWERS/2004/ANS01313.html. As befits a new style of IOL for a different purpose that was originally intended to treat cataracts, there have been numerous studies of late including those by Alio *et al.* (2005), Benedetti *et al.* (2005), Leccisotti and Fields (2005), Lifshitz *et al.* (2004), Lovisolo and Reinstein (2005), and Sridhar *et al.* (2005).

For traditional IOLs to treat cataracts, recent studies have focused on the materials used for an IOL as well as surgical and follow-up complications of various IOL designs. Examples of studies involving materials used for an IOL include a study of the use of PMMA by Frazer *et al.* (2005) as well as a comparison of PMMA, silicone and acrylics by Smith *et al.* (2005). Examples of studies involving surgical and postsurgical complications of an IOL include those by Cakmak *et al.* (2005), Collins *et al.* (2006), Martin and Sanders (2005), Oshika *et al.* (2005), Parsons *et al.* (2005), Tassignon *et al.* (2005), Tognetto *et al.* (2005), and Werner *et al.* (2006).

5 ARTIFICIAL AND REPLACEMENT CORNEA

The cornea is the outer coating of the human eye as can be visualized in Figures 40 and 41 and provides most of an eye's optical power. Together with the lens, the cornea refracts light and consequently helps the eye to focus. The cornea gives a larger contribution to the total refraction than the lens, but whereas the curvature of the lens can be adjusted to alter the focus based on accommodation, the curvature of the cornea is fixed. The cornea has sensitive nerve endings such that a touch of the cornea causes an involuntary reflex that closes the eyelid. As transparency is of prime importance, the cornea does not have blood vessels, but rather receives nutrients via diffusion from tears on the outside surface and the aqueous humor on the inside surface. The adult cornea has a diameter of approximately 12 mm (about one half inch) and a thickness of 0.5–0.7 mm in the center and 1.0–1.2 mm at the periphery. In humans, the refractive power of the cornea is approximately 45 D, which is approximately three-fourths of the eye's total refractive power. The cornea consists of five layers. From the outside to the inside they are as follows:

- *Corneal epithelium*: a thin epithelial layer of fast-growing and easily regenerated cells. Tears keep this layer moist.
- *Bowman's layer*: a tough layer that protects the corneal stroma. It consists of irregularly arranged collagen fibers.
- *Corneal stroma*: a thick, transparent middle layer responsible for most of the focusing power of the cornea. It consists of regularly arranged collagen fibers along with a few fibroblasts. If the stroma is damaged, for example, by injury or infection, it can lose its

transparency, causing vision problems. The corneal stroma consists of approximately 200 layers of type I collagen fibrils. The ordering of the fibrils is responsible for the transparency of the tissue.

- *Posterior limiting membrane (also called Descemet's membrane)*: a thin acellular layer that serves as the modified basement membrane of the corneal endothelium.
- *Corneal endothelium*: a simple scaly style epithelium, an inner lining acting as a barrier to prevent water inside the eyeball from moving into and hydrating the cornea, which would lead to blurred vision.

The cornea is composed mostly of dense connective tissue, similar to the surrounding sclera. However, the collagen fibers are arranged in a parallel pattern, allowing light waves to constructively interfere, letting the light pass through relatively uninhibited. The cornea is innervated by the long posterior ciliary nerves.

Various refractive eye surgery techniques, such as LASIK surgery, can change the shape of the cornea in order to reduce the need for glasses or otherwise improve the refractive state of the eye. In the techniques used today, parts of the cornea are removed with lasers. If the corneal stroma has developed opaque patches known as leukomas, a cornea of a deceased donor can be transplanted. Because there are few blood vessels in the cornea, there are also few problems with rejection of the new cornea. Replacement corneas are normally obtained from eye banks, which store donor corneas. There are also synthetic corneas in development. Most are merely plastic inserts, but there are also some made of plastics that encourage the eye tissue to grow into the synthetic cornea, making it a full replacement. In addition tissue-based artificial corneas are now in development.

5.1 Corneal Transplant

A corneal transplant is the replacement of a damaged cornea with an undamaged one. Corneal transplant is also called keratoplasty. A corneal transplant is not the same thing as laser-assisted *in situ* keratomileusis; LASIK involves reshaping, not removal, of the cornea, in order to correct focusing problems. Although the cornea is tough, due to its vulnerable location, it is quite susceptible to damage from accidents, bacteria, and particles in the air. The well-organized arrangement of the collagen is required in order for the cornea to remain transparent. Corneal transplantation is undertaken for blurry vision resulting from a cornea that no longer enables light to pass through uninterrupted, due to cloudiness or scarring. Reasons for opaqueness of the cornea may be

- degenerative disease;
- dystrophies (inherited diseases in which impurities gradually accumulate in the cornea);

- infection (infection of the cornea can also occur as a complication of other infections); and
- trauma.

In the procedure, the central region of the cornea is cut out with a trephine (an instrument resembling a cookie cutter). The replacement cornea is then inserted and sutured into place using extremely fine thread (narrower than a human hair). The patient is awake during the operation, which is conducted under local anesthesia. The eye is held open by special device. Surgery requires 30–90 min. After the surgery, eye drops are used to aid healing. Nevertheless, recovery from keratoplasty occurs extremely slowly; complete healing takes months to years, and so the sutures remain in place for at least several months. Complications include the possibility of rejection of the foreign tissue graft or bleeding or infection risks of surgery. The use of large amount of steroids following the operation improves the success rate. Nevertheless, about one in five patients rejects the new cornea. Over 40 000 corneal transplants are performed in the United States every year. Medicare reimbursement for a corneal transplant in one eye was about $1200 in 1997. Figure 55 depicts a cornea before and after corneal transplant surgery.

Figure 56 depicts the sequence of the surgical procedure for corneal transplantation.

There are numerous regional and local eye banks dedicated to receiving donor corneal tissue with proper storage until needed by a patient with a cloudy cornea. One can contact local eye banks by looking in the local yellow pages under "eye bank" or by searching yahoo.com under "eye bank, state name (or city name)." There will be links on how to become a donor including downloading a donor card. Physicians involved with corneal transplantation usually know how to access the local eye bank for donor tissue. Although blood-typing is not a necessity (as the

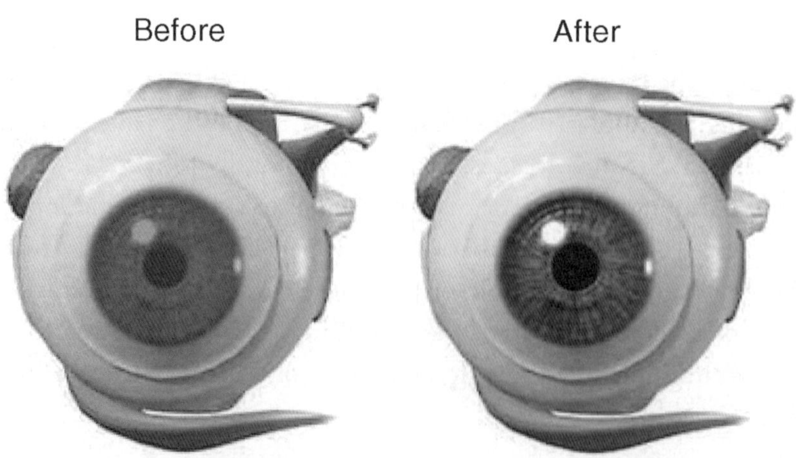

FIGURE 55: The cornea before and after transplant surgery. The cloudy cornea on the left is replaced with a clear transplant cornea on the right.

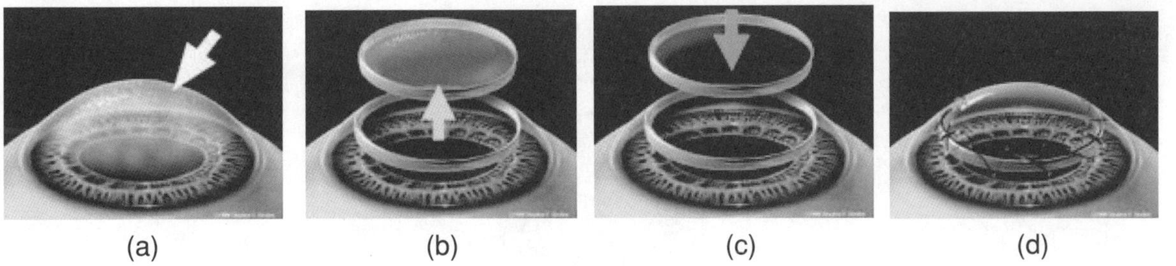

FIGURE 56: Corneal transplant surgery: (a) damaged cornea; (b) cornea removal; (c) replacement with donor cornea from eye bank or cadaver; and (d) suture replacement cornea in place.

corneal is avascular), it is often prudent to match blood type between donor and patient. The United Network for Organ Sharing, which is responsible for organ donation and transplantation for a variety of organs, does not handle eye tissue. This is done via eye banks at the local and regional level.

Replacement corneas that are either plastic or tissue based have been developed in recent years. An artificial (nontissue) cornea has been developed by AlphaCor produced by Argus Biomedical Corp. AlphaCor is a biocompatible, flexible, one-piece artificial cornea (keratoprosthesis) designed to replace a scarred or diseased native cornea. It is technically listed as a *hydrogel*. AlphaCor is designed for use in patients who have had multiple failed corneal transplants or in those patients in whom a donor graft is likely to fail. AlphaCor is available in two versions, to suit those with a natural lens (phakic) or artificial lens (pseudophakic) and for those without a lens (aphakic). AlphaCor is a one-piece convex disk consisting of a central transparent optic and an outer skirt that is entirely manufactured from poly(2-hydroxyethyl methacrylate) or PHEMA. The outer skirt is an opaque, high water content, PHEMA sponge. The porosity of the sponge encourages biointegration with host tissue and thus promotes retention of the implanted device. The central optic core is a transparent PHEMA gel providing a refractive power similar to that of the human cornea. The junctional zone between the skirt and the central optic is the interpenetrating polymer network (IPN). This is a permanent bond formed at the molecular level and is designed to prevent the down growth of cells around the optic, which can lead to the formation of retroprosthetic membranes, one of the major complications historically associated with artificial corneas. The AlphaCor device is shown in Figure 57.

The implantation of AlphaCor is a unique two-stage procedure. The first stage of the surgery involves the surgeon creating a pocket within the existing scarred cornea and inserting the device (after removing a portion of the scarred cornea first). This is usually performed under general anesthetic. Approximately 3 months after this initial surgery, a second, smaller procedure is performed. At this stage, the scarred tissue that is covering the AlphaCor is removed to allow the light to enter.

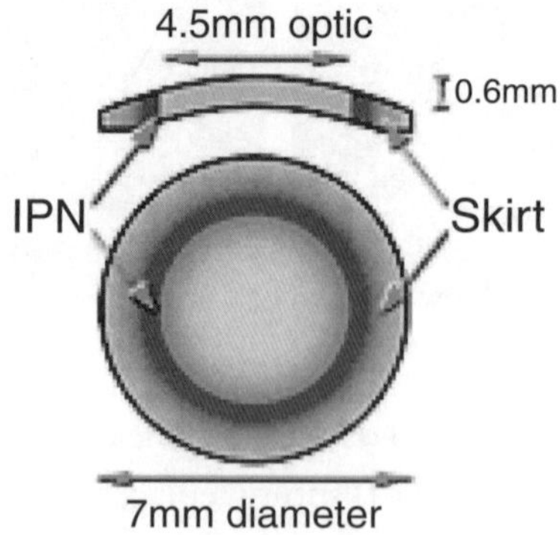

FIGURE 57: AlphaCor artificial cornea. Specifics of the outer skirt and the IPN are explained in the text.

Another type of artificial (nontissue) cornea is the Boston keratoprosthesis, manufactured by the Massachusetts Eye and Ear Hospital, which is affiliated to the Harvard Medical School. This device is a multipiece system as shown in Figure 58.

The Boston system appears to be similar to a bolt, washer and screw assembly, albeit biocompatible and sterilizable versions. The Boston keratoprosthesis has been under development since the 1960s and has gradually been perfected. It received FDA approval in 1992. About 600 implantations have been performed. It is one of the most commonly used keratoprosthesis in the United States. The keratoprosthesis is made of clear plastic with excellent tissue tolerance and optical properties. It consists of two parts but when fully assembled, it has the shape of a collar button (see Figure 58). The device is inserted into a corneal graft, which is then sutured into the patient's cloudy cornea. If the natural lens is in place, it is also removed. Finally, your

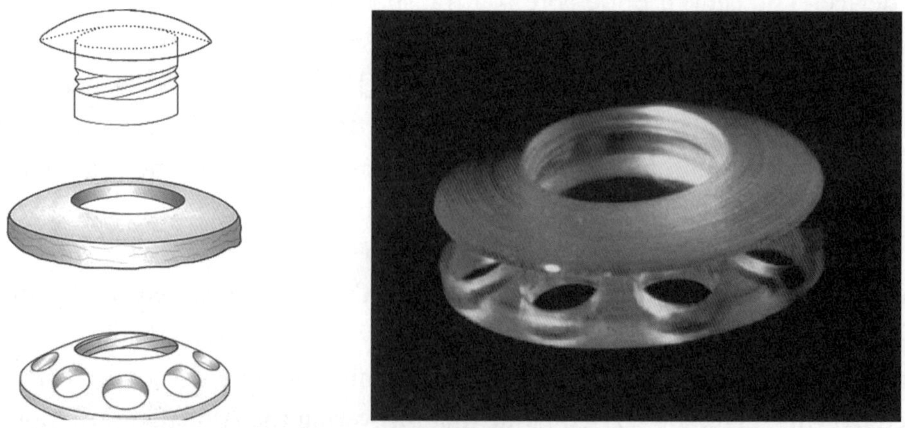

FIGURE 58: Boston keratoprosthesis consisting of three sections.

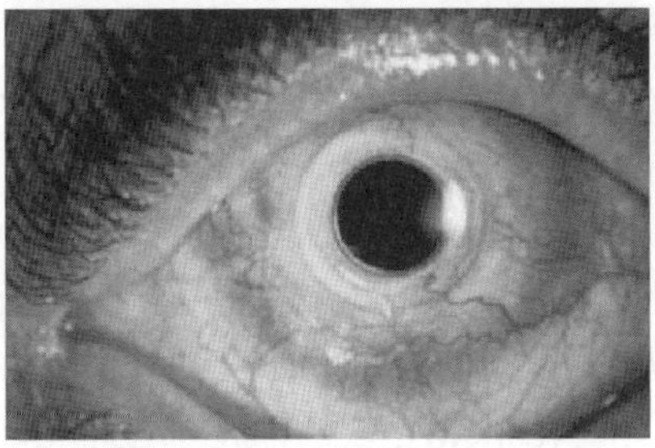

FIGURE 59: Boston keratoprosthesis 3 months after implantation.

physician may recommend that a soft contact lens may be applied to the surface. The Boston keratoprosthesis is shown implanted in a patient 3 months postsurgery in Figure 59.

Postoperative medication might be fluoroquinolone and vancomycin drops (14 mg/mL) once or twice daily. Prednisolone acetate 1% (a steroid) may also be used if needed. If pressure problems appear, steroids may be eliminated but the antibiotics should be used for life, on a daily basis. Soft contact lens should be worn around the clock on a long-term basis. The lens is very protective of the corneal tissue hydration.

Tissue-based artificial corneas consist of the following versions: 1) an artificial scaffold in which cells and nerves grow to form a natural corneal replacement, or 2) a tissue-engineered replacement that can replace a damaged cornea with a true tissue analog that will again grow into a finalized complete cornea. Scientists from the University of Ottawa created an artificial cornea that is grown around a "scaffold" of plastic and protein implanted into the eye. It regenerates the cells necessary to make a fully functioning cornea within a matter of weeks. So far the scaffold has been successfully tested only in pigs with corneal damage. Pigs given traditional corneal transplants showed no nerve regeneration in the weeks following transplantation. They reported, in the journal *Proceedings of the National Academy of Sciences*, that other corneal substitutes have been produced and tested, but that their implantable matrix performs as a physiologically functional tissue substitute and not simply as a prosthetic device (Griffith *et al.* 2002).

Another alternative is the true tissue-engineered replacement cornea. Human eye tissue has been grown in a laboratory and then successfully transplanted into patients. All the 14 patients in the University of California study had badly affected eyesight, but in 10 cases following treatment, vision was either restored or significantly improved. The scientists used the latest "bioengineering" techniques to grow corneal cells in the laboratory from a very small number supplied either by the patient, if there was one good eye, or by a related donor. Similar techniques are frequently used to "grow" new skin to lay over severe burns. The eye cells, called corneal stem cells, are naturally found just underneath the outside of cornea itself, which protects the delicate eye from damage. The stem cells mature into adult corneal cells that are needed to replace ageing corneal cells or repair injuries. Certain corneal injuries, such as burns—from fire, radiation, or chemicals—and some rare diseases and tumors can destroy the patient's stem cells. This means that the eye is no longer able to repair itself, and an accumulation of slight injuries destroys the eyesight. In these cases, a traditional corneal transplant, in which the outermost layer is taken from an organ donor, is simply not enough, as not enough stem cells are carried to replace those lost. The technique, perfected at the University of California Davis School of Medicine and Medical Center, harvests just a few stem cells under local anesthetic. These are converted into films just one cell thick in laboratory dishes, and placed on to a sterile membrane that gives them a framework on which to grow. A thicker layer is eventually produced, which is tough enough to be transplanted. The stem cells then mature into adult corneal cells so that sight can be restored. This work was initially reported by Schwab and Isserhoff (2000) followed by Han *et al.* (2002), and Schwab *et al.* (2000). Obviously, as was the case for the University of Ottowa study, the field of tissue and/or cellular based artificial corneas is still in its infancy. This is, however, the holy grail of corneal transplant, as is the case for organ transplant for other purposes—the ability to replace a failing organ with a replacement that looks, acts, and feels like the original.

REFERENCES

Aarts NL, Caffee CS. Manufacturer predicted and measured REAR values in adult hearing aid fitting: accuracy and clinical usefulness. *Int J Audiol* 2005 May;44(5):293–301.

Alio JL, Mulet ME, Gutierrez R, Galal A. Artisan toric phakic intraocular lens for correction of astigmatism. *J Refract Surg* 2005 Jul–Aug;21(4):324–31.

Beadle EA, McKinley DJ, Nikolopoulos TP, Brough J, O'Donoghue GM, Archbold SM. Long-term functional outcomes and academic-occupational status in implanted children after 10 to 14 years of cochlear implant use. *Otol Neurotol* 2005 Nov;26(6):1152–60.

Benedetti S, Casamenti V, Marcaccio L, Brogioni C, Assetto V. Correction of myopia of 7 to 24 diopters with the Artisan phakic intraocular lens: two-year follow-up. *J Refract Surg* 2005 Mar–Apr;21(2):116–26.

Bentler RA. Effectiveness of directional microphones and noise reduction schemes in hearing aids: a systematic review of the evidence. *J Am Acad Audiol* 2005 Jul–Aug;16(7): 473–84.

Berg AL, Herb A, Hurst M. Cochlear implants in children: ethics, informed consent, and parental decision making. *J Clin Ethics* 2005 Fall;16(3):239–50.

Biernath KR, Reefhuis J, Whitney CG, Mann EA, Costa P, Eichwald J, Boyle C. Bacterial meningitis among children with cochlear implants beyond 24 months after implantation. *Pediatrics* 2006 Jan 3.

Cakmak SS, Caca I, Unlu MK, Cakmak A, Olmez G, Sakalar YB. Surgical technique and postoperative complications in congenital cataract surgery. *Med Sci Monit* 2005 Dec 19;12(1):CR31–35.

Clark G, Shepherd R, Patrick J, Black R, Tong Y. Design and fabrication of the banded electrode array. *Ann NY Acad Sci* 1983;405:191–201.

Cohen-Mansfield J, Infeld DL. Hearing aids for nursing home residents: current policy and future needs. *Health Policy* 2005 Dec 30.

Collins JF, Gaster RN, Krol WF, VA Cooperative Cataract Study Group. Outcomes in patients having vitreous presentation during cataract surgery who lack capsular support for a nonsutured PC IOL. *Am J Ophthalmol* 2006 Jan;141(1):71–78.

Fabry D. Creating the evidence: lessons from cochlear implants. *J Am Acad Audiol* 2005 Jul–Aug;16(7):515–22.

Ferlito A, Arnold W, Rinaldo A, Niedermeyer HP, Bozorg Grayeli A, Devaney KO, McKenna MJ, Anniko M, Pulec JL, McCabe BF, van den Broek P, Shea JJ Jr. Viruses and otosclerosis: chance association or true causal link? *Acta Otolaryngol* 2003 Aug;123(6):741–6.

Flipsen P Jr, Colvard LG. Intelligibility of conversational speech produced by children with cochlear implants. *J Commun Disord* 2005 Dec 20.

Frazer RQ, Byron RT, Osborne PB, West KP. PMMA: an essential material in medicine and dentistry. *J Long Term Eff Med Implants* 2005;15(6):629–39.

Fu QJ, Chinchilla S, Nogaki G, Galvin JJ 3rd. Voice gender identification by cochlear implant users: the role of spectral and temporal resolution. *J Acoust Soc Am* 2005 Sep;118(3 Pt 1):1711–8.

Gordon KA, Tanaka S, Papsin BC. Atypical cortical responses underlie poor speech perception in children using cochlear implants. *Neuroreport* 2005 Dec 19;16(18):2041–5.

Griffith M, Hakim M, Shimmura S, Watsky MA, Li F, Carlsson D, Doillon CJ, Nakamura M, Suuronen E, Shinozaki N, Nakata K, Sheardown H. Artificial human corneas: scaffolds for transplantation and host regeneration. *Cornea* 2002 Oct;21(7 Suppl):S54–61.

Gustav Mueller H. Fitting hearing aids to adults using prescriptive methods: an evidence-based review of effectiveness. *J Am Acad Audiol* 2005 Jul–Aug;16(7):448–60.

Gustav Mueller H, Bentler RA. Fitting hearing aids using clinical measures of loudness discomfort levels: an evidence-based review of effectiveness. *J Am Acad Audiol* 2005 Jul–Aug;16(7):461–72.

Hachmeister JE. An abbreviated history of the ear: from Renaissance to present. *Yale J Biol Med* 2003 Mar 1;76(2):81–6.

Haensel J, Ilgner J, Chen YS, Thuermer C, Westhofen M. Speech perception in elderly patients following cochlear implantation. *Acta Otolaryngol* 2005 Dec;125(12):1272–6.

Han B, Schwab IR, Madsen TK, Isseroff RR. A fibrin-based bioengineered ocular surface with human corneal epithelial stem cells. *Cornea* 2002 Jul;21(5):505–10.

Hay-McCutcheon MJ, Pisoni DB, Kirk KI. Audiovisual speech perception in elderly cochlear implant recipients. *Laryngoscope* 2005 Oct;115(10):1887–94.

Heermann H. Johannes Kessel and the history of endaural surgery. *Arch Otolaryngol* 1969 Nov;90(5):652–8.

Henkin Y, Kaplan-Neeman R, Kronenberg J, Migirov L, Hildesheimer M, Muchnik C. Electrical stimulation levels and electrode impedance values in children using the Med-El Combi 40+ cochlear implant: a one year follow-up. *J Basic Clin Physiol Pharmacol* 2005;16(2–3):127–37.

Higgins MB, McCleary EA, Ide-Helvie DL, Carney AE. Speech and voice physiology of children who are hard of hearing. *Ear Hear* 2005 Dec;26(6):546–58.

Hochmair-Desoyer I, Hochmier E, Burian K. Design and fabrication of multiwire scala tympani electrodes. *Ann NY Acad Sci* 1983; 405:173–82.

Hong RS, Rubinstein JT. Conditioning pulse trains in cochlear implants: effects on loudness growth. *Otol Neurotol* 2006 Jan;27(1):50–56.

House HP, Wullstein H, Shea JJ Jr, Derlacki EL, Schuknecht H, Foller EP Jr, Juers Al. Techniques of stapes mobilization. *Arch Otolaryngol* 1960 Feb;71:338–53.

House W, Urban J. Long term results of electrode implantation and electronic stimulation of the cochlea in man. *Ann Otol Rhinol Laryngol* 1973;82:504–17.

Hyde M, Power D. Some ethical dimensions of cochlear implantation for deaf children and their families. *J Deaf Stud Deaf Educ* 2006 Winter;11(1):102–11.

Karosi T, Konya J, Petko M, Sziklai I. Histologic otosclerosis is associated with the presence of measles virus in the stapes footplate. *Otol Neurotol* 2005 Nov;26(6):1128–33.

Karosi T, Konya J, Szabo LZ, Sziklai I. Measles virus prevalence in otosclerotic stapes footplate samples. *Otol Neurotol* 2004 Jul;25(4):451–6.

Killion MC, Gudmundsen GI. Fitting hearing aids using clinical prefitting speech measures: an evidence-based review. *J Am Acad Audiol* 2005 Jul–Aug;16(7):439–47.

Laneau J, Wouters J, Moonen M. Improved music perception with explicit pitch coding in cochlear implants. *Audiol Neurootol* 2006 Jan–Feb;11(1):38–52.

Leccisotti A, Fields SV. Clinical results of ZSAL-4 angle-supported phakic intraocular lenses in 190 myopic eyes. *J Cataract Refract Surg* 2005 Feb;31(2):318–23.

Lempert J, Meltzer PE, Rambo JH, Wever EG. The effects of injury to the lateral semicircular canal. *Trans Am Acad Ophthalmol Otolaryngol* 1956 Sep–Oct;60(5):718–27.

Lewis MS, Valente M, Horn JE, Crandell C. The effect of hearing aids and frequency modulation technology on results from the communication profile for the hearing impaired. *J Am Acad Audiol* 2005 Apr;16(4):250–61.

Lifshitz T, Levy J, Aizenman I, Klemperer I, Levinger S. Artisan phakic intraocular lens for correcting high myopia. *Int Ophthalmol* 2004 Jul;25(4):233–8. Epub 2005 Sep 29.

Loizou PC. Introduction to cochlear implants. *IEEE Signal Processing Mag* 1998 Sep:101–130.

Lovisolo CF, Reinstein DZ. Phakic intraocular lenses. *Surv Ophthalmol* 2005 Nov–Dec;50(6):549–87.

Martin RG, Sanders DR. A comparison of higher order aberrations following implantation of four foldable intraocular lens designs. *J Refract Surg* 2005 Nov–Dec;21(6):716–21.

Merzenich M, White M. Cochlear implant – the interface problem, in *Functional Electrical Simulation: Applications in Neural Protheses* (F. Hambrecht and J. Reswick eds.). New York: Marcel Dekker, 1977, pp. 321–340.

Moore BC, Marriage J, Alcantara J, Glasberg BR. Comparison of two adaptive procedures for fitting a multi-channel compression hearing aid. *Int J Audiol* 2005 Jun;44(6):345–57.

Munson B, Nelson PB. Phonetic identification in quiet and in noise by listeners with cochlear implants. *J Acoust Soc Am* 2005 Oct;118(4):2607–17.

Myers EN, Ishiyama E, Heisse JW Jr. Histology of a successful stapes mobilization. *Ann Otol Rhinol Laryngol* 1970 Apr;79(2):321–30.

Myers EN, Myers D. Stapedectomy in advanced otosclerosis: a temporal bone report. *J Laryngol Otol* 1968 Jun;82(6):557–64.

Niedermeyer HP, Arnold W, Schuster M, Baumann C, Kramer J, Neubert WJ, Sedlmeier R. Persistent measles virus infection and otosclerosis. *Ann Otol Rhinol Laryngol* 2001 Oct;110(10):897–903.

Oshika T, Kawana K, Hiraoka T, Kaji Y, Kiuchi T. Ocular higher-order wavefront aberration caused by major tilting of intraocular lens. *Am J Ophthalmol* 2005 Oct;140(4):744–6.

Parsons C, Jones DS, Gorman SP. The intraocular lens: challenges in the prevention and therapy of infectious endophthalmitis and posterior capsular opacification. *Expert Rev Med Devices* 2005 Mar;2(2):161–73.

Perkins RC. Laser stepedotomy for otosclerosis. *Laryngoscope* 1980 Feb;90(2):228–40.

Reber MB, Kompis M. Acclimatization in first-time hearing aid users using three different fitting protocols. *Auris Nasus Larynx* 2005 Dec;32(4):345–51.

Reese JL, Hnath-Chisolm T. Recognition of hearing aid orientation content by first-time users. *Am J Audiol* 2005 Jun;14(1):94–104.

Ricketts TA, Hornsby BW. Sound quality measures for speech in noise through a commercial hearing aid implementing digital noise reduction. *J Am Acad Audiol* 2005 May;16(5):270–7.

Schuknecht HF, Frederick L. Jack (1861–1951). *Arch Otolaryngol* 1968 Mar;87(3):328–32.

Schwab IR, Isseroff RR. Bioengineered corneas – the promise and the challenge. *N Engl J Med.* 2000 Jul 13;343(2):136–8.

Schwab IR, Reyes M, Isseroff RR. Successful transplantation of bioengineered tissue replacements in patients with ocular surface disease. *Am J Ophthalmol* 2000 Oct;130(4):543–4.

Shampo MA, Kyle RA. Georg von Bekesy – audiology and the cochlea. *Mayo Clin Proc* 1993 Jul;68(7):706.

Shea JJ Jr. A personal history of stapedectomy. *Am J Otol* 1998 Sep;19(5 Suppl):S2–12.

Shea MC. Stapes fixation in chronic middle ear disease. *Laryngoscope* 1976 Feb;86(2):230–2.

Smith AF, Lafuma A, Berdeaux G, Berto P, Brueggenjuergen B, Magaz S, Auffarth GK, Brezin A, Caporossi A, Mendicute J. Cost-effectiveness analysis of PMMA, silicone, or acrylic intra-ocular lenses in cataract surgery in four European countries. *Ophthalmic Epidemiol* 2005 Oct;12(5):343–51.

Sridhar MS, Majji AB, Vaddavalli PK. Severe inflammation following iris fixated anterior chamber phakic intraocular lens for myopia. *Eye* 2005 Oct 21.

Stickney GS, Loizou PC, Mishra LN, Assmann PF, Shannon RV, Opie JM. Effects of electrode design and configuration on channel interactions. *Hear Res* 2005 Dec 7.

Stickney GS, Nie K, Zeng FG. Contribution of frequency modulation to speech recognition in noise. *J Acoust Soc Am* 2005 Oct;118(4):2412–20.

Tassignon MJ, De Groot V, Van Tenten Y. Searching the way out for posterior capsule opacification. *Verh K Acad Geneeskd Belg* 2005;67(4):277–88.

Tognetto D, Sanguinetti G, Ravalico G. Tissue reaction to hydrophilic intraocular lenses. *Expert Rev Med Devices* 2005 Jan;2(1):57–60.

Tomblin JB, Barker BA, Spencer LJ, Zhang X, Gantz BJ. The effect of age at cochlear implant initial stimulation on expressive language growth in infants and toddlers. *J Speech Lang Hear Res* 2005 Aug;48(4):853–67.

Tyler R. Open-set recognition with the 3M/Vienna single-channel cochlear implant. *Arch Otolaryngol Head Neck Surg* 1998;114:1123–6.

Uriarte M, Denzin L, Dunstan A, Sellars J, Hickson L. Measuring hearing aid outcomes using the Satisfaction with Amplification in Daily Life (SADL) questionnaire: Australian data. *J Am Acad Audiol* 2005 Jun;16(6):383–402.

van den Honert C, Stypulkowski P. Single fiber mapping of spatial excitation patterns in the electrically stimulated nerve. *Hear Res* 1987;29:195–206.

van Hooren SA, Anteunis LJ, Valentijn SA, Bosma H, Ponds RW, Jolles J, van Boxtel MP. Does cognitive function in older adults with hearing impairment improve by hearing aid use? *Int J Audiol* 2005 May;44(5):265–71.

Vermeire K, Brokx JP, Wuyts FL, Cochet E, Hofkens A, De Bodt M, Van de Heyning PH. Good speech recognition and quality-of-life scores after cochlear implantation in patients with DFNA9. *Otol Neurotol* 2006 Jan;27(1):44–49.

Vuorialho A, Sorri M, Nuojua I, Muhli A. Changes in hearing aid use over the past 20 years. *Eur Arch Otorhinolaryngol* 2005 Nov 9.

Waltzman SB, Cohen NL, Shapiro WH. The benefits of cochlear implantation in the geriatric population. *Otolaryngol Head Neck Surg* 1993 Apr;108(4):329–33.

Werner L, Hunter B, Stevens S, Chew JJ, Mamalis N. Role of silicon contamination on calcification of hydrophilic acrylic intraocular lenses. *Am J Ophthalmol* 2006 Jan;141(1):35–43.

Wilson B, Lawson D, Zerbi M. Advances in coding strategies for cochlear implants. *Adv Otolaryngol Head Neck Surg* 1995;9:105–29.

Xu L, Zwolan TA, Thompson CS, Pfingst BE. Efficacy of a cochlear implant simultaneous analog stimulation strategy coupled with a monopolar electrode configuration. *Ann Otol Rhinol Laryngol* 2005 Nov;114(11):886–93.

Author Biography

Gerald E. Miller is the Department Head of Biomedical Engineering at Virginia Commonwealth University as well as a Professor of Physiology, Professor of Cardiology, and a professor of Physical Medicine & Rehabilitation. The Biomedical Engineering Department is administered through the School of Engineering, but is situated within the Medical College of Virginia campus of the university, which contains the nations fourth largest academic medical complex. The Biomedical Engineering Department maintains graduate degrees leading to the M.S. and Ph.D. with 9 primary and 60 affiliate faculty and 65 graduate students as well as Virginia's first undergraduate BME program with an enrollment of 180. Dr. Miller received the B.S. in Aerospace Engineering from the Pennsylvania State University in 1971, the M.S. in Bioengineering from the Pennsylvania State University in 1975, and the Ph.D. in Bioengineering from the Pennsylvania State University in 1978. He is a member of Phi Kappa Phi as well as a Fellow of both ASME and AIMBE. His research activities include Rehabilitation Engineering, Physiological Fluid Mechanics, Artificial Internal Organs, Epilepsy Genesis, and the use of physiological signals in the control of mechanical systems. He has developed a multiple disk, centrifugal artificial ventricle, an early warning detection system for epileptic seizures, and has developed a protocol for prevention and treatment of decubitus ulcers. He has also developed a real time, hands free, PC based voice controller with infrared activated grippers for robots which can be easily implemented by disabled individuals. He has been an invited speaker at several international conferences in the area of fluid mechanics, artificial organs, and man-machine interfaces.

Artificial Organs

Gerald E. Miller
Virginia Commonwealth University

MORGAN & CLAYPOOL PUBLISHERS

ABSTRACT

The replacement or augmentation of failing human organs with artificial devices and systems has been an important element in health care for several decades. Such devices as kidney dialysis to augment failing kidneys, artificial heart valves to replace failing human valves, cardiac pacemakers to reestablish normal cardiac rhythm, and heart assist devices to augment a weakened human heart have assisted millions of patients in the previous 50 years and offers lifesaving technology for tens of thousands of patients each year. Significant advances in these biomedical technologies have continually occurred during this period, saving numerous lives with cutting edge technologies. Each of these artificial organ systems will be described in detail in separate sections of this lecture.

KEYWORDS

Artificial heart and ventricular assist, Artificial heart valve, Cardiac pacemaker, Dialysis, Human heart and heart surgery, Human kidney

Contents

CHAPTER 1

Artificial Heart Valves

The replacement or augmentation of failing human organs with artificial devices and systems has been an important element in health care for several decades. Such devices as kidney dialysis to augment failing kidneys, artificial heart valves to replace failing human valves, cardiac pacemakers to reestablish normal cardiac rhythm, and heart assist devices to augment a weakened human heart have assisted millions of patients in the previous 50 years and offers lifesaving technology for tens of thousands of patients each year. Significant advances in these biomedical technologies have continually occurred during this period, saving numerous lives with cutting edge technologies. Each of these artificial organ systems will be described in detail in separate sections below.

1.1 CARDIAC ANATOMY AND PATHOPHYSIOLOGY

The human heart consists of two pumping chambers (ventricles) and two filling chambers (atria). The heart is a pulsatile pump, operating via muscular contraction of both the ventricles and atria, and is designed to produce positive displacement of blood through two circulatory systems. The right ventricle pumps blood into the pulmonary circulation, where blood becomes oxygenated and the left ventricle pumps into the systemic circulation that allows oxygenated blood to reach tissues throughout the body, where oxygen is transported to these tissues.

In order for blood to flow in the proper direction from each of the ventricles as well as from each atrium to the associated ventricle, the heart contains heart valves to prevent backflow of blood. The atrioventricular valves (between the atria and the ventricles) prevent backflow into the atria when the ventricles contract during systole. The outflow valves (from each ventricle) are designed to prevent backflow during ventricular diastole. The human heart valves are shown in Figure 1, with the bileaflet mitral valve located between the left atrium and the left ventricle and the trileaflet aortic valve located between the left ventricle and the aorta (outflow tract). The trileaflet tricuspid valve is located between the right atrium and the right ventricle, while the trileaflet pulmonary valve is located between the right ventricle and the pulmonary artery. Each of these valves consists of an annulus from which valve leaflets project into the orifice opening. There are no valves located between the atria and the respective veins feeding blood into the atria. Thus, there is no valve from the vena cava to the right atrium, nor a valve from

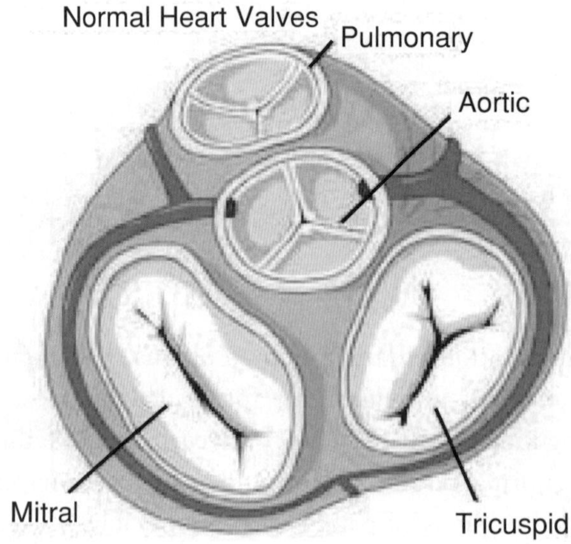

FIGURE 1: Human heart valves.

the pulmonary vein to the left atrium. All heart valves operate by means of a pressure gradient, moving in the direction of decreasing pressure at all times, whether toward the open or closed position.

Human heart valves are designed to close in an overlapping fashion with the valve leaflets abutting over each other in a fashion noted in Figure 2. The leaflets overlap to ensure adequate

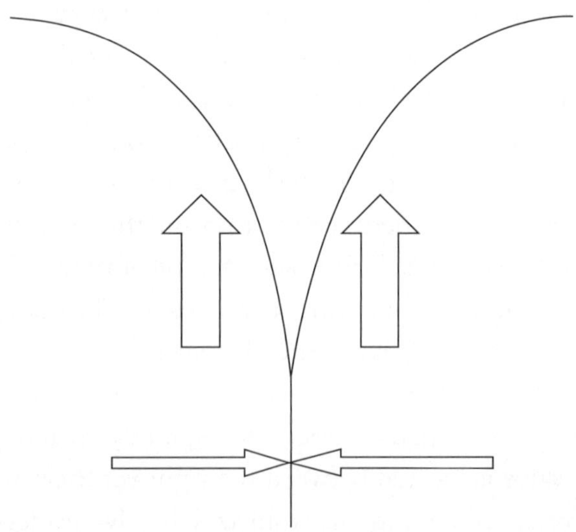

FIGURE 2: Human valve closure with overlapping valve leaflets showing pressure gradient acting upward against closed valve and lateral frictional force created by overlap of leaflets.

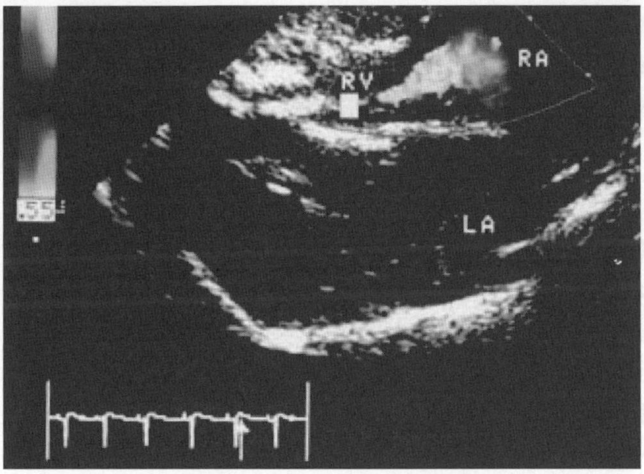

FIGURE 3: Doppler ultrasound image of a heart indicating jetting from the right ventricle toward the right atrium, resulting from reversing flow across the tricuspid valve.

valve closure with the vertical forces created by a pressure gradient across the valve being transferred (in part) to lateral (friction) forces. Thus, each valve leaflet is actually longer than one half of the radius of the valve orifice, which provides sufficient length for the leaflet overlap.

There are two potential sources of heart valve failure, each of which requires years to develop. One such mechanism is valve *stenosis* or narrowing of the valve orifice (best seen with the valve in an open position). Valve stenosis is often related to atherosclerosis and creates a narrowed outflow area. This not only reduces the area for blood flow, but also creates an added pressure drop due to the existence of a contraction in the flow pathway. This situation may be diagnosed with an echocardiogram that uses 2-D ultrasound and a Doppler flow probe. However, symptoms of such a narrowing would not normally appear until the orifice is significantly reduced (up to a 50% reduction in area). With the reduced area, it is possible that the flow across the valve orifice may become turbulent, which can also be visualized with Doppler ultrasound or even with the aid of a stethoscope. A color Doppler profile of a heart is shown in Figure 3.

Another source of valve failure is via weakening of the valve leaflets. In such a case, the leaflets begin to bulge upwards with a portion of the overlapping leaflets protruding above the valve orifice. Such an occurrence is termed valve *prolapse*. In the extreme case, the leaflets bulge so far that the overlap of the valve leaflets during closure is in jeopardy. If a gap appears between the leaflets in the (purportedly) closed position, there is a regurgitation of the blood backwards from its normal intended path (see Figure 3). Since the gap between the leaflets is initially small, the backflow is typically turbulent, which can be easily heard via a stethoscope or imaged via Doppler ultrasound. The sounds heard are called heart murmurs. In advanced heart valve failure from either mechanism, the use of a stethoscope followed by ultrasound (as needed) may indicate a need to replace the failing valve. The mitral valve and the aortic valve (both in the left heart) are the most prevalent valves to be replaced. This is not unusual as the left heart experiences greater pressures than the right heart.

1.2 PROSTHETIC HEART VALVES

As with any implantable device, issues related to biocompatibility, longevity, and function of replacement heart valves are important. Artificial heart valves have been clinically available for over 50 years with early designs consisting of either a caged ball or a tilting disk (within a caged or strut assembly). Early successful designs of each type include the Starr Edwards ball valve and the Bjork Shiley tilting disk valve. The ball valve, as seen in Figure 4, consists of a silastic or silicone rubber ball encased within a stainless steel cage and annular ring. The metallic ring (used for strength) was covered with a Dacron sewing ring. All materials are biocompatible with ample animal and clinical trials in support of these choices. The sewing ring allows attachment of the valve to the surrounding tissue of the original valve annulus.

The ball valve had an important advantage, at the time, in that it is structurally strong and durable. However, it closes with considerable force that can cause hemolysis and thrombosis, as seen in Figure 5. In addition, the ball acts as a central occluder with the valve in the open position, thus producing a greater pressure drop across the open valve that reduces the downstream

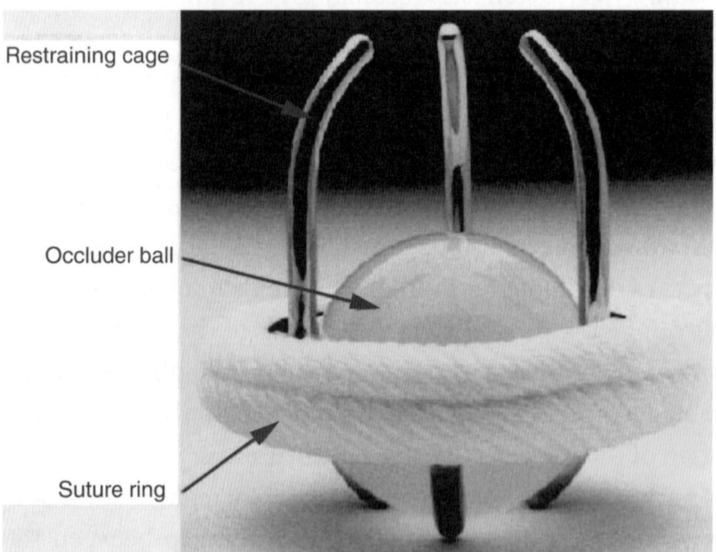

FIGURE 4: Prosthetic ball valve in the closed position showing with the ball sitting in the dacron-covered annular ring and the upper cage protecting the ball in the open position.

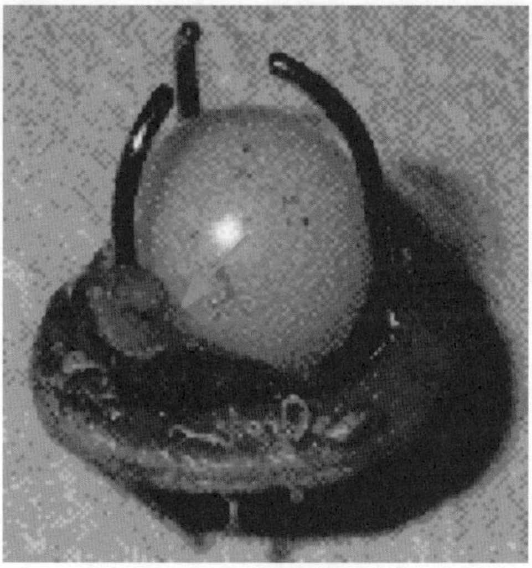

FIGURE 5: Thrombus formation at the seat of a ball valve as a result of the closing force of the ball against its seat.

pressure available to propel blood through a circulation. This type of valve also has a large vertical profile that makes implantation, particularly in the mitral position, somewhat difficult for many patients whose anatomical space is limited.

The tilting disk valve, as seen in Figure 6, opens to 60° as constrained by a stainless steel strut. The disk is constructed of pyrolytic carbon, given the trade name pyrolite.

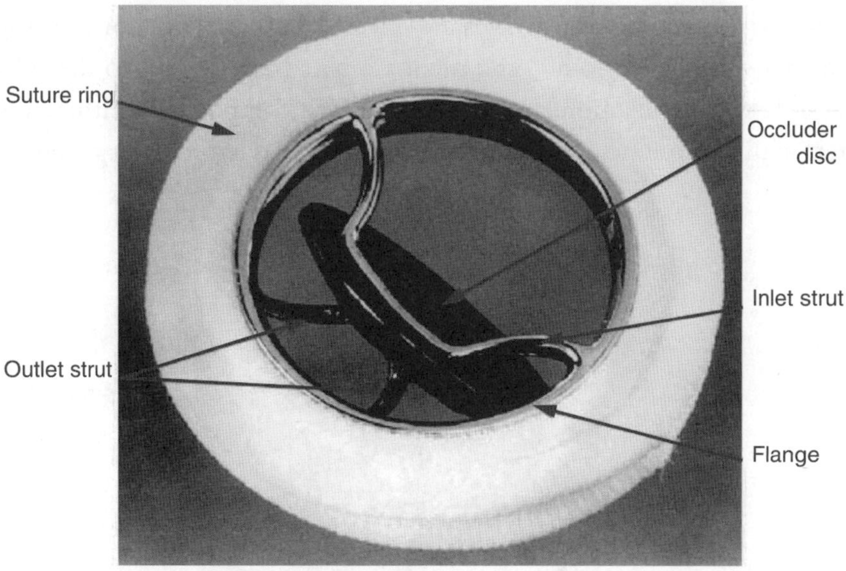

FIGURE 6: Tilting disk valve in the open position.

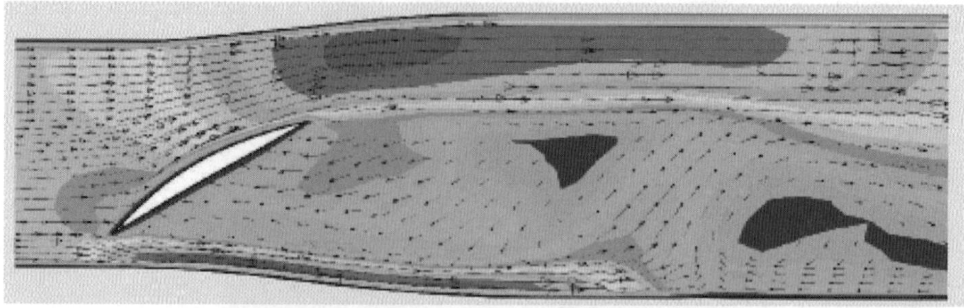

FIGURE 7: Flow patterns around a tilting disk valve generated by computational methods. Note the reversal of the velocity vector behind the tilted disk and the higher velocities at the periphery.

The tilting disk valve has an open profile that offers blood flow around the tilted disk as well as to the outer sides, which are between the disk and the blood chamber or artery. As such, it is far less of a central occluder to blood flow than the ball valve and has a smaller pressure drop across the open valve, which is also advantageous. This valve only opens to 60° due to the stainless steel strut assembly. The valve is not actually affixed to the struts. A flow pattern across the open valve is shown in Figure 7, which indicates a partial central occluder (the tilted disk) as well as peripheral flow around the disk. There is a bit of recirculation behind the disk and the peripheral flow is at relatively high velocities as compared to the central region. As is the case with most flow images, the red (or dark gray) colors in the image refer to higher velocities, the yellow (or lighter gray) refer to moderate velocities, the green (or lighter gray) refer to small velocities, and the blue (or very light gray) refer to extremely small (near zero) velocities. Arrows within the flow field indicate individual vector lines of fluid particles within the flow field.

Many modern prosthetic heart valves are bileaflet in nature as shown in Figure 8.

The bileaflet valves open to 80° and pivot via pins that connect to the valve leaflets and protrude into slots within the annular ring. The ring is composed of either stainless steel or titanium, often with a pyrolytic carbon coating. There is a dacron sewing ring attached to the

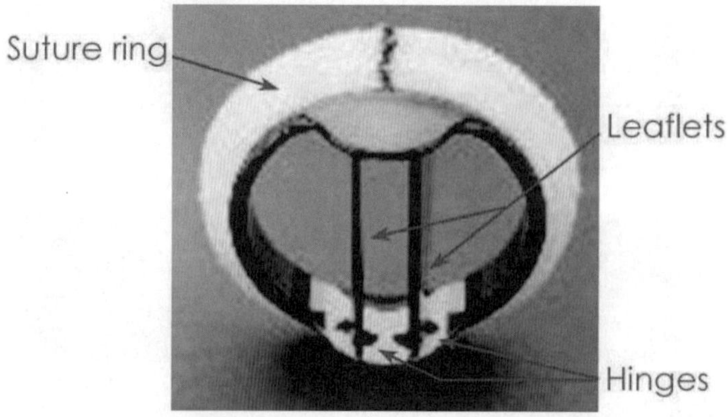

FIGURE 8: Bileaflet valve showing flanges where leaflet pins pivot within annular ring.

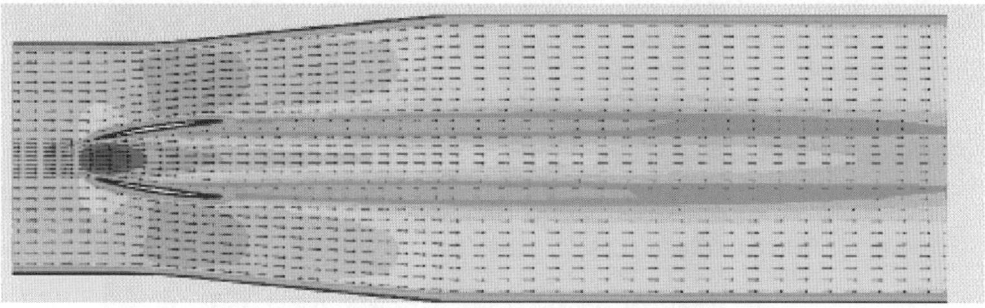

FIGURE 9: Flow through an open bileaflet valve with a large central flow region as well as a large peripheral flow region. Trailing edge vortices are seen emanating from the tips of the leaflet edges.

metal substrate for ease in attachment to the valve orifice. The leaflets are often composed of pyrolytic carbon for strength and durability. With two leaflets, the closing force is less than that of a single tilting disk valve, and certainly far less than that of a ball valve. As the pin assembly can be a site for thrombus formation, such valves normally have a washout zone near the pins to flush that area and avoid stagnant blood leading to thrombus formation. As such, this valve has mild regurgitation as a design element. Figure 9 depicts flow across an open bileaflet valve, indicating far more central flow than the tilting (monoleaflet) valve. There is no central occluder in the bileaflet valve as there was (partially) with the tilting (monoleaflet valve). Thus, there is a greater central flow region in this type of valve.

Ideally, prosthetic heart valves require low levels of hemolysis (red cell destruction), low levels of thrombosis, and a long life span (20–30 years). Most patients must receive anticoagulants indefinitely to deter thrombosis. Pyrolytic carbon and silicon rubber as valve or ball materials have been shown to be relatively antithrombogenic and their surfaces are often covered with proteins over time. Dacron polyester is sometimes used for valve leaflets. Such a material is more flexible than the pyrolite leaflets and thus closes with far less force and less hemolysis. However, such flexible leaflets also take longer to open and close, as they flex during the leaflet movement. An undue delay in valve closure produces a slight backflow, not unlike that of a natural human valve, which produces the incisura (dip) in the aortic pressure waveform. However, too much delay can adversely affect net forward flow, thus reducing net cardiac output.

Bioprosthetic valves, often a preferred alternative, are composed of natural valves from a pig, which has a cardiovascular system most similar to humans. These valves must be "treated" with glutaraldehyde to resist antigenicity or rejection of a foreign substance. As such, these biological valves are stiffer than natural heart valves due to the treatment process. In some cases, the bioprosthetic valve has a dacron sewing ring attached along its base and may even have a stiffening ring of stainless steel attached along its base as well. In some cases, dacron is attached along the outer face of the leaflets to provide greater strength to the leaflets. This does not

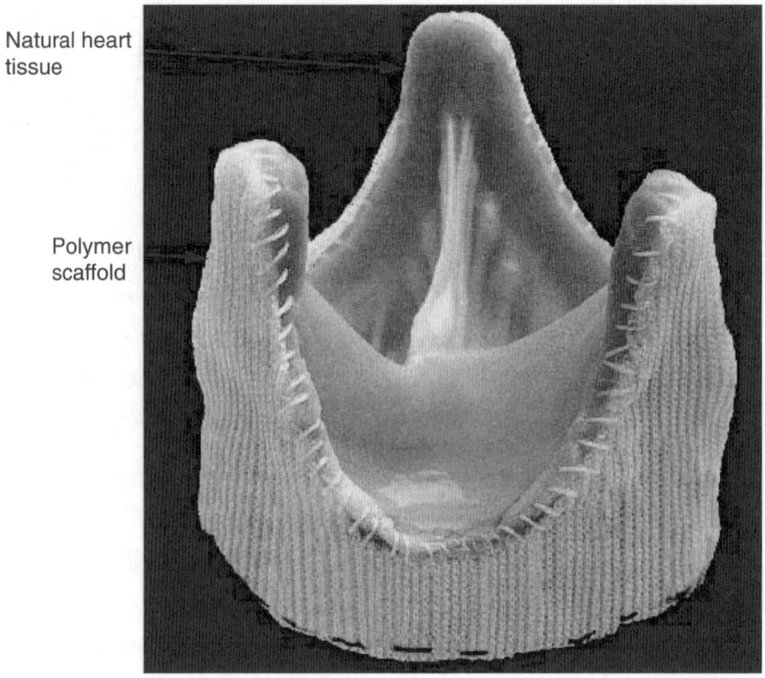

FIGURE 10: Bioprosthetic (pig) valve with dacron along the base and outer edges of the leaflets.

significantly add to the stiffness of the leaflets as compared to the glutaraldehyde treatment. Figure 10 depicts a bioprosthetic valve with a dacron covering.

Surgical implantation of prosthetic heart valves has evolved into an efficient and rapid procedure. The orifice diameter of the patient is evaluated with the aid of pulse echo ultrasound,

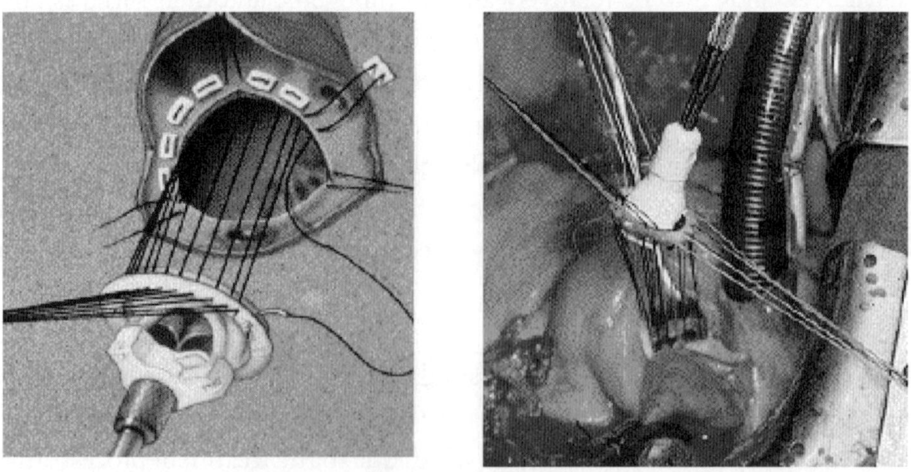

FIGURE 11: Prosthetic heart valve within valve holder with sutures preplaced within sewing ring and valve orifice.

the 2-D version often called echocardiography. The appropriately sized prosthetic valve is then placed into a valve holder and sutures are threaded through the sewing ring of the valve. The patient is placed on cardiopulmonary bypass (heart–lung machine) and the original valve is removed. The prosthetic valve is then placed into the orifice by one surgeon and another continually sews the sutures around the valve orifice using the curved needle attached to the sutures. The valve holder is then lowered into place, and the holder eventually depressed to release the valve—and the empty holder is then removed. The orifice is checked for potential leaks and additional sutures are placed on site as needed. The patient is then removed from bypass and the incision site closed as is the chest incision. The entire procedure often is completed within 1 h and the patient is normally on cardiopulmonary bypass for as little as 15 min. Figure 11 depicts a prosthetic heart valve within a valve holder with sutures in place.

1.3 EVALUATION OF PROSTHETIC VALVES

The performance and efficacy of prosthetic heart valves are normally evaluated by means of pressure gradients across the open valve, long-term durability, and prevalence of adverse physiological conditions such as calcification, stenosis, thrombosis, or hemolysis. Pressure gradients can be determined via direct pressure sensors or via Doppler ultrasound (indirect determination by means of flow/pressure relationship). Thrombosis, hemolysis, and calcification are determined via direct biochemical measures or via imaging modalities such as ultrasound. Such determinations are on mechanical test beds or blood flow loops in mock circulatory systems, or with animal models. Computer modeling techniques are also employed to evaluate valve performance using computational fluid dynamics. An excellent review of the fluid mechanics of prosthetic heart valves was produced by Yoganathan *et al.* (2004). Other groups involved with the evaluation of prosthetic and bioprosthetic heart valves include Labrosse *et al.* (2005), Jamieson *et al.* (2005), Medart *et al.* (2005), Pierrakos *et al.* (2004), Malouf *et al.* (2005), Goetze *et al.* (2004), Wu C *et al.* (2004), and Wu Y *et al.* (2004).

CHAPTER 2

Artificial Heart and Cardiac Assist Devices

2.1 CARDIAC ANATOMY AND PATHOPHYSIOLOGY

As was described above in the section dealing with heart valves, the human heart consists of two ventricles responsible for ejection of blood and two atria responsible for holding and filling the ventricles. The left heart pumps blood into the systemic circulation and the right heart into the pulmonary circulation. Figure 12 depicts the heart with all four chambers.

According to the National Center for Chronic Disease Prevention, 950 000 Americans die each year due to cardiovascular disease with over 61 million Americans suffering from this disease and over 6 million hospitalizations each year as a result. Major causes of cardiac-related pathologies include atherosclerosis in either the aorta or in coronary vessels such as depicted in Figure 13.

In many cases, coronary artery blockages and/or blockages in the aorta can be alleviated via surgery to remove the blockage, either through mechanical "reaming," via laser ablation, or via balloon catheter expansion. Often, stents are placed in a coronary vessel after removal of the blockage. In those instances where the blockage cannot be removed by such means, more extensive surgery can be performed to bypass the blockage with transplanted veins or via artificial vessels composed of biomedical polymers. This procedure is termed coronary artery bypass grafts or CABGs.

Often after such surgery, the patient's repaired heart is too weak to fully support the blood pumping requirements of the systemic circulation and will require some period of reduced load in order to recover. In other cases, the repaired heart cannot ever function at a normal level of 5 L/min at 95% oxygen saturation with a mean pressure of 100 mmHg. In still other cases, the heart cannot be adequately repaired and functions at a reduced level on a permanent basis. One approach to the latter cases would be to seek a heart transplant. However, given the vast number of potential cases in need of such a transplant and the inherent need for appropriate blood typing and tissue typing, there are insufficient numbers of donor hearts. The United Network for Organ Sharing notes that donor hearts number 2000–3000 per year with the stated need at least tenfold greater.

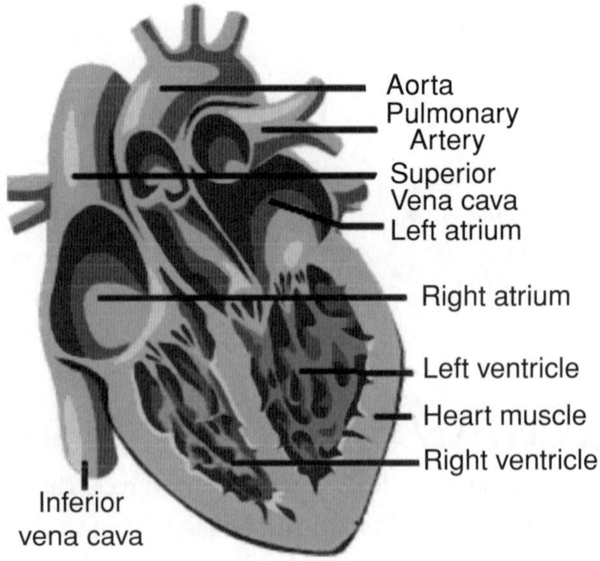

FIGURE 12: The human heart with four chambers: two atria and two ventricles.

As such, a stopgap mechanism is to employ a mechanical assist pumping system, called a ventricular assist device or VAD. This can be utilized in true "assist" mode for those patients recovering from open heart surgery and whose heart will recover full function within a few days. In those other cases where the patient's heart will not completely recover, or where surgical

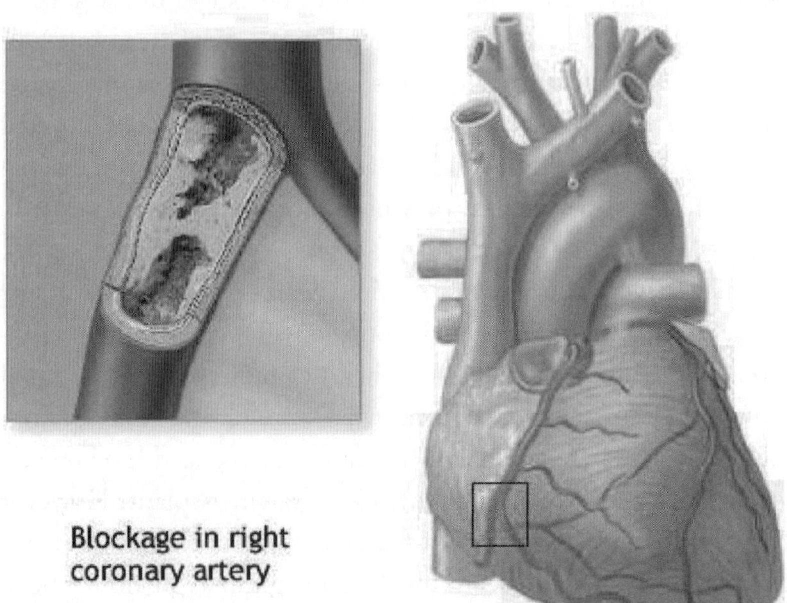

FIGURE 13: Coronary artery blockage with inset showing blockage and larger view showing location within the heart musculature.

intervention would not be beneficial, then this device can be utilized for a longer period until such time as a donor matching heart becomes available. In such cases, the device is termed as a "bridge to transplant." In the first two cases, the natural heart might pump as little as 2–3 L/min with the artificial heart pumping the remainder. When the left heart is being assisted, the device is termed an LVAD (left ventricular assist device). When the right heart is to be assisted, the device is termed an RVAD. For those cases where both hearts are being assisted, two devices are used in biventricular assist mode (BiVAD). In still other cases, when the patient's own heart cannot function even partially at the above level, it may be necessary to completely replace the failing heart with two pumps acting in tandem as a total artificial heart for end-stage heart failure.

2.2 HEART ASSIST TECHNOLOGY

Artificial hearts and heart assist devices have been in development and analysis for decades. The original pneumatically powered design, as typified by the Jarvik-3 and the Jarvik-7 pumps, consisted of a polycarbonate case surrounding a flexible polyurethane sac. The space between the two allowed high pressure air, designated as "systolic drive pressure" to collapse the blood containing sac and caused systolic ejection. A small vacuum pressure, designated as "diastolic pressure" pulled the sac toward the casing and caused diastolic filling. Artificial heart valves were employed for each artificial ventricle to prevent backflow during either portion of the cardiac cycle in a fashion similar to the natural heart. This sac type pump acted in a fashion similar to the natural heart in that it was inflow and inlet pressure sensitive (designated as preload sensitive), but not affected by the load against which the heart pumps (designated as not being afterload sensitive). The Jarvik pump is shown in Figure 14, with the air lines from the outer casing which are then connected to a drive system that provides the alternating high and low pressure to the blood sac. The two pressures, the systolic duration, and beat frequency can all be controlled and varied from a drive console.

Although pneumatically driven pumps are still utilized clinically, many original such designs have now incorporated electrically powered cam drive systems that move a piston/diaphragm arrangement to eject blood during systole and then, when the piston is pulled back toward the drive unit, initiates diastolic filling. Such blood pumps are still pulsatile and thus need valves to eliminate backflow and regurgitation. A biventricular device of this type is shown in Figure 15, with no pneumatic air lines emanating from the pump as was the case for the sac-type design.

Power for the electrically operated LVADs or BiVADs is achieved with external, rechargeable batteries as is shown in Figure 16. Transcutaneous energy transmission occurs via a pair of coils—one placed over the skin and one placed below the skin, the latter attached to the VAD. The VAD is connected to the apex (base) of the left ventricle (or to the base of both ventricles in biventricular support) with the aid of an inlet cannula. This orientation allows gravity to

FIGURE 14: The Jarvik-7 pneumatically driven artificial heart showing two ventricles and the air lines which lead to the spacing between the outer casing and the inner flexible blood sac.

assist emptying of the ventricle into the device, which offloads the ventricle in part. The VAD then pumps a portion of the cardiac output through a cannula that connects to the aorta via an end-to-side anastomosis.

Unlike the pneumatically driven blood pumps, where the air from the pneumatic drive unit is vented to the atmosphere, the electrically driven pumps are normally closed systems.

FIGURE 15: Biventricular electrically operated pusher plate design for a blood pump with four valves similar to that of the natural heart.

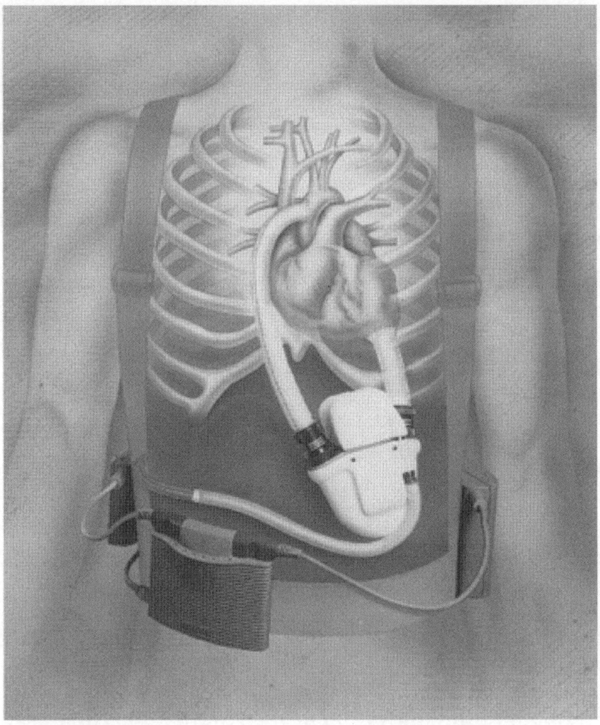

FIGURE 16: Novacor electrically powered LVAD with external battery pack and connecting cannula from apex of left ventricle and to aorta.

As such, when the diaphragm moves back toward the internal drive system, there can be a period where the air within the drive chamber is compressed, resulting in an increase in local pressure. As the diaphragm moves away from the drive unit (during systole), the air in this chamber becomes rarified and the pressure is reduced. Rather than exist with changing pressures within the drive system, early versions of electrically powered pumps contained a flexible sac, a compliance chamber, whereby the excess air in the drive system is shunted to the compliance chamber, thus maintaining a relatively constant pressure within the pump drive chamber. As the diaphragm moves forward during systole, the air is shunted from the compliance chamber into the drive chamber of the pump and vice versa during diastole. Later versions of electrically operated pumps utilize a venting system to avoid the need for a compliance chamber, which is an added component that takes up space and adds to the complications of the device operation.

The use of positive displacement pumps, such as the sac or pusher plate blood pumps, offers several advantages. These include the use of pulsatile flow similar to that of the natural heart as well as an insensitivity to afterload—i.e., the device can pump against any reasonable load, even a hypertensive load. Issues with such pumps include the sensitivity to preload and

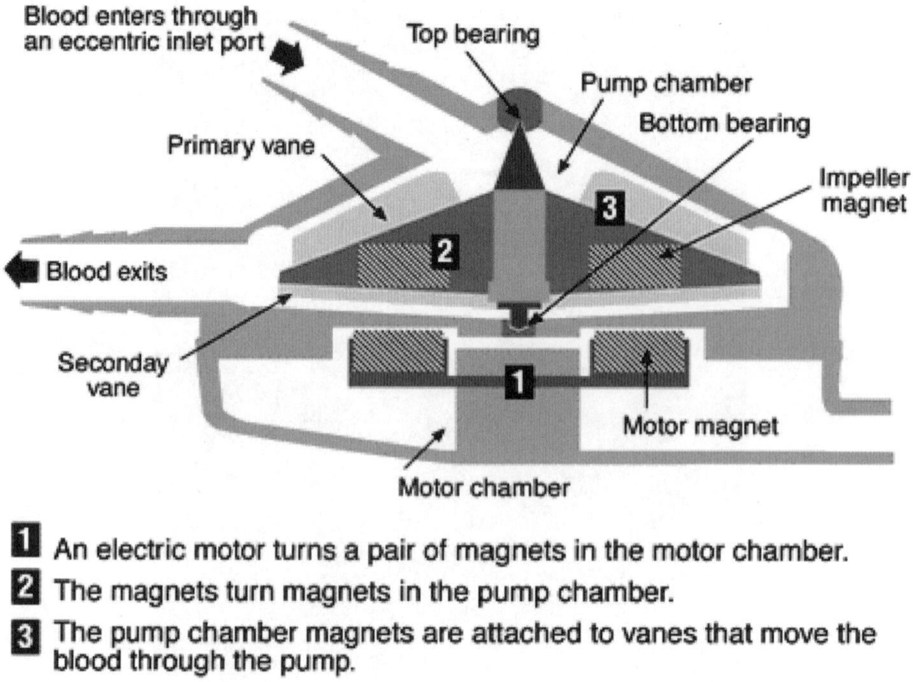

FIGURE 17: Centrifugal blood pump with inlet at top center and outlet at bottom left. Spinning rotor produces increased pressure at the periphery.

problems associated with artificial heart valves such as calcification, hemolysis, thrombosis, and mechanical failure. It was these issues that initiated the later use of rotary blood pumps as assist and/or bridge to transplant devices. Such rotary devices often require less power than their positive displacement counterparts and do not utilize valves, as they produce steady flow. Such rotary blood pumps are configured as either a centrifugal pump or as an axial pump.

The centrifugal blood pump is configured with an inlet to the pump from the top center and an outlet radially at the periphery. This arrangement is not unlike a tornado in flow design and the flow of such a device is similar—low pressure and velocity in the center with higher pressure and velocity at the periphery. Centrifugal pumps are electrically powered and utilize magnets to spin the rotor and, oftentimes, suspend the drive assembly as seen in Figure 17.

Another centrifugal design was produced by Medtronic and consists of nested cones as seen in Figure 18. Still another centrifugal design consists of a series of parallel disks spinning about an axis in unison as seen in Figure 19.

Centrifugal blood pumps have several advantages including a lack of valve-related problems, a relatively lower power drain (than positive displacement pumps), and an ability to pump at higher rates by spinning the rotor faster. Without a displacing drive chamber, there is no need

FIGURE 18: Medtronic BP-80 centrifugal blood pump with nested cones and magnetic drive.

for a compliance chamber as there is for some positive displacement pumps. The centrifugal pumps are placed below the left ventricle in a fashion similar to the positive displacement pumps with the outlet cannula extending toward the aorta as is the case for the pulsatile pump connections. Centrifugal pumps are sized similar to their pulsatile analogs and usually can produce 5–7 L/min with an increase in pressure of 100 mmHg at rotation rates of 1500–3000 rpms. Drawbacks to such pumping systems include the use of steady flow rather than pulsatile flow, as well as being significantly affected by afterload. In fact, it is possible for such pumps to spin, but produce no forward flow, if the outlet pressure generated by the pump is lower than the afterload pressure. This is known as "deadheading" and can be remedied by spinning the rotor faster to increase the outlet pressure of the pump.

The other versions of rotary pumps are the smaller axial flow pumps. Unlike the centrifugal pumps, the axial flow pumps have their inlet forward and their outlet directly behind, much as a jet engine is configured. One advantage of such an arrangement is that the flow is more stable in the flow-through arrangement. In addition, the axial pumps, being smaller than their centrifugal counterparts, can be fitted within the base of the ventricle, thus eliminating the inlet cannula. This type of pump can also be placed within the aorta to allow for flow through the device from the left ventricle, and on toward the remainder of the systemic circulation with an added pressure and flow generated from the pump. As such pumps are smaller and have lesser blood contacting area, they require higher rotation rates than their centrifugal counterparts—often reaching 15 000–25 000 rpms. It was originally thought that such rotation rates would produce harmful shearing stresses to blood elements. However, the extremely short exposure time of blood within the device allows only partial deformation of blood elements without permanent damage occurring. A typical axial flow pump is shown in Figure 20.

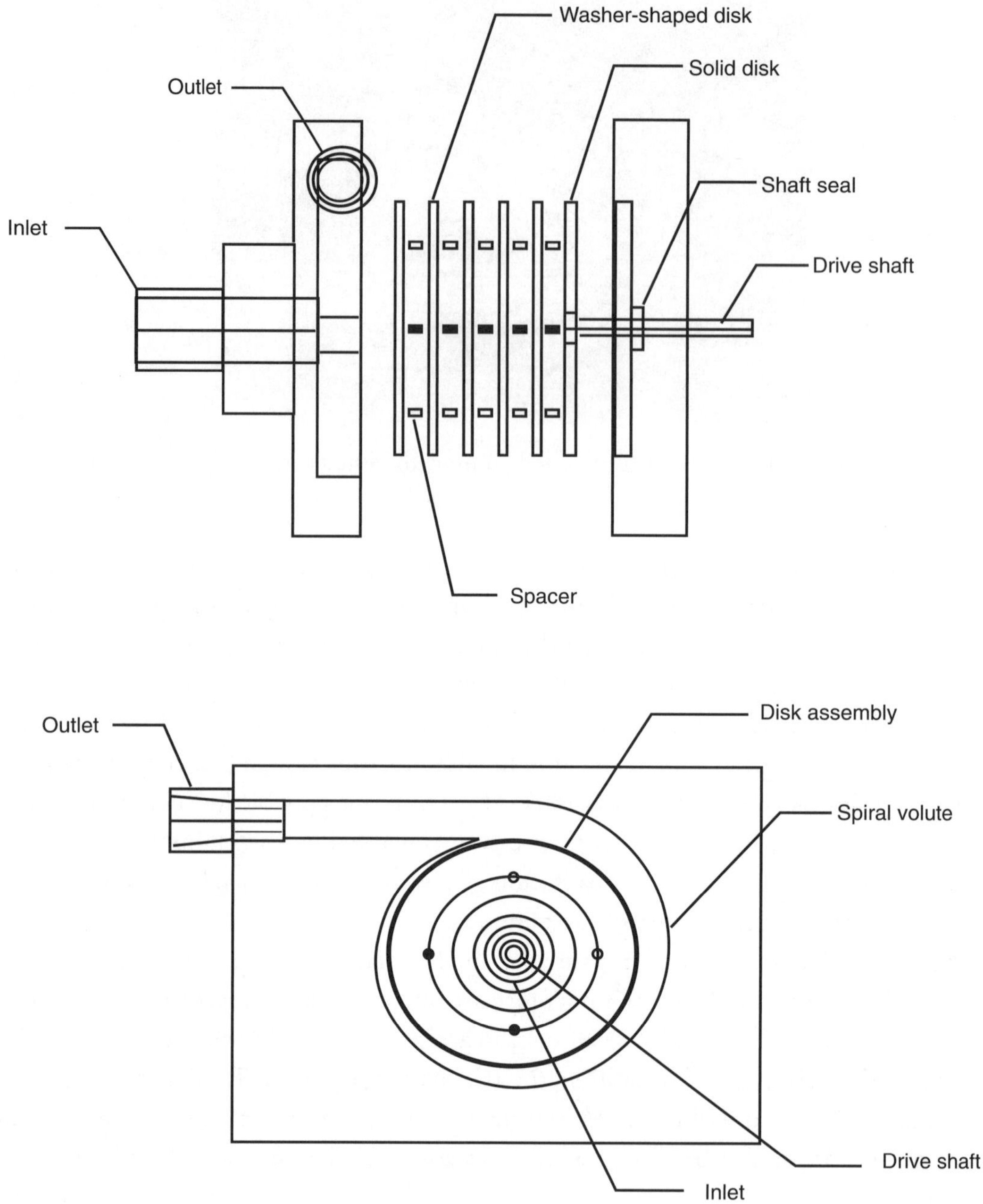

FIGURE 19: Multiple disk centrifugal blood pump based upon the Tesla Turbine design.

The axial flow pump is not only magnetically driven, but requires magnetic suspension. The rotor is magnetically suspended inside the stator, often with blood immersed bearings. A front and rear diffuser are utilized to direct blood flow toward the space between the rotor and the stator. As with their centrifugal pump counterparts, axial flow pumps are preload

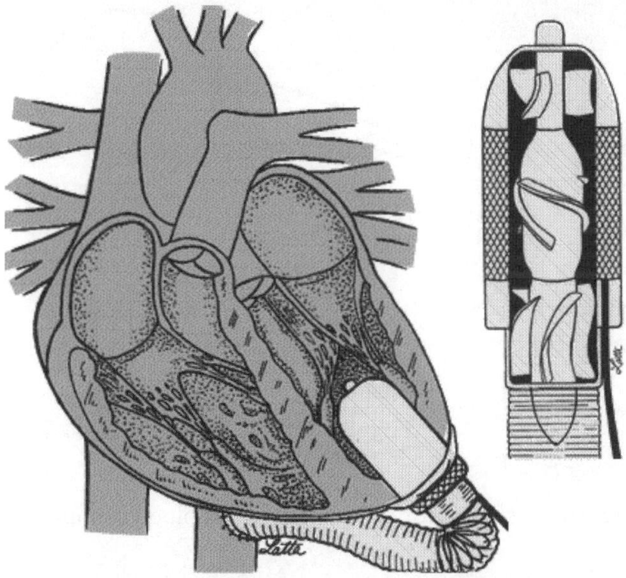

FIGURE 20: Axial flow blood pump configured within the base of the left ventricle. Flow through the device is completely axial with forward and rear diffusers to direct flow to the outer sections between the rotor and the stator.

insensitive and afterload sensitive. As such, they are relatively unaffected by inlet conditions. Axial flow pumps, being smaller in size, do not normally produce flow rates as large as centrifugal pumps, and are not normally utilized for long-term bridge-to-transplant support. However, being of a more compact size, it is far easier to configure such pumps than the larger centrifugal versions.

2.3 EVALUATION OF BLOOD PUMPS

Blood pumps are usually evaluated either experimentally on the bench (designated *in vitro*) or in animal models (*in vivo*) or in clinical studies (*in situ*). There are also computational models using finite element techniques, collectively termed "computational fluid dynamics." Experimental studies *in vitro* (on the bench) employ a mock circulatory system that mimics the human circulation in terms of the physical load against which the device must pump blood. In the bench top study, a blood analog is often employed consisting of a glycerin–water mixture. A mock circulatory system is shown in Figure 21 and consists of a compliance element simulating aortic compliance, a resistor simulating peripheral resistance, and a venous reservoir or compliance element simulating either venous inlet pressure or venous compliance.

Experimental studies of the fluid mechanics within blood pumps and inside the attached cannula (both input and output cannulae) can be typically conducted by laser Doppler anemometry (LDA) and/or particle image velocimetry. Both techniques require that the test fluid is transparent and that the associated flow field is within a transparent section. Collectively,

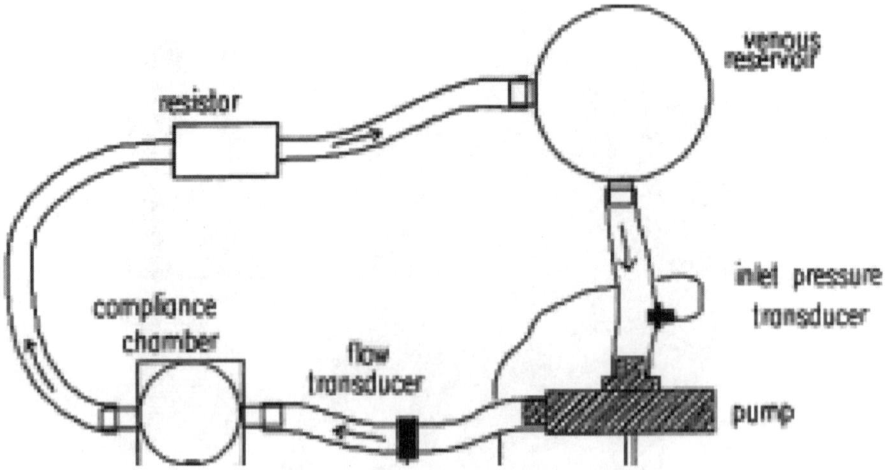

FIGURE 21: Mock circulatory system for *in vitro* testing of blood pumps.

this field is known as *flow visualization*. A particle image velocimetry (PIV) system requires the use of neutrally buoyant particles that are fluorescent so as to be easily illuminated by a dual laser system. A PIV system employs a CCD camera and dual lasers to illuminate the particles at two time intervals closely spaced (within 30 ms) with the tracks between particles calculated by the supporting computer. A typical PIV system is shown in Figure 22.

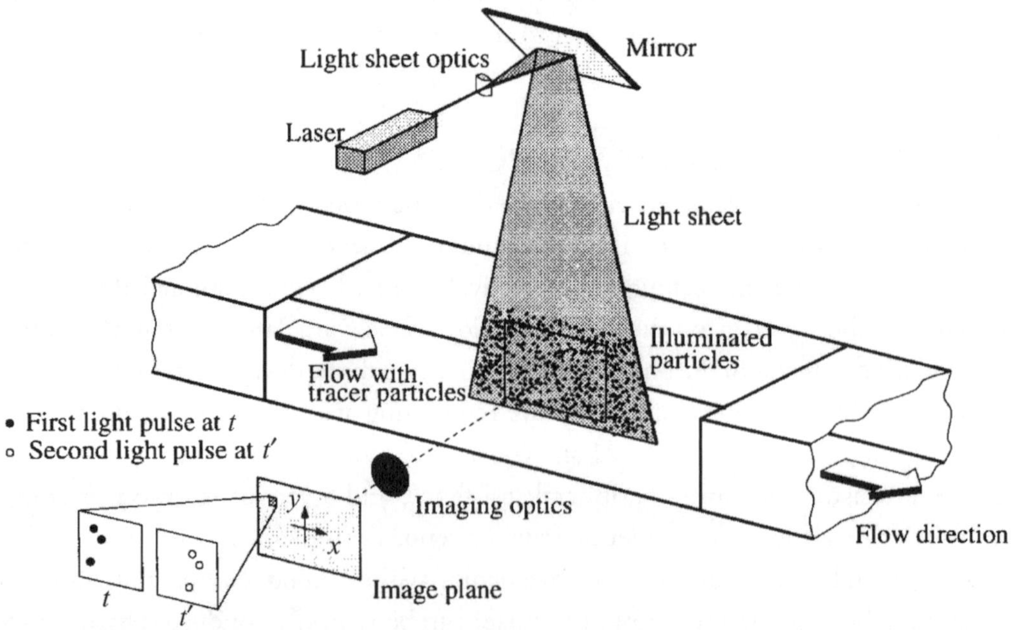

FIGURE 22: Typical PIV system with laser-illuminated particles within the flow field.

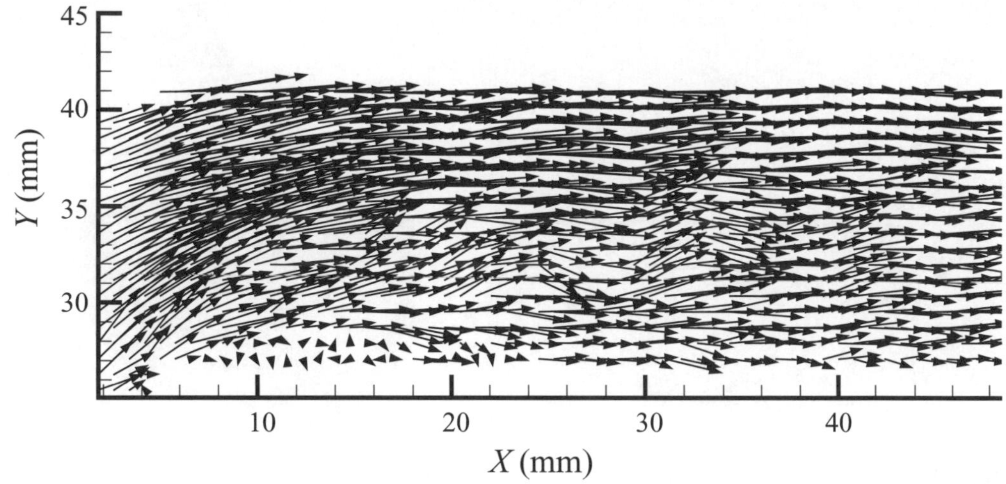

FIGURE 23: PIV vector mapping (snapshot) of particles within a flow field at one instance in time (over a 30-ms interval).

A snapshot of the fluid dynamics of the flow field generated by PIV-illuminated particles is shown in Figure 23. Such snapshots are generated 30 or more times per second and 100 of such images can be averaged to produce a flow profile indicating overall flow trends. Such an average flow profile is shown in Figure 24.

In addition to flow visualization techniques, such as LDA or PIV, it is common to measure volume flow rate with either a Doppler ultrasound flow sensor or an electromagnetic

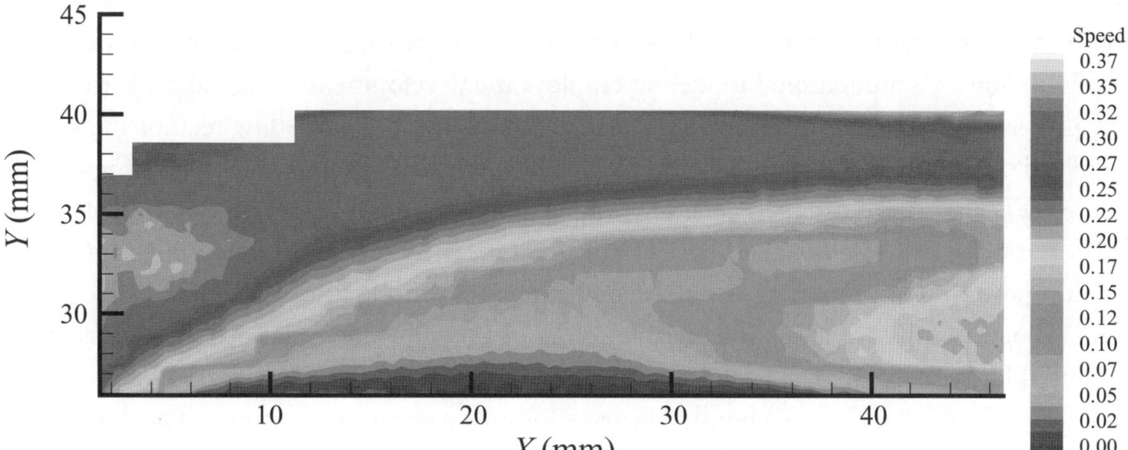

FIGURE 24: Time averaged (contour) plots from sequential vector plots of a PIV-measured flow field. Blue zone indicates low velocity (near lower wall) and red zone indicates higher velocity at outer (top) wall. Slight reversal of flow in yellow is seen at the far right.

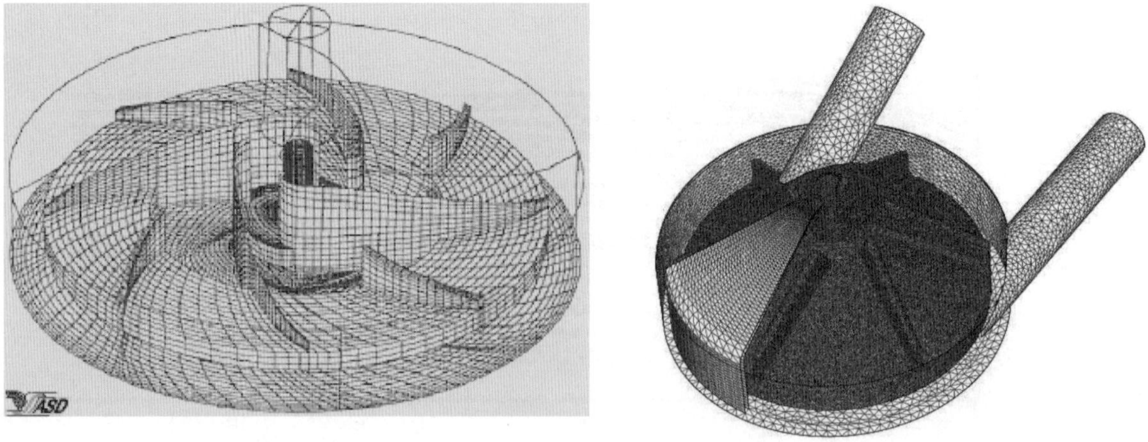

FIGURE 25: Grids generated by computational fluid mechanics algorithms to represent flow boundaries and computational accuracy steps.

flow sensor. It is also common to measure pressure within the flow field at several locations including at the inlet to the pump, at the outlet to the pump, within the compliance chamber of the mock circulatory system, and within the pumping chamber itself. Standard dome-type pressure transducers or solid-state transducers are employed. The former are less expensive, but have a limited frequency response of 100 Hz while the latter are more sensitive and accurate, can be mounted flush within a flow field conduit, but are also far more expensive and fragile.

In many cases, it may be appropriate to model the flow field by computational methods, particularly when a variety of configurations must be examined. The use of computational fluid mechanics (CFD) is often advantageous as a means of verifying experimental data as well as a method of avoiding the need to make expensive *in vitro* experiments with numerous physical configurations. Computational modeling employs the development of digital grids that depict the physical boundaries of the flow field within which the pressure–flow relationships can be analyzed. A typical grid pattern generated within a centrifugal pump is shown in Figure 25.

CFD algorithms can then generate simulated flow field information consisting of pressure and/or velocity fields within the physical grid framework as seen in Figure 26. It is possible to superimpose PIV or other flow visualization techniques onto the grid system as well.

Blood pumps are also evaluated in terms of pressure flow relations as well as for the potential for hemolysis. A typical pressure–flow relationship for a centrifugal LVAD is shown in Figure 27, which indicates a family of curves for various rpms of the pump. The data indicate that the flow is reduced at increasing afterload pressures, which is typical of centrifugal pumps. Similar performance curves of pressure versus flow are typically utilized for all LVAD designs including positive displacement pumps. Alterations in design of the pumps are often based upon the performance results noted in this figure. Hemolysis data is shown in Figure 28 and

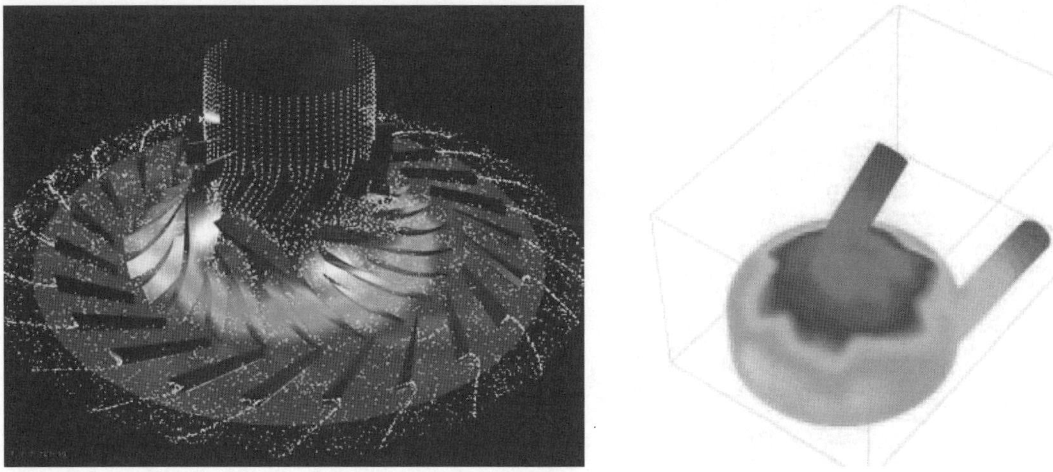

FIGURE 26: CFD-generated flow fields within designated grid system. Left figure indicates flow field data within the device while right figure indicates stress data on walls of the device.

is generated with the pump attached to a modified mock circulatory system with whole blood used within the circulation. Hemolysis can be evaluated as either measured evidence of plasma free hemoglobin or by an index of hemolysis which is based in part on a relative scale of plasma free hemoglobin.

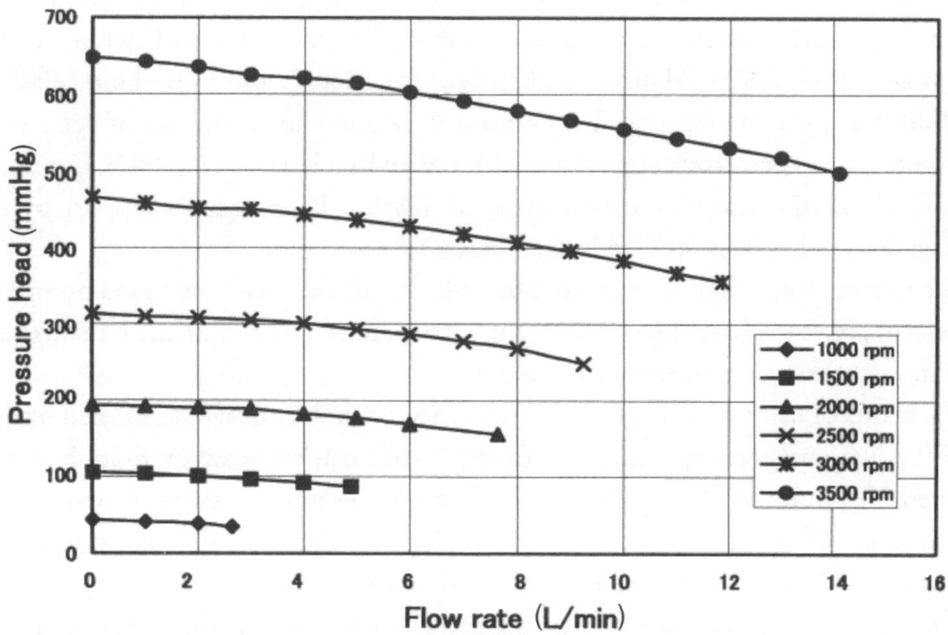

FIGURE 27: Performance curves for a centrifugal blood pump indicating pressure–flow relationships for varying pump rpms. Flow is reduced at increasing pressures (afterloads).

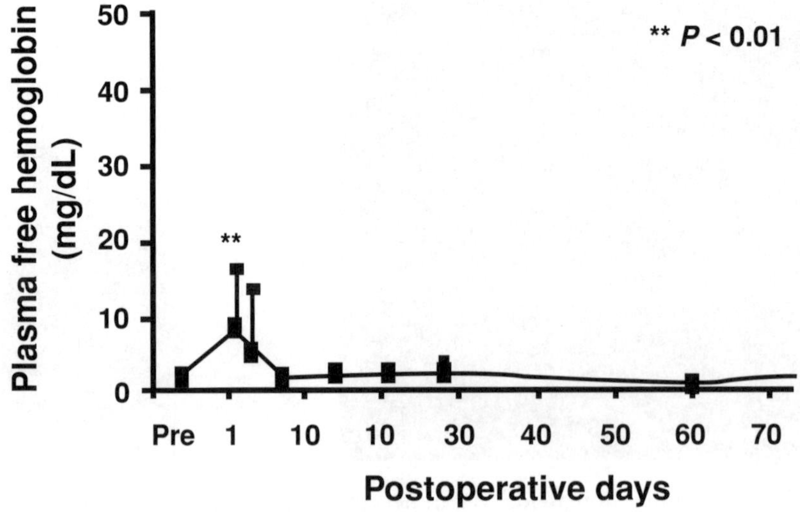

FIGURE 28: Hemolysis data for an LVAD showing levels of plasma free hemoglobin postoperative (postimplantation of pump). After an initial spike, the hemolysis data is quite low, which is typical of many LVAD designs.

Studies of the efficacies of various VAD technologies have been conducted by numerous investigators over the previous 50 years. Of late, several investigators have analyzed the role of ventricular assist devices as either a bridge to transplantation or as a long-term destination therapy including Mehra (2004), Birks *et al.* (2004), Westaby (2004), Kherani *et al.* (2004), Stevenson and Rose (2003), Matsuda and Matsumiya (2003), and Wheeldon (2003), among others. Additionally, there have been significant developments in the use of ventricular assist technology for use in pediatric applications as noted by Reinhartz *et al.* (2002), Duncan (2002), Deiwick *et al.* (2005), and Throckmorton *et al.* (2002). Biventricular support by means of mechanical devices has been studied by Samuels (2004).

The role of steady flow as generated by centrifugal and axial flow blood pumps has been debated for many years. Saito (2004), Mesana (2004), Myers *et al.* (2003), Ichikawa and Nose (2002), among others have examined this issue.

The status of using ventricular and circulatory assist devices was examined by a working group of the National, Heart, Lung, and Blood Institute as reported by Reinlib *et al.* (2003). The United Network for Organ Sharing (UNOS) has examined issues related to the use of ventricular assist technology as it pertains to the selection of patients for heart transplants. Does the use of the VAD lessen the patient's urgent need for a transplant, thus lowering their status "in line" while waiting for a transplant? Or does it increase their needs? Morgan *et al.* (2004) report on the latest findings from UNOS on that topic. Deng and Naka (2002) provide an overview of the state of the art for mechanical circulatory support as do Nemeh and Smedira (2003).

Velocity measurements within ventricular assist devices via flow visualization techniques have been reported by Yamane *et al.* (2004), Tsukiya *et al.* (2002), Manning and Miller (2002), Day *et al.* (2002), Wu *et al.* (1999), Mulder *et al.* (1997), Kerrigan *et al.* (1996), and Miller *et al.* (1995). Computational fluid dynamics (CFD) techniques have been employed by numerous investigators to analyze various design configurations of circulatory assist devices including Song *et al.* (2004a,b), Okamoto *et al.* (2003), Curtas *et al.* (2002), Anderson *et al.* (2000), and Pinotti and Rosa (1995), among others.

CHAPTER 3
Cardiac Pacemakers

3.1 CARDIAC ELECTROPHYSIOLOGY

The heart weighs between 7 and 15 ounces (200–425 g) and is a little larger than the size of your fist. By the end of a long life, a person's heart may have beat (expanded and contracted) more than 3.5 billion times. In fact, each day, the average heart beats 100 000 times, pumping about 2000 gallons (7571 L) of blood. Electrical impulses from your heart muscle (the myocardium) cause your heart to contract, and are created by the movement of ions (principally potassium) across membranes in a fashion called depolarization of cells. This depolarization is propagated along pathways as the ion transport continues. This electrical signal begins in the sinoatrial (SA) node, located at the top of the right atrium. The SA node is sometimes called the heart's "natural pacemaker" with the average heart beat occurring 72 beats/min. An electrical impulse from this natural pacemaker travels through the muscle fibers of the atria and ventricles, causing them to contract. Although the SA node sends electrical impulses at a certain rate, your heart rate may still change depending on physical demands, stress, or hormonal factors. The initiating electrical signal from the SA node travels down a preferred pathway of specialized conducting myocardial tissue toward the atrioventricular (AV) node, which is the electrical connecting point from the atria to the ventricles. After a slight transmission delay (designed to allow the atria to contract and fill the ventricles before the ventricles contract), the specialized conduction pathway continues into the ventricles. There are two bundles of fibers, called the left and right bundle branches of His, culminating in purkinje fibers, that promote rapid transmission of electrical current through both ventricles. The electrical pathways are shown in Figure 29.

3.2 THE ELECTROCARDIOGRAM

The waves of depolarization that spread through the heart during each cardiac cycle generate electrical currents, which in turn spread through the body's interstitial fluids and eventually up to the body's surface. Recording electrodes, placed on the surface of the skin on opposite sides of the heart, are used to detect such electrical potentials. These signals are filtered, amplified, and recorded as a measure of the underlying cardiac electrical activity. The record that results from this procedure is termed an electrocardiogram (ECG or EKG). An ECG is an important

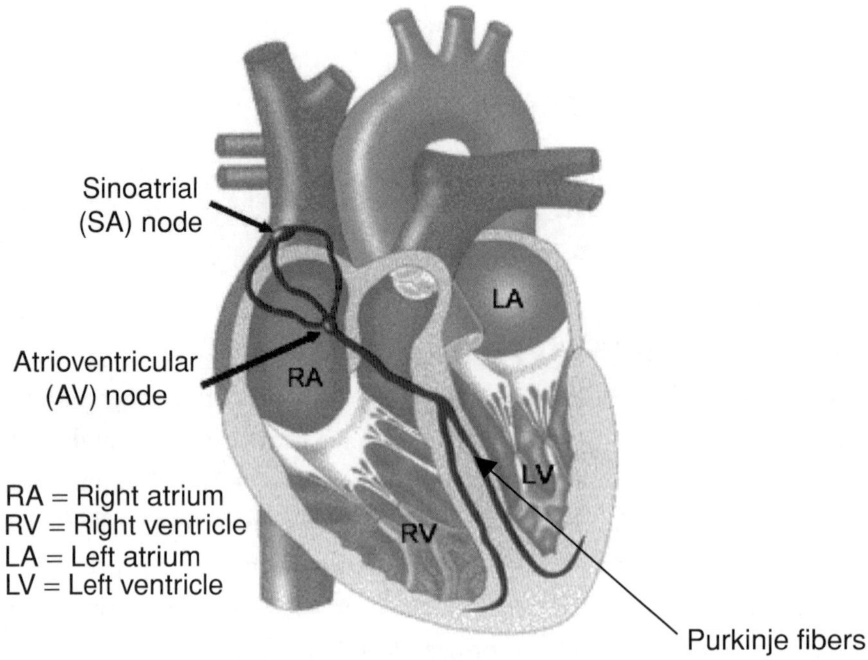

FIGURE 29: Electrical excitatory pathways in the heart.

diagnostic tool to determine whether any cardiac malfunction may have an underlying electrical reason. A typical ECG waveform is seen in Figure 30. The normal ECG consists of three basic features: a P wave, a QRS complex, and a T wave. On some patient waveforms, the QRS complex is seen as three separate waves.

The electrical currents produced as the atrial muscle cells depolarize prior to contraction generate the P wave. This is followed in time by the QRS complex and results from currents

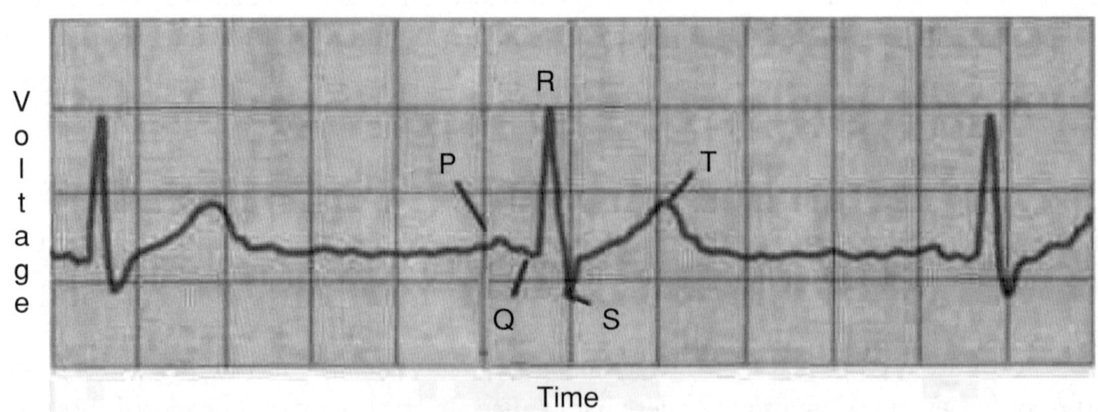

FIGURE 30: Typical ECG showing P, QRS, and T waves.

generated as the ventricles depolarize prior to their contraction. There is a slight time delay between these waves resulting from the delay produced at the AV node, as stated above. Thus, both the P wave and the QRS complex represent depolarization waves. Following ventricular contraction, the ventricle muscle cells repolarize (reverse polarity and reverse ionic transport), and these reversing electrical currents produce the T wave that typically occurs 0.25–0.35 s following ventricular depolarization.

Repolarization of the atria is masked by the QRS complex, and is thus not normally seen on the standard ECG. The intervals between these waves and the width and shape of these waves on the ECG are useful diagnostic measures of cardiac function, as is the overall frequency of the cardiac cycle. The interval between the beginning of the P wave and the beginning of the QRS complex is designated by the PQ interval. Since the Q wave is often absent, it is sometimes termed the PR interval and represents the duration of time (normally about 0.16 s) between the onset of atrial contraction and the onset of ventricular contraction. In patients with certain heart diseases including scarred and/or inflamed tissue, this may lead to a lengthening of the PR interval because more time is required for the depolarization wave to spread through the atrial myocardium and the AV node. The time required for the generation of the QRS complex is termed the QRS duration and represents the amount of time needed for ventricular depolarization. The QT interval extends from the beginning of the QRS complex to the end of the T wave and represents the time required for ventricular contraction and repolarization. Typically, the normal QT interval lasts for approximately 0.35 s. The ST interval represents the time required for the ventricles to repolarize and extends from the S wave to the end of the T wave.

The term *tachycardia* denotes an overall fast heart rate, typically at more than 100 beats/min. Tachycardia may result from fever, from stimulation of cardiac sympathetic nerves, from certain hormones or drugs, or by weakening of the heart muscle itself. When the myocardium is unable to pump blood effectively, homeostatic reflexes are activated that subsequently increase the heart rate. The term *bradycardia* denotes a slow heart rate of less than 60 beats/min and is a condition common to athletes, whose enlarged hearts pump a greater stroke volume per heart beat than that of a nonathlete. In addition, bradycardia may also result from decreased body temperature, due to certain drugs, or via stimulation of the heart by its parasympathetic nerve fibers originating from the vagus nerve. Bradycardia may also occur in patients with atherosclerotic lesions of the carotid sinus region of the carotid arteries. The term *arrythmia* refers to any electrical abnormality in the ECG including rate-related conditions as well as conduction delays or even complete blockages.

The SA node is the natural pacemaker of the heart, providing a stimulating source that initiates the electrical depolarization cycle at approximately 72 beats/min. Should there be a *block* of the excitation wave between the SA node and the AV node, then the origin of the excitation for the remainder of the heart becomes the AV node, whose normal rhythmic rate is slower than that of the SA node (approximately 55 beats/min). The remainder of the excitatory pathway (AV node to purkinje fibers) remains intact. Should there be a partial or total block between the AV node and the purkinje fibers, then alternate excitatory sources within the ventricular muscle assume the excitatory role, albeit at a much lower rate (30 beats/min). These are readily evident in the ECG waveforms as a reduced heart rate, as widening of the waves within the ECG, and/or as changes in the shape of the waves.

Premature contractions can occur within a normal rhythmic cardiac excitation (called sinus rhythm) due to altered excitatory events within an atrium or a ventricle. Premature atrial contractions (PACs or PABs for premature atrial beats) are atrial beats that occur too early due to an abnormal electrical signal. Often, things such as caffeine, alcohol, medications (especially decongestants), certain medical conditions such as hyperthyroidism, anemia, and hypertension, and stress can trigger PACs. Some people may feel a fluttering in their hearts when experiencing PACs, whereas others have no symptoms. PACs are benign and may require no treatment. Premature ventricular contractions (PVCs or PVBs for premature ventricular beats) are early ventricular contractions that occur when the ventricles contract out of sequence with normal heart rhythm. Though they are generally benign and usually do not require treatment, PVCs may result in more serious arrhythmias in those with heart disease or a history of tachycardia. For these people, antiarrhythmic drugs and an implantable cardioverter defibrillator (ICD) may be prescribed. PVCs most often occur spontaneously; however, like PACs, they can also be triggered by caffeine, alcohol, medications (especially decongestants), certain medical conditions such as hyperthyroidism, anemia, and hypertension, and stress. PACs and PVCs are readily observed as added waves within the ECG. An *ectopic* or premature atrial beat produces an added P wave within the standard ECG waveform. A PVC produces an added R wave or QRS segment within the normal ECG waveform. Repeated PAC or PCV events may indicate a serious condition that will require intervention.

Fibrillation is caused when the heart muscle begins to quiver, or fibrillate, continually and cannot contract normally. When a heart is in a state of fibrillation, there is no synchronization between the atria and the ventricles, nor any standard heart beat/contraction. This may cause the patient to experience a racing sensation—and sometimes discomfort in the chest—and/or to feel light-headed or faint. Ventricular fibrillation (VF or V Fib) is a life-threatening arrhythmia that necessitates immediate treatment with an external defibrillator, an internal defibrillator (ICD), antiarrhythmic drugs, or VT ablation. Fibrillation is easily recognized on the ECG waveform as a loss of the QRS segment replaced by a noisy, lower amplitude signal. Atrial fibrillation (AF or A Fib) is a very fast, uncontrolled atrial heart rate caused by rapidly fired signals. During an episode of AF, the atrial heart rate may exceed 350 beats/min. Not all of these beats reach the ventricles, so the ventricular rate is not this high. However, the ventricular rate is often higher than normal and can also be erratic, exceeding 100 beats/min. Sometimes an impulse will circle the atria, triggering atrial flutter, which is similar to AF. Alone, AF is rarely serious, but if a patient has high blood pressure, valvular disease, or heart muscle damage, AF can increase the risk of stroke or heart failure.

There are several treatments for AF, including medication, ablation, and an electrical therapy called cardioversion. Electrical cardioversion converts the heart rate back to normal sinus rhythm through the use of a controlled electrical shock that excites all the heart cells at once, allowing the SA node to resume its role as the heart's natural pacemaker. If medication and electrical cardioversion do not improve the AF, the physician may recommend cardiac ablation to prevent conduction of abnormal electrical impulses from the atria to the ventricles, with implantation of a permanent pacemaker to control heart rate. Ventricular fibrillation (VF) is a chaotic heart rate resulting from multiple areas of the ventricles attempting to control the heart's rhythm. Ventricular fibrillation can occur spontaneously (generally caused by heart disease) or when ventricular tachycardia has persisted too long. When the ventricles fibrillate, they cannot contract normally, hence, they cannot effectively pump blood. The instant VF begins, effective blood pumping stops. VF quickly becomes more erratic, resulting in sudden cardiac arrest or sudden cardiac death. This arrhythmia must be corrected immediately via a shock from an external defibrillator or an internal device (ICD). The defibrillator can stop the chaotic electrical activity and restores normal rhythm by depolarizing the entire heart with the idea that normal sinus rhythm will be restored if the SA node fires first and initiates the normal excitation process.

3.3 CARDIAC PACEMAKER

An artificial cardiac pacemaker is an implantable device that produces an excitatory wave at an appropriate site within the heart: a) at the SA node to replace a failing natural SA node (or correct atrial fibrillation) and generate a normal sinus rhythm, b) at the AV node where a malfunctioning AV node exists, or c) within one or both ventricles to continue excitation beyond a partial or total heart block (or to correct ventricular fibrillation). A cardiac pacemaker consists of 1) an electrical device, the pulse generator, which produces the excitatory signal, 2) the wires, called the leads, which connect the pulse generator to the cardiac tissues, 3) a battery system (housed within the pulse generator) to power the pulse generator, and 4) a programmable segment (also housed within the pulse generator) to evaluate the heart excitatory function and send the appropriate excitatory segment from the pulse generator. A schematic diagram of a cardiac pacemaker and the leads is shown in Figure 31.

The batteries within the pulse generator "can" are typically lithium iodide and have a life span of approximately 10–12 years. The pulse generator casing is typically titanium which is welded and hermetically sealed. An epoxy top is attached from which the leads are attached to the pulse generator. This allows for changes in the leads without disrupting the pulse generator case. A typical pulse generator is small in size, often less than an ounce in weight, less than two inches wide, and a quarter-inch thin. Thus, the device is roughly the size of two silver dollars stacked on top of one another. Once implanted in the upper chest, just below the skin

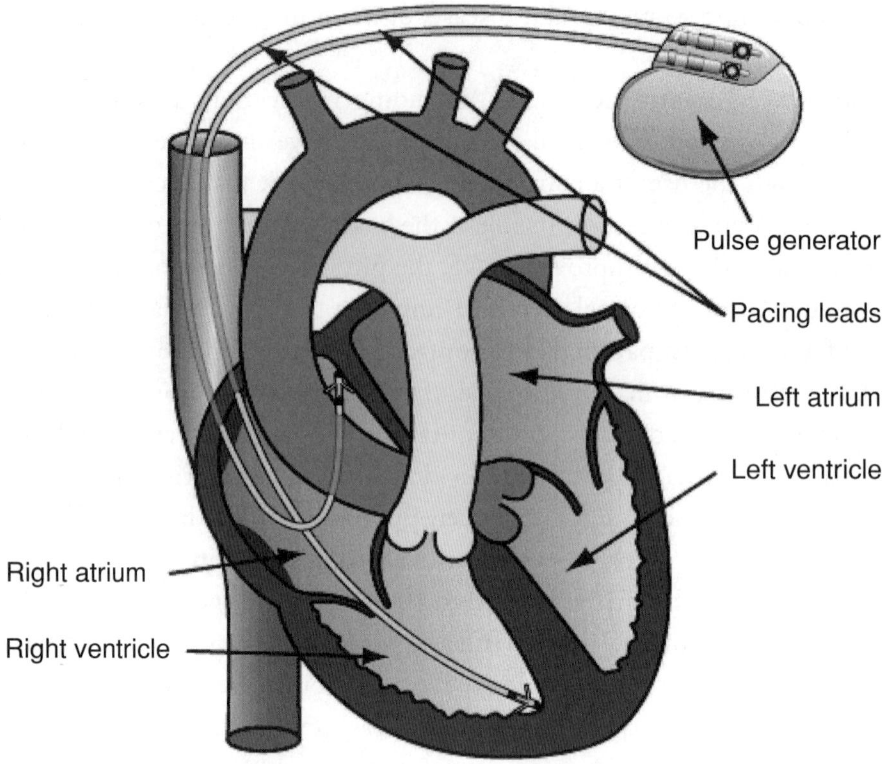

FIGURE 31: Schematic diagram of an implantable cardiac pacemaker.

near the collarbone, the pacemaker's presence is nearly invisible to the eye. When the batteries have become depleted, the entire pulse generator case is replaced. The surgical implantation procedure will be explained in more detail below. A pulse generator is seen in Figure 32, with the leads connected to the "can" seen in Figure 33.

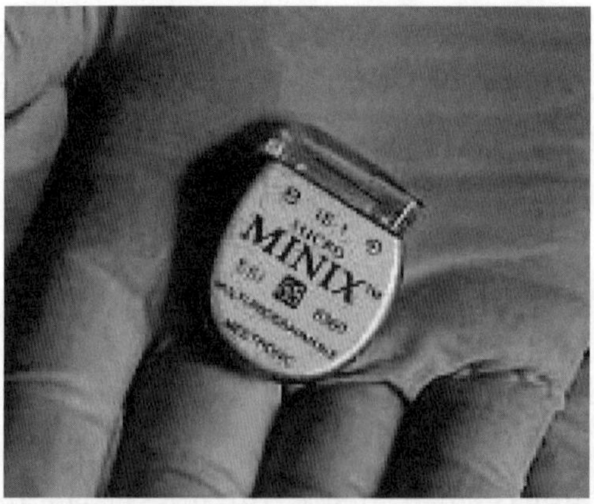

FIGURE 32: Typical pacemaker pulse generator with epoxy top for connection to leads.

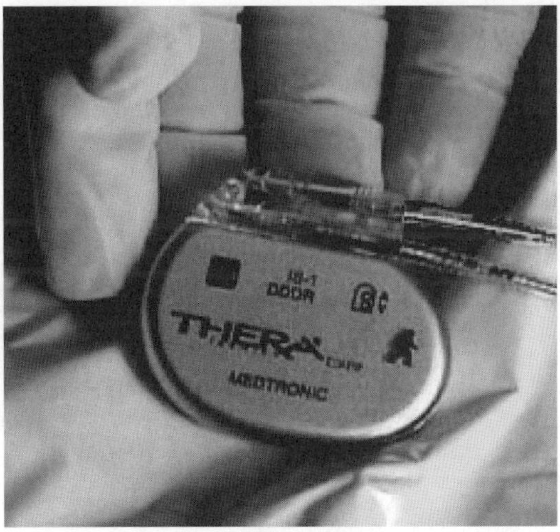

FIGURE 33: Pulse generator with leads attached.

The pacing lead, which connects to the pulse generator header, is a flexible insulated wire with an electrode tip. This tip, which can be inserted through a vein into the heart, carries impulses from the pulse generator to the heart, stimulating the heart to beat. It also carries information from the heart back to the pulse generator, which the physician accesses via a special programmer or into the pulse generator circuit embedded program. Leads that are inserted via a vein into the heart have tines on the end as shown in Figure 34 or have a screw end, which can be threaded into the cardiac muscle as shown in Figure 35.

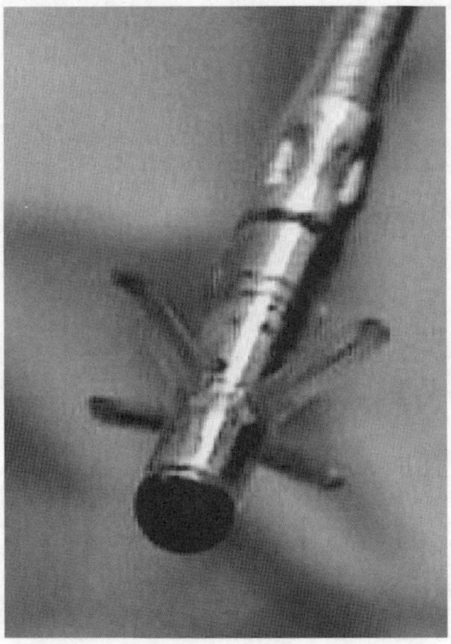

FIGURE 34: Pacemaker lead with tines on end for embedding into heart tissue.

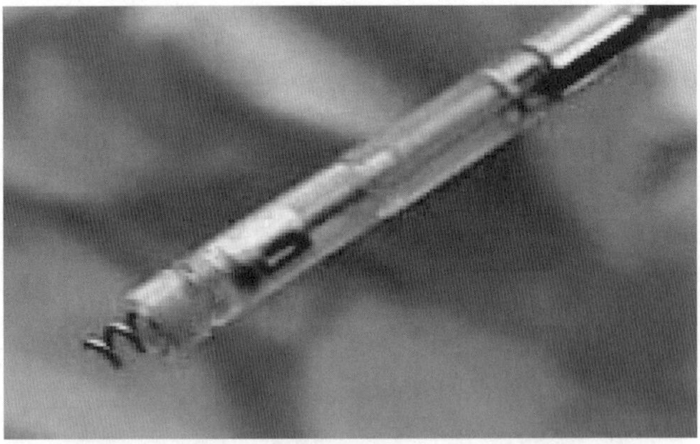

FIGURE 35: Pacemaker lead with screw end for implantation into heart tissue.

The leads shown in Figures 34 and 35 are both designated as endocardial or transvenous leads. The tined lead shown in Figure 34 is a passive fixation device where the tines become lodged in the trabeculae (fibrous meshwork) of the heart. The helix or screw-ended lead shown in Figure 35 is an active fixation device that extends into the endocardial tissue (interior surface of the heart). Such a lead can be positioned anywhere in the heart chamber. The other category of lead is the myocardial or epicardial lead which is secured to the outer surface of the heart as shown in Figure 36. Such leads are connected to cardiac tissue via sutures or screwed into tissue as was the case with the endocardial lead.

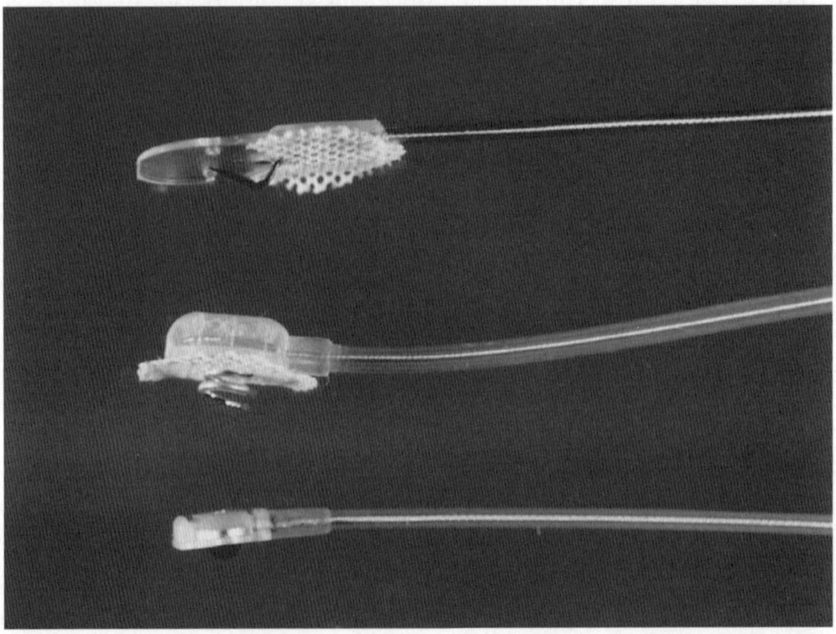

FIGURE 36: Myocardial pacemaker leads which are sewn into exterior cardiac tissue via sutures or screwed into the exterior surface.

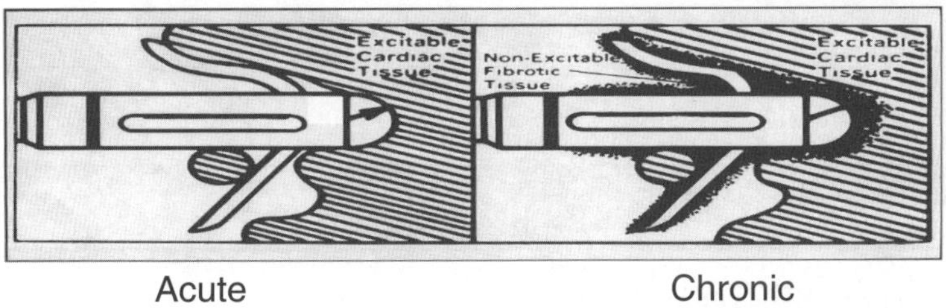

FIGURE 37: Fibrous capsule surrounding a tined lead as a result of an inflammatory response.

Once leads have been implanted, there is often a tissue reaction that creates a fibrotic capsule around the lead, as shown in Figure 37. Such a capsule can increase the resistance to the stimulating pulse to cardiac tissue, which would require an increase in the amplitude of the original pulse from the pacemaker. This would deplete the batteries at a rate faster than normal. In order to avoid such a circumstance, some leads contain a steroid that eludes into the surrounding tissue to reduce the onset of an inflammatory response. This is seen in Figure 38.

Leads can be either unipolar or bipolar. Unipolar leads, as shown in Figure 39, are single leads inside of an insulating sheath, which send the impulse signal from the pulse generator with the returning electrical signal traveling through the tissues to the pacemaker. The unipolar lead normally has a smaller diameter than the bipolar lead and exhibits a greater tendency for pacing artifacts on the surface EKG, since the returning electrical signal travels through tissue.

The bipolar lead, as shown in Figure 40, is a coaxial lead with a larger overall diameter. The returning signal (after stimulating cardiac muscle) pathway is through the second embedded lead, rather than through tissues. As a result, the bipolar lead has a lower susceptibility to affect

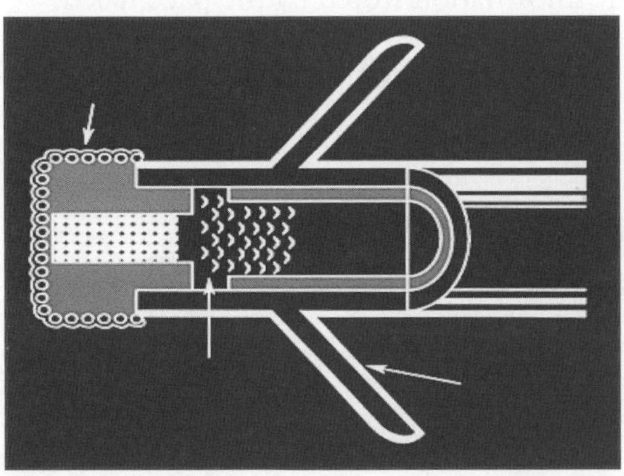

FIGURE 38: Steroid eluding lead designed to deter inflammatory response and fibrous capsule formation.

FIGURE 39: Unipolar pacemaker lead.

the EKG and is less susceptible to external interference, such as from ambient electromagnetic "noise."

The insulation on pacemaker leads is either silicone or polyurethane. Silicone is inert, biocompatible, biostable, repairable with medical adhesive, and has a long and successful history in biomedical applications. Polyurethane is also biocompatible, but additionally has a higher tear strength, lower friction coefficient, and a smaller required thickness.

Pacemaker systems that are fixed rate often utilize a single lead that is placed in the atrium (if there is an SA nodal problem) or the ventricle for AV nodal or bundle branch problems. For circumstances where there is a need for a demand style pacemaker, there must be two leads placed within the heart as seen in Figure 41.

The sensing electrode is placed within the atrium to sense any variation in normal sinus rhythm, such as what might occur during exercise of increased heart activity. The stimulating electrode might be placed within the ventricle and the rate of stimulation is dependent on the sensed SA nodal rate. Modern demand pacemakers thus have four main functions:

- Stimulate cardiac depolarization
- Sense intrinsic cardiac function
- Respond to increased metabolic demand by providing rate responsive pacing
- Provide diagnostic information stored by the pacemaker.

The output of the pacemaker, which stimulates cardiac tissue, can be visualized on the surface EKG as is seen in Figure 42.

FIGURE 40: Bipolar pacemaker lead.

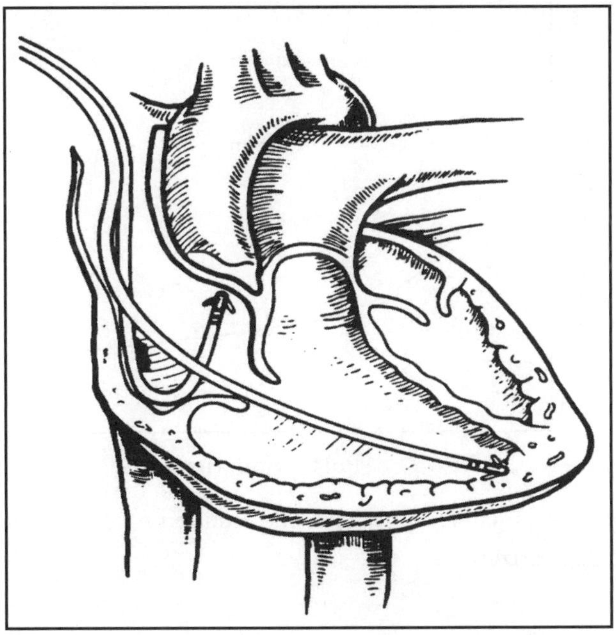

FIGURE 41: Demand style pacemaker with two leads: one for sensing and one for stimulation.

A typical stimulating pulse from the pulse generator must be large enough to cause depolarization (i.e., to "capture" the heart) and must be sufficient to provide an appropriate pacing safety margin. The pulse width must be long enough for depolarization to disperse to the surrounding tissue. As such, pulse generation waveforms are typically evaluated on a strength duration curve as noted in Figure 43.

Once a pacemaker elicits a stimulating pulse and the heart muscle depolarizes, there exists a refractory period whereby another stimulating pulse would not elicit further depolarization.

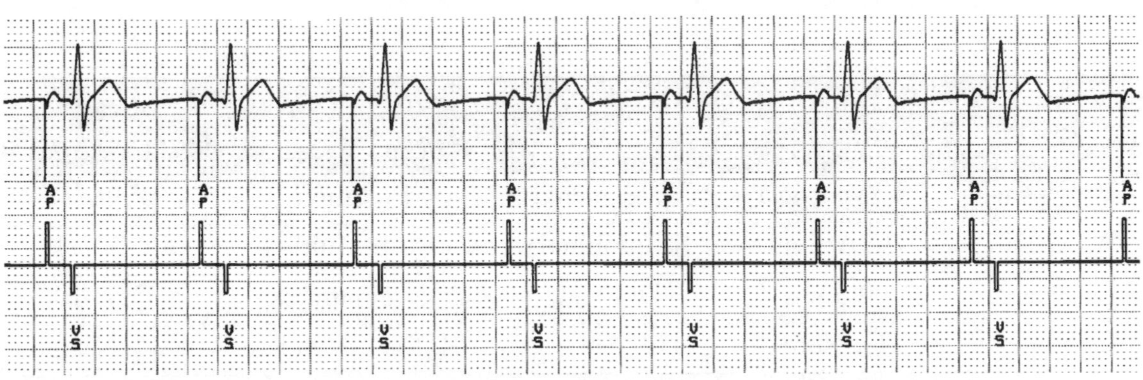

FIGURE 42: Surface EKG with pacemaker stimulation pulses noted within the EKG waveform as well as on the axis below.

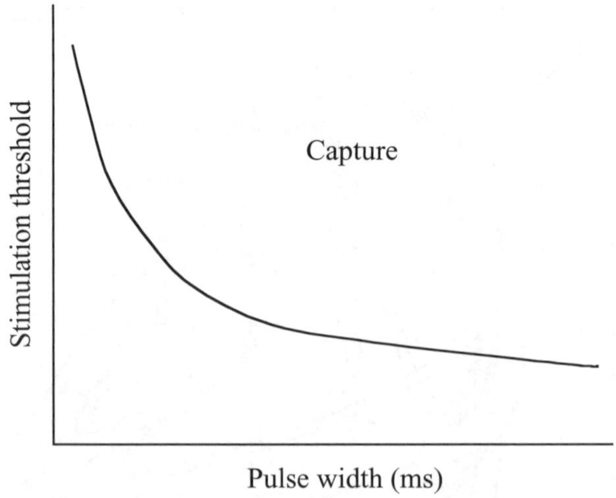

FIGURE 43: Strength–duration curve to evaluate the required necessary stimulating pulse to achieve "capture," i.e., produce a depolarization.

This same refractory period exists during normal cardiac electrical stimulation, such that any ectopic wave would not elicit a second depolarization. Repolarization would normally need to occur before a second pulse would produce another depolarizing wavefront. As was noted above, modern pacemakers often employ a sensing electrode in one heart chamber and a second stimulating electrode in another chamber. A system of classification has been established for pacemakers which is used to describe the various parameters that can be incorporated into modern pacemakers, as seen in Figure 44.

Referring to this classification system, a VOO pacemaker (see chart) stimulates the ventricle with no atrial or ventricular sensing (nor any response to sensing). This would be a fixed rate pacemaker that stimulated the ventricle regardless of the patient's intrinsic cardiac

CHAMBER PACED	CHAMBER SENSED	RESPONSE TO SENSING	PROGRAM FUNCTION	FUNCTION
V: Ventricle	V: Ventricle	T: Triggered	P: Simple	P: Pace
A: Atrium	A: Atrium	I: Inhibited	M: Multi	S: Shock
D: Dual	D: Dual	D: Dual	C: Communicating	D: Dual
O: None	O: None	O: None	R: Rate modulate	O: None
S: Single	S: Single	O: None		

FIGURE 44: Pacemaker classification system.

electrophysiology. A VVI pacemaker (see chart) senses ventricular electrical activity and paces the ventricle if there is a lack of normal intrinsic activity sensed. If normal ventricular excitatory pulses are sensed, then the pacemaker output is inhibited (the I) in the pacemaker classification. A VVIR pacemaker is identical to the VVI pacemaker except for the added "R" term indicating program function (R = rate). This pacemaker changes the stimulation rate of firing when the intrinsic ventricular stimulation warrants it. Similarly, a VAIR pacemaker senses atrial activity and stimulates the ventricles. The pacemaker pulse rate is affected by the atrial activity. If the patient's metabolic activity alters the SA nodal firing rate, then the pacemaker senses this and increases the ventricular firing rate. If there is a normal sinus rhythm, then the pacemaker output is inhibited. Similarly, a VDIR pacemaker senses intrinsic activity in both chambers with the pacemaker capable of delivering a variable rate.

Originally, pacemakers of the 1950s were external devices with the leads implanted into the heart. The first transistorized, wearable, battery powered pacemaker developed by Wilson Greatbach in 1957, and the first totally implanted pacemaker was in 1960. The battery life was approximately 18 months. In the mid-1960s, transvenous leads were developed that could be introduced through a vein, rather than require opening the chest to attach the lead. Original designs for pacemakers incorporated a fixed rate. In the mid-1960s, the first "demand" pacemakers were developed that sensed when there was a missed beat and then initiated a pulse to excite the cardiac muscle, while being inactive during normal sinus rhythm. The screw tip and tined leads were developed in the 1970s, and in 1975, the longer lasting lithium iodide battery was developed to replace the mercury–zinc battery. At that same time, the titanium case and epoxy top was developed, which better shielded the pacemaker from outside electromagnetic disturbances, such as a microwave oven. In that same time span, the first programmable pacemakers were developed as were the first dual chamber pacemakers. In the 1990s, the pacemakers became smaller with advances in batteries and in microelectronics. To date, over 2 million cardiac pacemakers have been successfully implanted.

3.4 PACEMAKER IMPLANTATION

A schematic diagram depicting the surgical implantation of a cardiac pacemaker is shown in Figure 45. The pacemaker pulse generator with its epoxy top is implanted just below one of the collarbones. The leads are then inserted through a vein leading to the heart or via minimally invasive surgery to affix them to the exterior of the heart.

The use of the easily accessible site below the collarbone allows the pulse generator with its battery pack to be easily replaced once the batteries have become depleted. By using a local anesthetic, a small opening in the skin and underlying thin tissues can be created in order to remove the pulse generator. Since the leads are connected to the epoxy top of the pulse generator by means of a force fit, then the leads can be easily removed and snapped

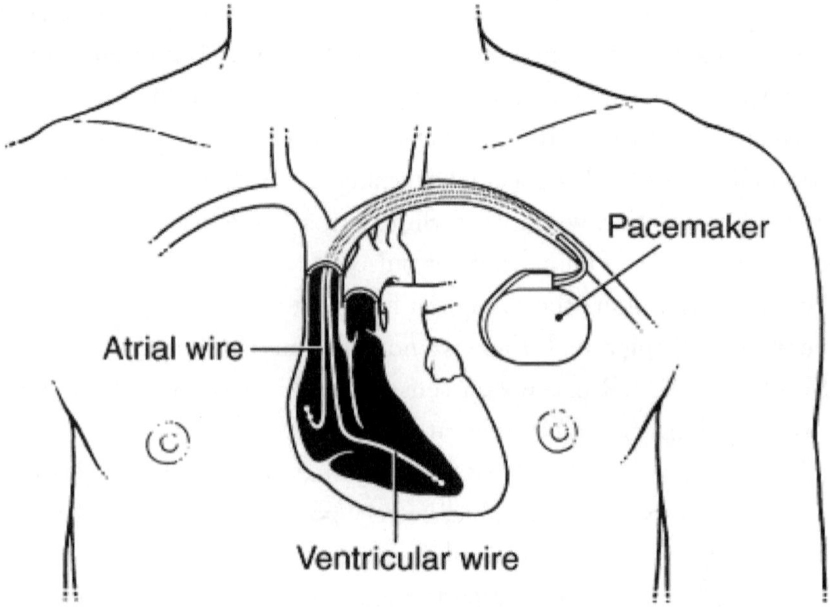

FIGURE 45: Implantation of a cardiac pacemaker pulse generator below a collarbone with the leads threaded through a nearby vein toward the heart.

into the replacement pulse generator (and battery pack) within the new epoxy top associated with the new generator. The time "off" of pacing is thus extremely short—the time it takes to pull the lead(s) out of one epoxy top and into another. The new pulse generator is then placed in the site of the old unit, and the small incision is then closed. The entire replacement procedure often takes as little as 30 min from open to close. As the modern batteries for cardiac pacemakers can last as long as 12 years, replacement of the pulse generator is infrequent.

3.5 CARDIOVERTER

If the patient is susceptible to ventricular fibrillation, then a standard cardiac pacemaker may not be able to reestablish sinus rhythm. As with the case when an individual suffers a "heart attack," where VF occurs, it is necessary to convert the patient by completely depolarizing the entire heart by means of a defibrillator. Most individuals are familiar with external defibrillators that use large paddles and a large current density to provide an overwhelming depolarization to the heart. The premise is that when the heart then completely repolarizes, that the SA node will fire before any other site within the heart and reestablish sinus rhythm. The large paddles and large current density are necessary for external defibrillation, as the skin and exterior tissues provide a large resistance to electrical current flow.

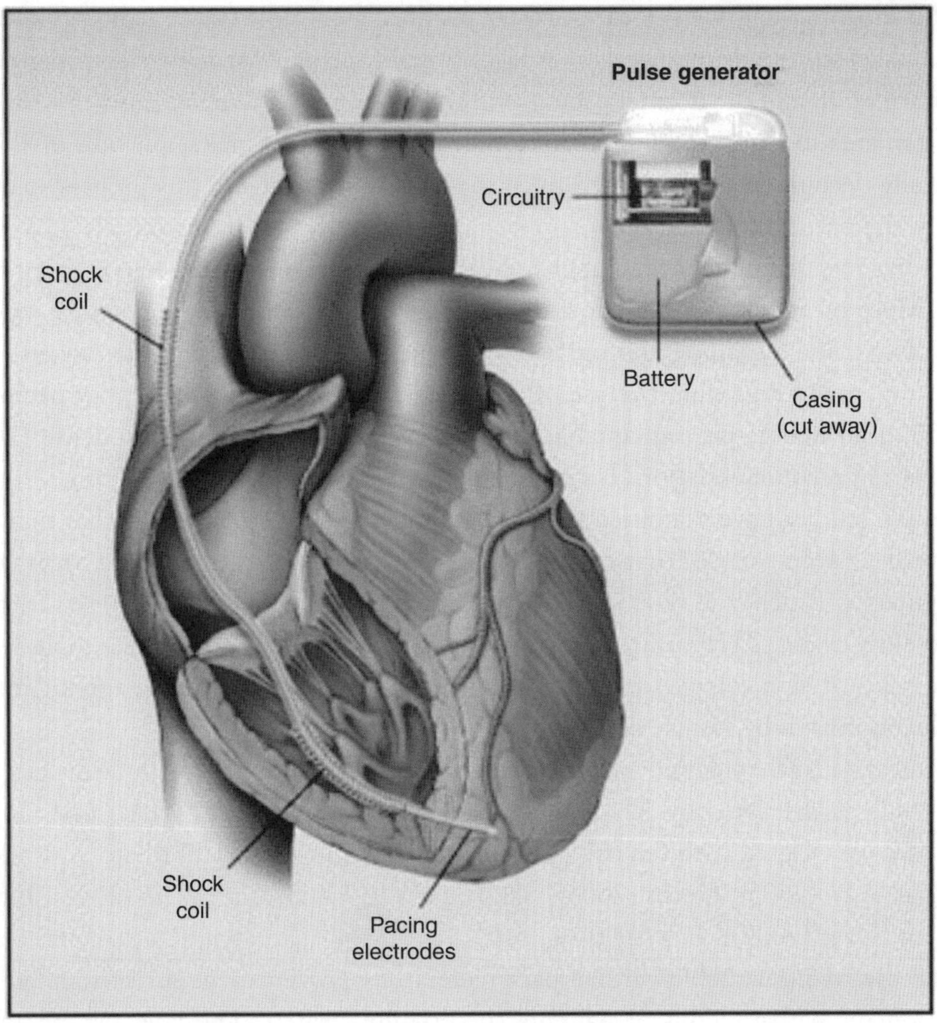

FIGURE 46: Dual pacemaker and cardioverter within the same case. Note the shocking coil which surrounds the pacing lead.

However, if the defibrillator were placed deeper within the body, then the amount of current necessary to defibrillate the heart is far less. As a result, the device is far smaller. For those patients who may require defibrillation on a regular enough basis, an implantable version can be placed in the same site as a cardiac pacemaker—below the collarbone with the leads threaded through a vein toward the heart. For those patients who require frequent cardiac pacing, but are also less frequently susceptible to ventricular fibrillation, then a combination unit of pacemaker and implantable cardioversion device (ICD) is implanted as shown in Figure 46. This device is sometimes called an implantable cardiac defibrillator (ICD).

The shocking coil for the cardioverter (defibrillator) unit (within the same housing as the pacemaker) surrounds the pacemaker lead. Should the pacer sensing electrodes detect ventricular fibrillation and the normal pacing protocol does not reestablish sinus rhythm, then the cardioverter fires to completely depolarize the heart. The pacemaker portion of the housing is electrically decoupled and insulated from the cardioverter during this process so as to avoid damage to the pacemaker circuitry.

Studies regarding the development of cardiac pacemakers have been ongoing for several decades. More recent studies include a discussion of permanent transvenous pacing by Petrie (2005), a discussion regarding appropriate selection of a pacemaker by Ellenbogen and Wood (2005), a discussion regarding the use of single chamber versus dual chamber pacing by Toff *et al.* (2005), an update on implantable pacemakers by Woodruff and Prudente (2005), and indications and recommendations for pacemaker therapy by Gregoratos (2005).

Several studies have examined the role of interfering external factors that may adversely affect a cardiac pacemaker. Of particular note has been the effect of an MRI on pacemaker activity as reported by Irnich *et al.* (2005), Vahlhaus (2005), and Del Ojo *et al.* (2005). Other external factors recently examined regarding interference with pacemaker performance included electromagnetic fields of high voltage lines by Scholten *et al.* (2005), and electromagnetic interference from induction ovens by Hirose *et al.* (2005).

Clinical issues that may adversely affect pacemaker performance include dislodged and/or fibrosed leads as discussed by Khan *et al.* (2005), Shimada *et al.* (2002), and Cutler *et al.* (1997). Issues associated with the use of implantable cardioverter defibrillators were examined by Nazarian *et al.* (2005), Gersh (2004), Sanders *et al.* (2005), Alter *et al.* (2005), Bryant *et al.* (2005), and Silver (2005).

The use of implantable cardiac pacemakers has proven to be an extremely successful "artificial organ" with millions of patients successfully treated. The later use of implantable defibrillators and dual pacemaker/defibrillators has also been successful, albeit with far less cases to date than with pacemakers alone.

CHAPTER 4
Dialysis

The human kidneys are responsible for continually cleansing blood of metabolic waste products (such as urea, uric acid, creatinine), excess ions, and excess water. The resultant filtrate is then transported to the ureter to be removed as urine. Approximately 20% of the blood supply is routed to the kidneys at any time in order to provide a continual means of removing potentially toxic materials from the blood. The water removal from the kidneys serves as a secondary means of controlling blood pressure, as total blood volume is a factor in venous return, cardiac output, and thus arterial pressure. There are two kidneys, each consisting of a network of millions of individual mass transfer elements called *nephrons*. Transport occurs across the surface of the nephron. As is the case with the lungs, which have similar elements called alveoli, the nephrons provide a means of greatly expanding the surface area available for mass transfer within a constrained volume. The theory is similar to that of a single large sphere as compared to millions of smaller spheres within the same volume. The net surface area of the large sphere is much smaller than that of the combined surface area of the numerous small spheres. Thus, mass transfer of metabolic waste products occurs by branching blood into smaller and smaller channels until the actual mass transfer occurs at the small end point—the nephron. Again, this is similar to the trachea branching into the bronchi and eventually to the alveoli in the lungs.

4.1 THE NEPHRON AND MASS TRANSFER

The kidney and its most elemental mass transfer unit, the nephron, are shown in Figure 47. Unlike most organs within the human body, the kidneys are supplied with blood directly from an arteriole, which branches from the renal artery. The use of an arteriole allows for control of the source hydrostatic pressure to each nephron (by means of vasoconstriction or vasodilation) and provides a constant source of pressure to each mass transfer unit.

The nephron consists of the glomerulus, a relatively porous membrane with pores diameters of 50 Å and pore lengths of 500 Å. These pore diameters allow a large amount of fluid to enter into the nephron tubule system, while excluding (by size) such blood elements as cells, large proteins, large sugars, etc. The glomerular filtration rate (GFR), the amount of fluid flow through all of these membranes, equals 125–150 mL/min. As the total blood flow to the kidneys

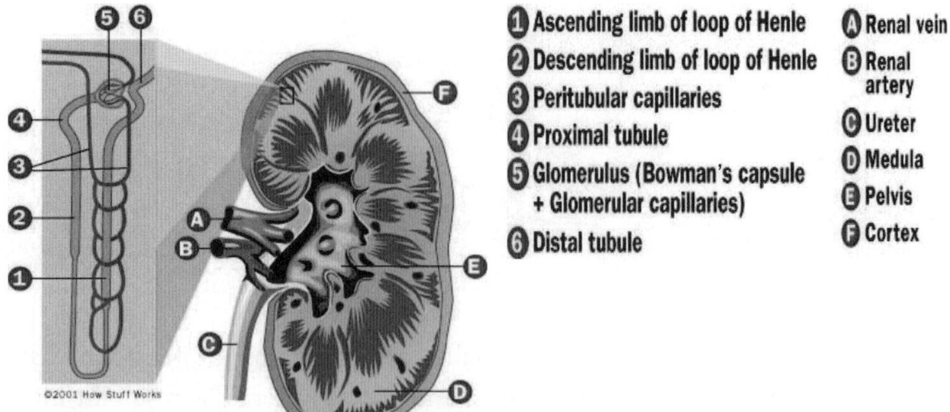

FIGURE 47: The human kidney and the nephron (shown in detail to the left).

equals 1000–1200 mL/min (20–25% of total blood flow), and plasma (the liquid portion of blood) represents approximately one half of the renal blood flow (for a hematocrit of 45% plus white cells and platelets), then the GFR represents approximately 25% of plasma renal flow (150 of 600 mL/min). This means that 25% of the liquid portion of the blood is leaving the blood as it travels into the tubules of the nephron. Luckily, only about 1–2 mL/min eventually finds its way to the ureter, which indicates that the remainder is returned to the blood supply. This occurs through reabsorption of the filtrate into blood vessels that parallel each nephron, the *vasa recta* or *peritubular capillaries* as is seen in Figure 47. The glomerular filtration rate is controlled by a hydrostatic pressure gradient as well as an osmotic pressure gradient (from the blood side to the filtrate side across the glomerulus). The hydrostatic pressure gradient is a relatively large 50 mmHg (for so short of a "pore" length) and the reverse osmotic pressure gradient is 25 mmHg. The osmotic pressure gradient is reversed from the hydrostatic pressure gradient since cells and large proteins on the blood side do not travel across the pores, and exert an osmotic pressure from the filtrate side to the blood side. The net 25 mmHg pressure gradient that pushed fluid across the glomerulus is still quite large and results in a significant transport of fluid as was described above.

The filtrate (the name of the fluid once it enters the tubules of the nephron) then follows the tubule pathway from the proximal tubule, through the loop of Henle, then through the distal tubule and the collecting duct, eventually ending in the ureter. Along the way, through a combination of concentration-driven passive diffusion as well as active transport, positive ions are transported across the tubules into the surrounding extracellular fluid. Negative ions follow the same pathway, primarily by means of a charge balance as well as passive diffusion. Water follows along the same pathway by means of a concentration-driven osmosis. All of these constituents are then absorbed by the vasa recta to return to the blood supply.

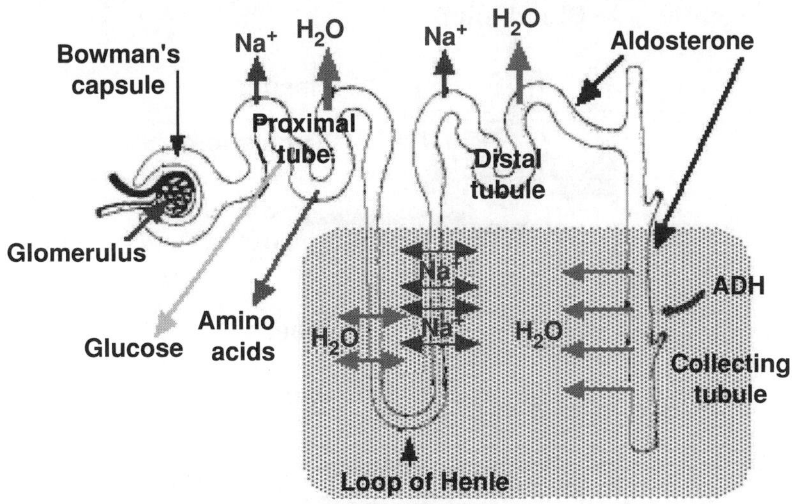

FIGURE 48: Movement of ions and water across the tubule system of the nephron as controlled by the hormones: aldosterone and ADH.

The loop of Henle operates by a countercurrent mechanism whereby the downward and upward sections are both contributing to the extracellular osmolarity, which is very high at the bottom of each loop and lessens toward the top of each loop. This occurs as ions are transported in both directions within each section of the loop, with most of the ion transport out of the loop at the bottom and into the loop at the top. The counter current mechanism allows the ascending loop to return fluid (and ions) into the descending loop and eventually into the extracellular fluid at the bottom of both loops. This fluid and ion mixture eventually finds its way back to the vasa recta.

The entire process of fluid flow and mass transfer across the glomerulus and through the tubule system is designed to transport a significant amount of fluid into the tubules with the bulk of it returning to the vasa recta. This process allows for selective retention of water and ions inside the tubules as controlled by hormones, which in turn allows for selective concentration of urine as is seen in Figure 48.

This approach—moving alot of fluid into the tubules to allow for a variable amount to be returned, cannot be duplicated by means of dialysis—the artificial cleansing of blood. Dialysis utilizes a standard transport of ions, metabolic wastes and water across a semipermeable membrane with this fluid mixture traveling across the membrane by means of simple diffusion, and not returning to the blood, as it does in the natural kidney.

4.2 DIALYSIS PROCEDURE AND THE DIALYSIS SYSTEM

Dialysis is the artificial cleansing of blood to remove the same components as those removed by the natural kidney: metabolic waste products such as urea, ureic acid, and creatinine; excess ions; and excess water. This is accomplished by means of concentration gradient-driven diffusion for

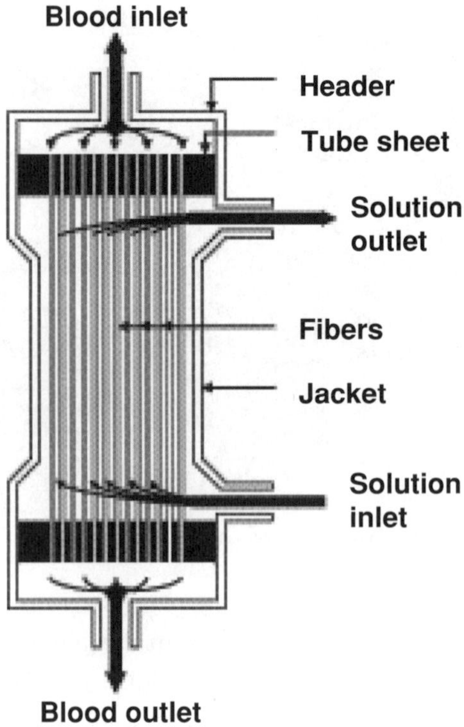

FIGURE 49: Capillary tube dialysis cartridge.

the first two components and by a pressure gradient for water. This process occurs within a capillary tube dialysis cartridge as is shown in Figure 49.

The cartridge is a polycarbonate canister containing 11 000 minute capillary tubes, each with pores small enough to allow transport of the three major blood components listed above, while too small to allow blood cells, large proteins, etc. to cross. In this fashion, the capillary tube pores are not unlike the glomerulus pores in the natural kidney nephron. The cartridge has upper and lower ports for blood to enter and exit the canister. Surrounding the capillary tubes is a fluid, called *dialysate*, which bathes the capillary tubes as serves as the recipient of the wastes, ions, and excess water which leaves the blood from within the capillary tubes. The dialysate is pumped from one side port of the cartridge and leaves from the other side port. The capillary tubes are typically composed of cellulose, which has been proven to be a biocompatible material. At the top and bottom of the capillary tube pack is a polyurethane "potting" section which appears to be a yellowish mass. In fact, this compound serves to briefly pool the incoming blood from the large inlet port so that it can more readily enter the minute capillary tubes. The reverse is true at the bottom of the capillary tubes where the blood enters the potting compound and then into the large outlet port. Dialysis cartridges come in many sizes and are matched to the size of the patient—from children to large adults as can be seen in Figure 50.

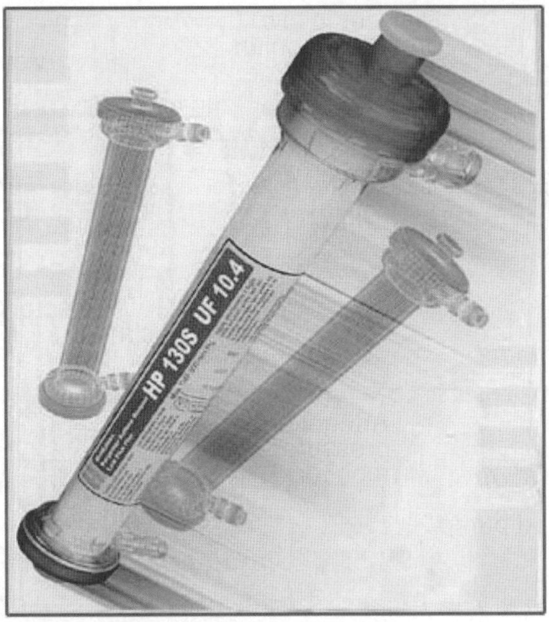

FIGURE 50: Capillary tube dialysis cartridges.

Dialysate is similar to clean blood, in that it has no waste products and has the normal level of ions. The typical dialysate mixture consists of the following normal level of blood ions:

NaCl	5.8 g/L	Na	132 meq/L
$NaHO_3$	4.5 g/L	K	2.0 meq/L
KCl	0.15 g/L	Cl	105 meq/L
$CaCl_2$	0.18 g/L	HCO_3	33 meq/L
$MgCl_2$	0.15 g/L	Ca	2.5 meq/L
Glucose	2.0 g/L	Mg	1.5 meq/L

The glucose is used to provide an osmotic gradient to assist in water transport from blood to dialysate. Blood with a high concentration of wastes and ions enters into the top of the dialysis cartridge with clean dialysate entering in the side port nearby. Along the length of the dialysis cartridge, simple diffusion takes place with wastes moving from the source of high concentration (blood) to the point of zero/low concentration (dialysate) across the capillary pores. Similarly, ion transport occurs from a high concentration (in the blood) to one of low concentration (in the dialysate). Water is transported from blood to dialysate by means of a concentration gradient of glucose assisted by a pressure gradient. A typical blood flow rate through the capillary tubes (as a whole) is 200 mL/min with the dialysate flow rate at 500 mL/min. The higher flow rate of dialysate ensures that dialysate with newly acquired waste products are quickly dispelled from

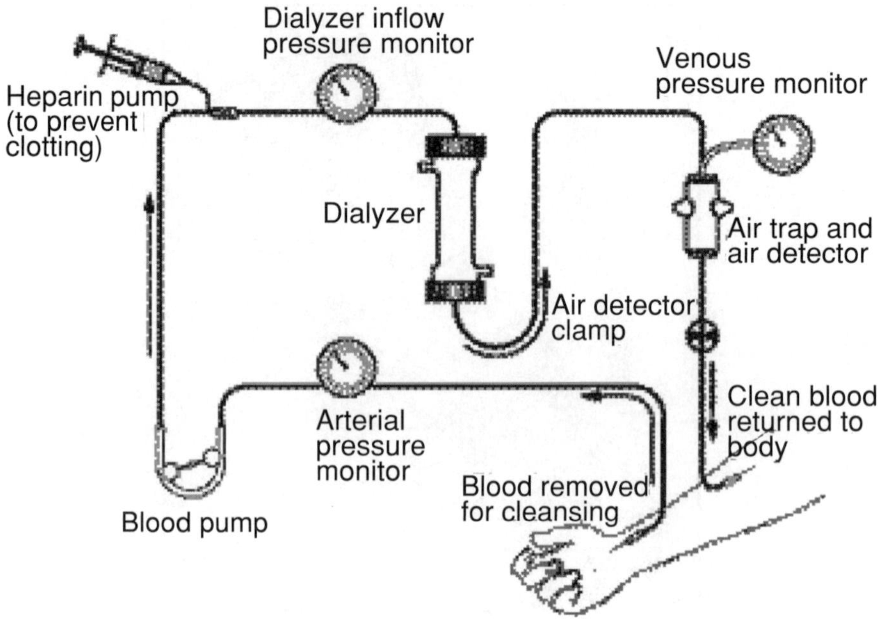

FIGURE 51: Typical dialysis system.

the cartridge and replaced with clean dialysate, thus maintaining a large concentration gradient for mass transfer.

A typical dialysis system is shown in Figure 51. Blood access from the body is connected from needles to tubing, which is routed throughout the machine and into the dialysis cartridge. Blood from the bottom of the dialysis cartridge is then routed through the remainder of the tubing and onwards to return to the body. The dialysis machine consists of various sensors and monitors along with two key elements—a roller pump which pushes blood slowly along a tubing pathway (to avoid stagnation and resultant clotting) and bubble traps to allow any ambient air from remaining in the blood to cause an air embolism.

The dialysis cartridge and tubing set are disposed of following dialysis, although in some dialysis centers, the cartridge itself is cleaned and may be reused. This latter issue will be discussed in more detail in a subsequent section.

Blood is accessed from the radial artery in the forearm and returns to the cephalic vein. Typical chronic dialysis patients undergo dialysis three times per week for 4 h/session. This results in numerous insertions of needles into the forearm. Although the skin becomes tough after time, the underlying blood vessels do not. As a result, chronic dialysis patients often undergo a minor procedure whereby an arteriovenous graft is placed below the skin connecting the radial artery to the cephalic vein as is seen in Figure 52.

The graft not only protects the blood vessels from repeated needle insertions, but also connects the high pressure artery to the low pressure vein—keeping the vein from collapsing.

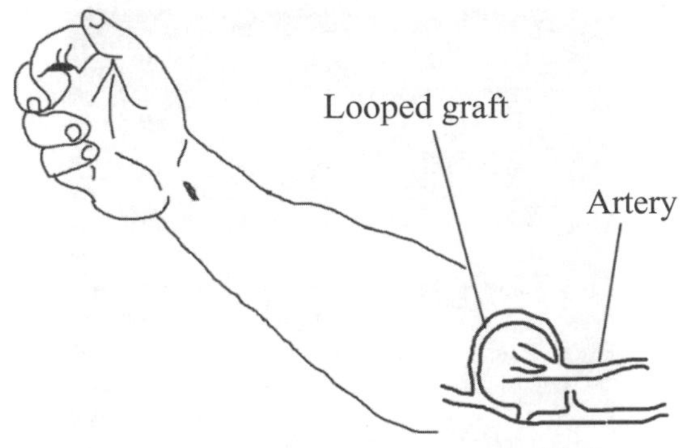

FIGURE 52: Arteriovenous graft in the forearm for chronic dialysis patients.

The pressure gradient from the artery to the vein also provides a gradient to propel blood through the tubing set and the dialysis cartridge. Thus, the roller pump within the dialysis system merely provides a boost for blood flow rather than having to provide the sole means of blood flow through the system.

A typical dialysis tubing set is shown in Figure 53 and includes bubble traps and sufficient tubing to connect from the artery/vein access needles to the dialysis machine, the dialysis cartridge, and back again to the forearm.

A typical dialysis machine is shown in Figure 54 with slots for the dialysis cartridge and bubble traps as well as an embedded roller pump and blood sensors.

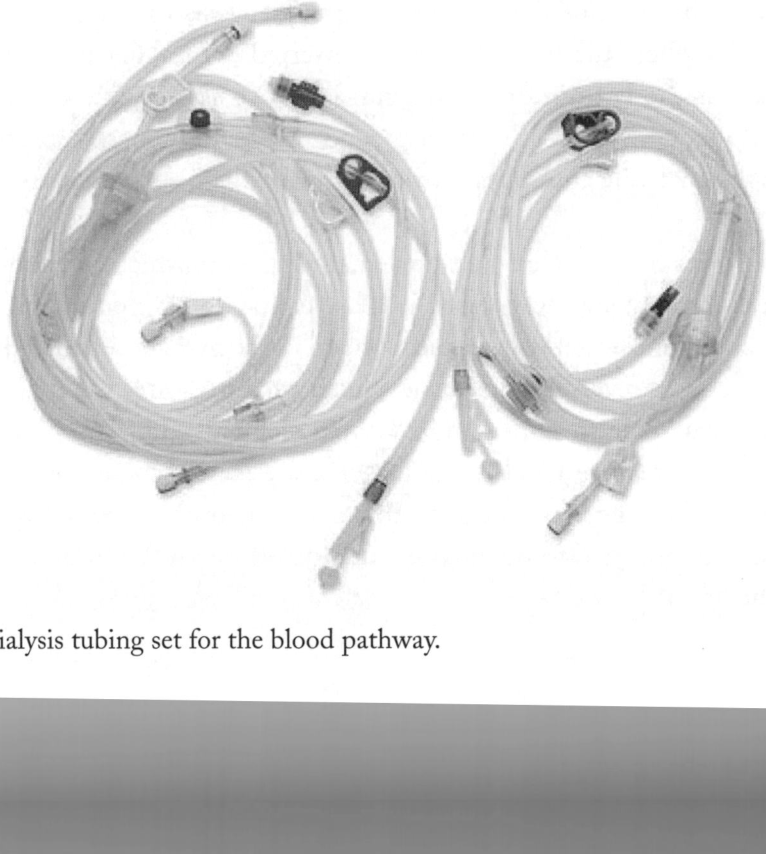

FIGURE 53: Dialysis tubing set for the blood pathway.

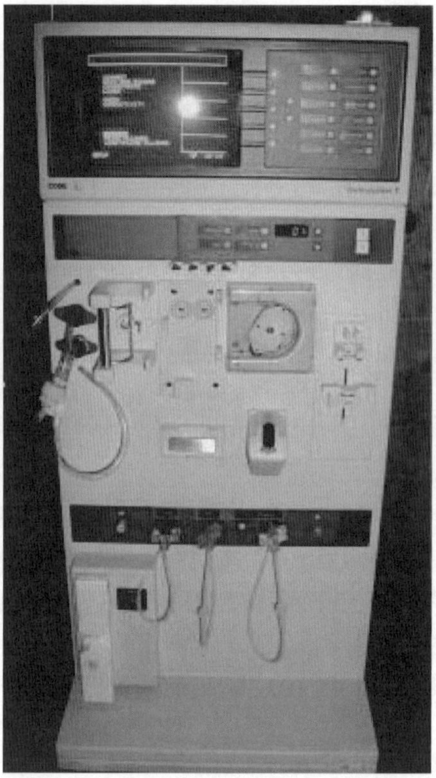

FIGURE 54: Dialysis machine.

The left-hand color-coded (blue and red) tubing is for the dialysate (clean and "dirty" ports). The slot to the right of this tubing is for the dialysis cartridge. To the right of the cartridge location is where the bubble traps are inserted. The roller pump is to the right of the bubble traps. At the bottom of the machine are two tubes with a tray below them. A dialysate concentrate jug is placed on the tray and one of the tubes is inserted into the jug. The machine draws dialysate concentrate from the jug, which is mixed with processed/treated water.

A typical dialysis center includes a water treatment room which converts municipal water into that which is clean enough to be used in close contact with blood. This water treatment includes a sediment filter, a water softener section, an ion exchanger section, an ultraviolet light to destroy bacteria, and a reverse osmosis unit which back-pressures water across an extremely fine (small pore) membrane to remove microscopic elements. The resulting treated water is sent to various patient treatment stations and each dialysis machine is connected to the water port via a hose located on the back of the machine. The mixture of treated water and dialysate concentrate is then transported to the dialysis cartridge through the front, color-coded tubing as was seen to the left of Figure 54.

A typical dialysis machine is approximately 60 inches high, 17 inches wide, 22 inches deep, and weights 190 pounds. It is on rollers to allow placement near the patient. The typical front-mounted screen displays the time on dialysis, time remaining, target water loss (in kg), water uptake rate (in kg/h), and blood pressure and/or heart rate.

4.3 HISTORY OF DIALYSIS

Thomas Graham, Professor of Chemistry at Anderson's University in Glasgow, coined the term dialysis in 1861. He noticed that crystalloids were able to diffuse through vegetable parchment coated with albumin (which acted as a semipermeable membrane). He called this "dialysis." Using this method, he was able to extract urea from urine. In 1913, Abel, Rowntree, Turner, and colleague constructed the first artificial kidney. They used hirudin, produced from leeches obtained from Parisian barbers, as an anticoagulant. They passed animal blood from an arterial cannula through celloidin tubes that were contained in a glass "jacket." The glass jacket was filled with saline or artificial serum. They coined the term "artificial kidney." Blood was returned into the vein of the animal via another cannula. The inventors wrote, "this apparatus might be applied to human beings suffering from certain toxic states, especially if due to kidney damage, in the hope of tiding a patient over a dangerous chemical emergency." The apparatus was never used to treat a patient (Robson JS, 1978). George Haas from Germany performed the first successful human dialysis in autumn 1924. The dialysis was performed on a patient with terminal uremia "because this was a condition against which the doctor stands otherwise powerless." The dialysis lasted for 15 min, and no complications occurred.

The first practical human hemodialysis machine was developed by WJ Kolff and H Berk from the Netherlands in 1943. This rotating drum artificial kidney consisted of 30–40 m of cellophane tubing in a stationary 100-L tank. It was Kolff who made clinicians and experimentalists interested in the treatment of uremia, and this machine delivered the effective hemodialysis treatments. This rotating drum machine is seen in Figure 55.

In 1946, Nils Alwall produced the first dialyzer with controllable ultrafiltration. It consisted of 10–11 m of cellophane tubing wrapped around a stationary, vertical drum made of a metal screen—resembling a rotating drum device stood on its end. In 1956, Kolff and Watschinger developed the principles of the Alwall machine to develop the "twin coil" artificial kidney (Figures), a modification of the "pressure cooker" dialyzer developed by Inouye and Engelberg in 1952. The first patients treated by dialysis were all believed to have acute renal failure. The methods in use for getting adequate flows of blood into the machine exhausted veins and arteries very quickly, and only a few dialysis treatments could be undertaken. The development of methods to use blood vessels repeatedly while preserving them made it possible to contemplate keeping a few patients alive for longer periods even though they had permanent

FIGURE 55: Kolff rotating drum dialysis machine—the first practical hemodialysis system.

renal failure. The arteriovenous shunt was the key development. The first substantial program for dialysis of patients with chronic renal failure began in Seattle in the same year.

Home hemodialysis was introduced to overcome the difficulties in providing adequate facilities in hospitals for the increasing number of patients being put forward for treatment. If a relative provided help for the patient, it could be carried out without the use of doctors, nurses or hospital premises, extending the number of patients that could be treated, as well as being better for the patient. However, in 1965, at the American Society of Artificial Internal Organs meeting, reports of home hemodialysis of four patients in Boston and two in Seattle were supplemented by a report of two patients treated at home in London (Shaldon). All reported success and plans to expand their programs.

Today, hundreds of thousands of chronic dialysis patients undergo routine, periodic dialysis three times per week at local dialysis centers located throughout the nation. Dialysis represents one of the most successful organ replacement systems with millions of patients treated successfully for partial or total renal failure. Diabetes and heart disease remain principal causes for renal failure with alcohol and substance abuse also accounting for numerous cases.

4.4 DIALYZER CARTRIDGE REUSE

Many local dialysis centers are privately operated facilities. Although dialysis conducted in hospitals for hospital patients (acute dialysis centers) results in single use of dialysis cartridges, many private facilities reuse cartridges. This entails the rinsing of the cartridge following patient dialysis, after which the cartridge is treated with a sterilant such as formaldehyde, gluteraldehyde, or renalin. Up to four cartridges are placed into a cleaning machine that provides several rinse steps and introduction of the sterilant. The patient's name is written on the cartridge and the cartridge is placed into a bin where it will be stored until that patient needs it for the next session.

The cartridge is then rinsed several times before used for that patient. Although there is a capital expense in purchasing the cleaning machine along with supplies and manpower needed to rinse and sterilize the cartridges, there is a net modest savings rather than purchasing a new cartridge for every patient session. The cartridges cost $10–12 each and the dialysis facility is reimbursed at a set rate for each patient session by insurance, Medicare, or Medicaid. If there is a small net savings, even after expenses associated with cleaning and storing the cartridges, then the dialysis facility would show a large profit, given the number of patient sessions conducted each year. A typical dialysis center might have 20 patient stations. Each patient undergoes dialysis three times per week for 4 h/session. If the center runs an MWF morning, MWF afternoon, T-Th-S morning and T-Th-S afternoon patient cohort, then there are a total of 240 sessions/week (3 sessions for each of 20 stations for each of four patient cohorts). As such, even a modest savings per session adds up to substantial overall savings and profit.

Dialyzer-reuse machines are shown in Figure 56.

Although cost issues are often at the forefront of the rationale to reuse dialyzer cartridges, there has been a considerable debate regarding the efficacy and safety when reusing cartridges. One health-related rationale for the reuse of cartridges is the "first use syndrome" which refers

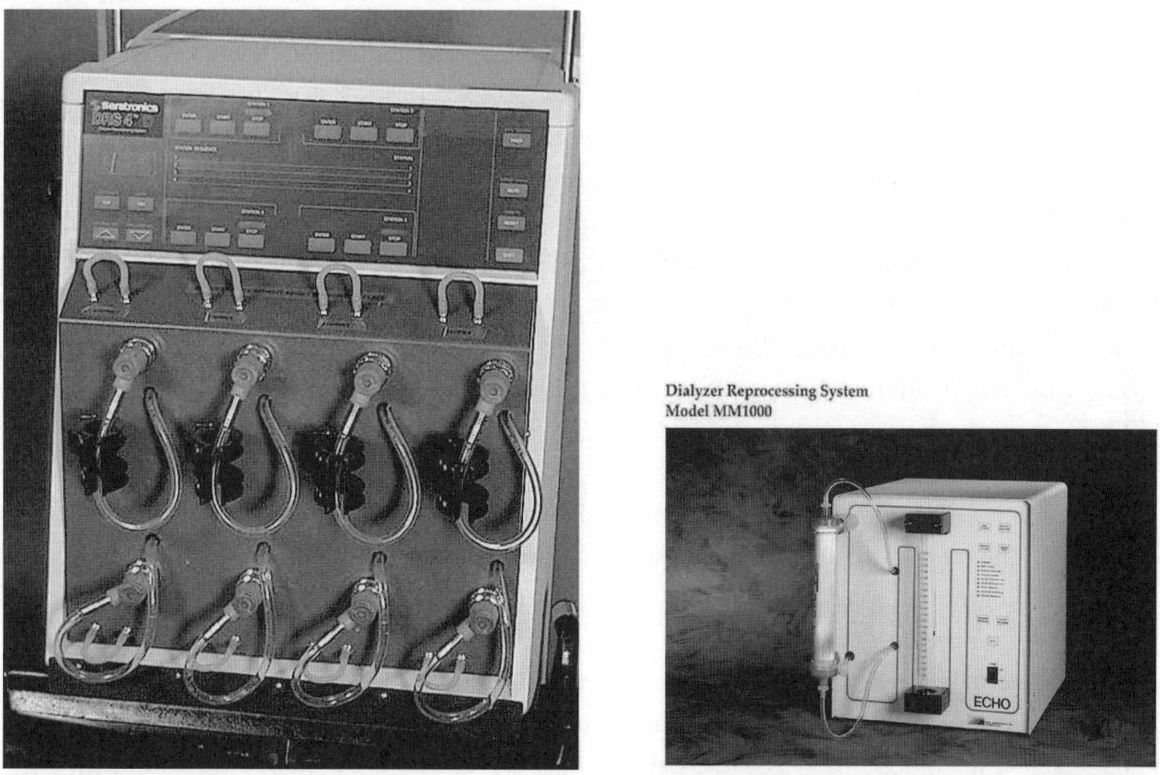

FIGURE 56: Dialyzer-reuse machines: a four-cartridge system on the left and a single-cartridge unit on the right.

to residual manufacturing byproducts still inside the cartridge when it is unpacked from its wrapping. By reusing cartridges, the uses after the first use no longer suffer from this problem. On the other hand, the health-related arguments against reuse are that a) there is residual sterilant (formaldehyde, etc.) even after rinsing that might harm the patient over time, and b) the clearance of wastes and ions from the dialysis cartridge is reduced over time as blood byproducts clog some of the capillary tubes. Many dialysis cartridges may be reused up to 25 times per patient before the clearance levels fall below 80% of the maximum, which is the standard cutoff before the cartridge must be discarded. The Association for the Advancement of Medical Instrumentation (AAMI) sets standards for cartridge clearance levels. Numerous studies have examined the various factors regarding reuse of dialyzer cartridges including Fan *et al.* (2005), Robinson and Feldman (2005), Szathmary *et al.* (2004), Narsipur (2004), Stragier (2003), Ward and Ouseph (2003), Rahmati *et al.* (2003), and Parks (2002), among others. The issues regarding reuse continue to be debated, although the prevalence of reuse is high among the vast majority of chronic, privately operated dialysis centers.

As dialysis is utilized by hundreds of thousands of patients and the process affects blood chemistry and overall health, there have been numerous studies on the process itself, on disease states that require dialysis, on the various techniques and technologies regarding dialysis. In particular, there are issues associated with cardiovascular disease that impact hemodialysis and vice versa. Studies that have examined this link are numerous and include Familoni *et al.* (2005), Saxena and Panhotra (2005), Di Benedetto *et al.* (2005), Iorio *et al.* (2005), and Ronco and Tetta (2005), among others. Issues associated with blood access and the use of AV fistulas have been studied by Dember *et al.* (2005), Wijnen *et al.* (2005), and Peirce *et al.* (2005), among others. Issues related to blood chemistry and general health considerations have been examined by Panichi *et al.* (2005), Piccoli *et al.* (2005), Prado *et al.* (2005), Gusella *et al.* (2005), Lee *et al.* (2005), and Kiss *et al.* (2005), among others. In addition, there have been studies that have examined the effect of missed dialysis sessions in patient health as well as the desire of patients to quit dialysis including those by Gee (2005), Davison and Jhangri (2005), and Unruh *et al.* (2005), among others.

Dialysis remains a popular and cost-effective means of augmenting reduced kidney function and is a viable alternative to kidney transplantation. The latter approach can be costly and the numbers of available donor kidneys that are properly blood typed, tissue typed, and in a nearby geographic zone are relatively few in number. Although there are continuing issues related to the health of dialysis patients, particularly those with cardiovascular disease, the use of dialysis remains steady and there is no projection that it will decrease in the near future.

The future of dialysis may be in the development of a miniaturized, implantable system. Nissenson *et al.* (2005a,b) at the UCLA Medical School have developed a nanotechnology-based artificial nephron system that employs two membranes operating in series within one very small cartridge. The first membrane mimics the function of the glomerulus, allowing substances up to the size of albumin (MW 69 000). The second membrane mimics the function of the renal tubules, selectively reclaiming designated solutes, in a manner similar to that of the natural kidney. No dialysate is used in this device. As such, this miniaturized system is closer to that of an actual nephron than to currently employed dialysis systems.

References

Aagaard J. The Carbomedics aortic heart valve prosthesis: a review. *J Cardiovasc Surg (Torino)* 2004 Dec;45(6):531–4.

Alter P, Waldhans S, Plachta E, Moosdorf R, Grimm W. Complications of implantable cardioverter defibrillator therapy in 440 consecutive patients. *Pacing Clin Electrophysiol* 2005 Sep;28(9):926–32.

Anderson J, Wood HG, Allaire PE, Olsen DB. Numerical analysis of blood flow in the clearance regions of a continuous flow artificial heart pump. *Artif Organs* 2000 Jun;24(6): 492–500.

Arcidiacono G, Corvi A, Severi T. Functional analysis of bioprosthetic heart valves. *J Biomech* 2005 Jul;38(7):1483–90. Epub 2004 Nov 11.

Bach DS, Sakwa MP, Goldbach M, Petracek MR, Emery RW, Mohr FW. Hemodynamics and early clinical performance of the St. Jude Medical Regent mechanical aortic valve. *Ann Thorac Surg* 2002 Dec;74(6):2003–9; discussion 2009.

Baldwin JT, Tarbell JM, Deutsch S, Geselowitz DB. Mean flow velocity patterns within a ventricular assist device. *ASAIO Trans* 1989 Jul–Sep;35(3):429–33.

Bech-Hanssen O, Gjertsson P, Houltz E, Wranne B, Ask P, Loyd D, Caidahl K. Net pressure gradients in aortic prosthetic valves can be estimated by Doppler. *J Am Soc Echocardiogr* 2003 Aug;16(8):858–66.

Bernacca GM, McColl JH, Wheatley DJ. Comparison of prosthetic valve hydrodynamic function: objective testing using statistical multilevel modeling. *J Heart Valve Dis* 2004 May;13(3):467–77.

Birks EJ, Yacoub MH, Banner NR, Khaghani A. The role of bridge to transplantation: should LVAD patients be transplanted? *Curr Opin Cardiol* 2004 Mar;19(2):148–53.

Bolno PB, Kresh JY. Physiologic and hemodynamic basis of ventricular assist devices. *Cardiol Clin* 2003 Feb;21(1):15–27.

Botzenhardt F, Eichinger WB, Bleiziffer S, Guenzinger R, Wagner IM, Bauernschmitt R, Lange R. Hemodynamic comparison of bioprostheses for complete supra-annular position in patients with small aortic annulus. *J Am Coll Cardiol* 2005 Jun 21;45(12):2054–60.

Botzenhardt F, Gansera B, Kemkes BM. Mid-term hemodynamic and clinical results of the stented porcine medtronic mosaic valve in aortic position. *Thorac Cardiovasc Surg* 2004 Feb;52(1):34–41.

Brodell GK, Wilkoff BL. A novel approach to determining the cause of pacemaker lead failure. *Cleve Clin J Med* 1992 Jan–Feb;59(1):91–2.

Bryant J, Brodin H, Loveman E, Payne E, Clegg A. The clinical and cost-effectiveness of implantable cardioverter defibrillators: a systematic review. *Health Technol Assess* 2005 Sep;9(36):1–150, iii.

Cook WL, Jassal SV. Prevalence of falls among seniors maintained on hemodialysis. *Int Urol Nephrol* 2005;37(3):649–52.

Curtas AR, Wood HG, Allaire PE, McDaniel JC, Day SW, Olsen DB. Computational fluid dynamics modeling of impeller designs for the HeartQuest left ventricular assist device. *ASAIO J* 2002 Sep–Oct;48(5):552–61.

Cutler NG, Karpawich PP, Cavitt D, Hakimi M, Walters HL. Steroid-eluting epicardial pacing electrodes: six year experience of pacing thresholds in a growing pediatric population. *Pacing Clin Electrophysiol* 1997 Dec;20(12 Pt 1):2943–8.

Davis PK, Pae WE Jr, Pierce WS. Toward an implantable artificial heart. Experimental and clinical experience at The Pennsylvania State University. *Invest Radiol* 1989 Jan;24(1): 81–7.

Davison SN, Jhangri GS. The impact of chronic pain on depression, sleep, and the desire to withdraw from dialysis in hemodialysis patients. *J Pain Symptom Manage* 2005 Nov;30(5): 465–73.

Day SW, McDaniel JC, Wood HG, Allaire PE, Landrot N, Curtas A. Particle image velocimetry measurements of blood velocity in a continuous flow ventricular assist device. *ASAIO J* 2001 Jul–Aug;47(4):406–11.

Day SW, McDaniel JC, Wood HG, Allaire PE, Song X, Lemire PP, Miles SD. A prototype HeartQuest ventricular assist device for particle image velocimetry measurements. *Artif Organs* 2002 Nov;26(11):1002–5.

Deiwick M, Hoffmeier A, Tjan TD, Krasemann T, Schmid C, Scheld HH. Heart failure in children—mechanical assistance. *Thorac Cardiovasc Surg* 2005 Feb;53(Suppl 2): S135–40.

Del Ojo JL, Moya F, Villalba J, Sanz O, Pavon R, Garcia D, Pastor L. Is magnetic resonance imaging safe in cardiac pacemaker recipients? *Pacing Clin Electrophysiol* 2005 Apr;28(4):274–8.

Dember LM, Kaufman JS, Beck GJ, Dixon BS, Gassman JJ, Grene T, Himmelfarb J, Hunsicker LG, Kusek JW, Lawson JH, Middloton JP, Radeva M, Schwab SJ, Whitting JF, Feldman HI; DAC Study Group. Design of the dialysis access consortium (DAC) clopidogrel prevention of early AV fistula thrombosis trial. *Clin Trials* 2005;2(5):413–22.

Deng MC, Naka Y. Mechanical circulatory support devices—state of the art. *Heart Fail Monit* 2002;2(4):120–8.

Di Benedetto A, Marcelli D, D'Andrea A, Cice G, Cappabianca F, Pacchiano G, D'Amato R, Oggero AR, Bonanno D, Pergamo O, Calabro R. Risk factors and underlying cardiovascular diseases in incident ESRD patients. *J Nephrol* 2005 Sep–Oct;18(5):592–8.

Dumont K, Vierendeels JA, Segers P, Van Nooten GJ, Verdonck PR. Predicting ATS open pivot heart valve performance with computational fluid dynamics. *J Heart Valve Dis* 2005 May;14(3):393–9.

Duncan BW. Mechanical circulatory support for infants and children with cardiac disease. *Ann Thorac Surg* 2002 May;73(5):1670–7.

Ellenbogen KA, Wood MA. Pacemaker selection—the changing definition of physiologic pacing. *N Engl J Med* 2005 Jul 14;353(2):202–4.

Familoni OB, Alebiosu CO, Avodele OE. Effects and outcome of haemodialysis on QT intervals and QT dispersion in patients with chronic kidney disease. *Cardiovasc J S. Afr* 2005 Nov 30:1–4. [Epub ahead of print]

Fan Q, Liu J, Ebben JP, Collins AJ. Reuse-associated mortality in incident hemodialysis patients in the United States, 2000 to 2001. *Am J Kidney Dis* 2005 Oct;46(4):661–8.

Gasparini G, Curnis A, Gulizia M, Occhetta E, Corrado A, Bontempi L, Mascioli G, Maura Francese G, Bortnik M, Magnani A, Di Gregorio F, Barbetta A, Raviele A. Rate-responsive pacing regulated by cardiac haemodynamics. *Europace* 2005 May;7(3): 234–41.

Gee MD. Thought and discussion among patients about stopping dialysis. *Nephrol News Issues* 2005 Nov;19(12):80–1, 83–4.

Gersh B. Do implantable cardioverter defibrillators lower mortality risk in patients with cardiomyopathy? *Nat Clin Pract Cardiovasc Med* 2004 Nov;1(1):16–17.

Goetze S, Brechtken J, Agler DA, Thomas JD, Sabik JF 3rd, Jaber WA. In vivo short-term Doppler hemodynamic profiles of 189 Carpentier-Edwards Perimount pericardial bioprosthetic valves in the mitral position. *J Am Soc Echocardiogr* 2004 Sep;17(9):981–7.

Gregoratos G. Indications and recommendations for pacemaker therapy. *Am Fam Physician* 2005 Apr 15;71(8):1563–70.

Grigioni M, Daniele C, Del Gaudio C, Morbiducci U, Balducci A, D'Avenio G, Barbaro V. Three-dimensional numeric simulation of flow through an aortic bileaflet valve in a realistic model of aortic root. *ASAIO J* 2005 May–Jun;51(3):176–83.

Gura MT. Implantable cardioverter defibrillator therapy. *J Cardiovasc Nurs* 2005 Jul–Aug;20(4):276–87.

Gusella M, Rebeschini M, Cartei G, Ferrazzi E, Ferrari M, Padrini R. Effect of hemodialysis on the metabolic clearance of 5-Fluorouracil in a patient with end-stage renal failure. *Ther Drug Monit* 2005 Dec;27(6):816–18.

Hirose M, Hida M, Sato E, Kokubo K, Nie M, Kobayashi H. Electromagnetic interference of implantable unipolar cardiac pacemakers by an induction oven. *Pacing Clin Electrophysiol* 2005 Jun;28(6):540–8.

Hirsch DJ, Cooper JR Jr. Cardiac failure and left ventricular assist devices. *Anesthesiol Clin North America* 2003 Sep;21(3):625–38.

Ho TH, Van Nguyen P, Phan PK, Pham VN. Up to nine-years' experience with the Allcarbon prosthetic heart valve. *J Heart Valve Dis* 2005 Jul;14(4):512–17.

Huang CY, Tuan TC, Lee WS, Cheng CM, Lin SJ, Kong CW. Long-term efficacy and stability of atrial sensing in VDD pacing. *Clin Cardiol* 2005 Apr;28(4):203–7.

Ichikawa S, Nose Y. Centrifugal blood pumps for various clinical needs. *Artif Organs* 2002 Nov;26(11):916–18.

Iorio L, Violi F, Simonelli R, Nacca RG, Caliendo A, De Santo LS. Temporary dialysis treatments for heart failure in chronic kidney disease. *Semin Nephrol* 2005 Nov;25(6):408–12.

Irnich W, Irnich B, Bartsch C, Stertmann WA, Gufler H, Weiler G. Do we need pacemakers resistant to magnetic resonance imaging? *Europace* 2005 Jul;7(4):353–65.

Jamieson WR, von Lipinski O, Miyagishima RT, Burr LH, Janusz MT, Ling H, Fradet GJ, Chan F, Germann E. Performance of bioprostheses and mechanical prostheses assessed by composites of valve-related complications to 15 years after mitral valve replacement. *J Thorac Cardiovasc Surg* 2005 Jun;129(6):1301–8.

Kadish A, Mehra M. Heart failure devices: implantable cardioverter-defibrillators and biventricular pacing therapy. *Circulation* 2005 Jun 21;111(24):3327–35.

Kerrigan JP, Yamazaki K, Meyer RK, Mori T, Otake Y, Outa E, Umezu M, Borovetz HS, Kormos RL, Griffith BP, Koyanagi H, Antaki JF. High-resolution fluorescent particle-tracking flow visualization within an intraventricular axial flow left ventricular assist device. *Artif Organs* 1996 Jun;20(6):534–40.

Khan EM, Voudouris AA, Hood RE, Shorofsky SR. Repositioning of a dislodged and fibrosed ventricular lead. *J Interv Card Electrophysiol* 2005 Jun;13(1):55–7.

Kherani AR, Maybaum S, Oz MC. Ventricular assist devices as a bridge to transplant or recovery. *Cardiology* 2004;101(1–3):93–103.

Kirsch ME, Tzvetkov B, Vermes E, Pouzet B, Sauvat S, Loisance D. Clinical and hemodynamic performance of the 19-mm medtronic mosaic bioprosthesis. *J Heart Valve Dis* 2005 May;14(3):433–9.

Kiss E, Keusch G, Zanetti M, Jung T, Schwarz A, Schocke M, Jaschke W, Czermak BV. Dialysis-related amyloidosis revisited. *AJR Am J Roentgenol* 2005 Dec;185(6):1460–7.

Kleine P, Abdel-Rahman U, Klesius AA, Scherer M, Simon A, Moritz A. Comparison of hemodynamic performance of Medtronic Hall 21 mm versus St. Jude Medical 23 mm prostheses in pigs. *J Heart Valve Dis* 2002 Nov;11(6):857–63.

Koertke H, Seifert D, Drewek-Platena S, Koerfer R. Hemodynamic performance of the Medtronic Advantage prosthetic heart valve in the aortic position: echocardiographic evaluation at one year. *J Heart Valve Dis* 2003 May;12(3):348–53.

Labrosse MR, Beller CJ, Robicsek F, Thubrikar MJ. Geometric modeling of functional trileaflet aortic valves: development and clinical applications. *J Biomech* 2005 Sep 29. [Epub ahead of print]

Labrousse LM, Choukroun E, Serena D, Billes MA, Madonna F, Deville C. Prospective study of hemodynamic performances of standard ATS and AP-ATS valves. *J Heart Valve Dis* 2003 May;12(3):341–7.

Lamson TC, Ojan OS, Geselowitz DB, Tarbell JM. A two-phase fluid volume compensation chamber for an electric ventricular assist device. *Artif Organs* 1990 Aug;14(4):270–7.

Lee CT, Hsu CY, Lam KK, Lin CR, Chen JB. Inflammatory markers and hepatocyte growth factor in sustained hemodialysis hypotension. *Artif Organs* 2005 Dec;29(12):980–3.

Maisel WH. Cardiovascular device development: lessons learned from pacemaker and implantable cardioverter-defibrillator therapy. *Am J Ther* 2005 Mar–Apr;12(2):183–5.

Malouf JF, Ballo M, Connolly HM, Hodge DO, Herges RM, Mullany CJ, Miller FA. Doppler echocardiography of 119 normal-functioning St Jude Medical mitral valve prostheses: a comprehensive assessment including time–velocity integral ratio and prosthesis performance index. *J Am Soc Echocardiogr* 2005 Mar;18(3):252–6.

Manning KB, Miller GE. Flow through an outlet cannula of a rotary ventricular assist device. *Artif Organs* 2002 Aug;26(8):714–23.

Matsuda H, Matsumiya G. Current status of left ventricular assist devices: the role in bridge to heart transplantation and future perspectives. *J Artif Organs* 2003;6(3):157–61.

Medart D, Schmitz C, Rau G, Reul H. Design and in vitro performance of a novel bileaflet mechanical heart valve prosthesis. *Int J Artif Organs* 2005 Mar;28(3):256–63.

Mehra MR. Ventricular assist devices: destination therapy or just another stop on the road? *Curr Heart Fail Rep* 2004 Apr–May;1(1):36–40.

Mesana TG. Rotary blood pumps for cardiac assistance: a "must"? *Artif Organs* 2004 Feb;28(2):218–25.

Miller GE, Madigan M, Fink R. A preliminary flow visualization study in a multiple disk centrifugal artificial ventricle. *Artif Organs* 1995 Jul;19(7):680–4.

Miller LW. Patient selection for the use of ventricular assist devices as a bridge to transplantation. *Ann Thorac Surg* 2003 Jun;75(6 Suppl):S66–71.

Mitamura Y, Okamoto E, Hirano A, Mikami T. Development of an implantable motor-driven assist pump system. *IEEE Trans Biomed Eng* 1990 Feb;37(2):146–56.

Moffatt-Bruce SD, Jamieson WR. Long-term performance of prostheses in mitral valve replacement. *J Cardiovasc Surg (Torino)* 2004 Oct;45(5):427–47.

Morgan JA, Mazzeo PA, Flannery MR, Oz MC, Naka Y. Effects of changes in UNOS policy regarding left ventricular assist devices. *J Heart Lung Transplant* 2004 May;23(5): 620–2.

Mulder MM, Hansen AC, Mohammad SF, Olsen DB. In vitro investigation of the St. Jude Medical Isoflow centrifugal pump: flow visualization and hemolysis studies. *Artif Organs* 1997 Aug;21(8):947–53.

Myers TJ, Robertson K, Pool T, Shah N, Gregoric I, Frazier OH. Continuous flow pumps and total artificial hearts: management issues. *Ann Thorac Surg* 2003 Jun;75(6 Suppl):S79–85.

Narsipur SS. Measurement of fiber bundle volume in reprocessed dialyzers. *Clin Nephrol* 2004 Feb;61(2):130–3.

Nazarian S, Maisel WH, Miles JS, Tsang S, Stevenson LW, Stevenson WG. Impact of implantable cardioverter defibrillators on survival and recurrent hospitalization in advanced heart failure. *Am Heart J* 2005 Nov;150(5):955–60.

Nemeh HW, Smedira NG. Mechanical treatment of heart failure: the growing role of LVADs and artificial hearts. *Cleve Clin J Med* 2003 Mar;70(3):223–6, 229–33.

Nissenson AR, Ronco C, Pergamit G, Edelstein M, Watts R. The human nephron filter: toward a continuously functioning, implantable artificial nephron system. *Blood Purif* 2005a;23(4):269–74. Epub 2005 May 20.

Nissenson AR, Ronco C, Pergamit G, Edelstein M, Watts, R. Continuously functioning artificial nephron system: the promise of nanotechnology. *Hemodial Int* 2005b Jul;9(3):210–17.

Noiseux N, Khairy P, Fournier A, Vobecky SJ. Thirty years of experience with epicardial pacing in children. *Cardiol Young* 2004 Oct;14(5):512–19.

Okamoto E, Hashimoto T, Mitamura Y. Design of a miniature implantable left ventricular assist device using CAD/CAM technology. *J Artif Organs* 2003;6(3):162–7.

Ostberg BN, Ostberg GN. Total artificial heart design with integrated electric motor. *Med Biol Eng Comput* 1987 May;25(3):345–6.

Padeletti L, Barold SS. Digital technology for cardiac pacing. *Am J Cardiol* 2005 Feb 15;95(4):479–82.

Panichi V, Rizza GM, Taccola D, Paoletti S, Mantuano E, Migliori M, Grangioni S, Flippi C, Carpi A. C-reactive protein in patients on chronic hemodialysis with different techniques and different membranes. *Biomed Pharmacother* 2005 Oct 25. [Epub ahead of print]

Parks MS. Practical considerations in converting to Renalin 100 for dialyzer reprocessing. *Nephrol News Issues* 2002 Nov;16(12):55, 59–62, 64.

Peirce RM, Funaki B, Van Ha TG, Lorenz JM. Percutaneous declotting of virgin femoral hemodialysis grafts. *AJR Am J Roentgenol* 2005 Dec;185(6):1615–19.

Petrie JP. Permanent transvenous cardiac pacing. *Clin Tech Small Anim Pract* 2005 Aug;20(3):164–72.

Phillips SJ. Selecting the best heart valve for your patient: mechanical or tissue. *Am Heart Hosp J* 2004 Summer;2(3):149–52.

Piccoli GB, Amgnano A, Perrotta L, Piccoli G. Daily dialysis, nocturnal dialysis and randomized controlled trials: are we asking the right questions? *Kidney Int* 2005 Dec;68(6):2913–14.

Pierce WS, Rosenberg G, Snyder AJ, Pae WE Jr, Donachy JH, Waldhausen JA. An electric artificial heart for clinical use. *Ann Surg* 1990 Sep;212(3):339–43; discussion 343–4.

Pierrakos O, Vlachos PP, Telionis DP. Time-resolved DPIV analysis of vortex dynamics in a left ventricular model through bileaflet mechanical and porcine heart valve prostheses. *J Biomech Eng* 2004 Dec;126(6):714–26.

Prado M, Roa LM, Palma A, Milan JA. Double target comparison of blood-side methods for measuring the hemodialysis dose. *Kidney Int* 2005 Dec;68(6):2863–76.

Rahmati MA, Rahmati S, Hoenich N, Ronco C, Kaysen GA, Levin R, Levin NW. On-line clearance: a useful tool for monitoring the effectiveness of the reuse procedure. *ASAIO J* 2003 Sep–Oct;49(5):543–6.

Reinhartz O, Stiller B, Eilers R, Farrar DJ. Current clinical status of pulsatile pediatric circulatory support. *ASAIO J* 2002 Sep–Oct;48(5):455–9.

Reinlib L, Abraham W; Working Group of the National, Heart, Lung, and Blood Institute. Recovery from heart failure with circulatory assist: a working group of the National, Heart, Lung, and Blood Institute. *J Card Fail* 2003 Dec;9(6):459–63.

Rezaian GR, Aghasadeghi K, Kojuri J. Evaluation of the hemodynamic performance of St. Jude mitral prostheses: a pilot study by dobutamine-stress Doppler echocardiography. *Angiology* 2005 Jan–Feb;56(1):81–6.

Richenbacher WE, Pae WE Jr, Magovern JA, Rosenberg G, Snyder AJ, Pierce WS. Roller screw electric motor ventricular assist device. *ASAIO Trans* 1986 Jul–Sep;32(1):46–8.

Roberts PR. Follow up and optimisation of cardiac pacing. *Heart* 2005 Sep;91(9):1229–34.

Robinson BM, Feldman HI. Dialyzer reuse and patient outcomes: what do we know now? *Semin Dial* 2005 May–Jun;18(3):175–9.

Ronco C, Tetta C. Dialysis patients and cardiovascular problems: can technology solve the complex equation? *Expert Rev Med Devices* 2005 Nov;2(6):681–7.

Rosenberg G, Snyder A, Weiss W, Landis DL, Geselowitz DB, Pierce WS. A cam-type electric motor-driven left ventricular assist device. *J Biomech Eng* 1982 Aug;104(3):214–20.

Rosenberg G, Snyder AJ, Landis DL, Geselowitz DB, Donachy JH, Pierce WS. An electric motor-driven total artificial heart: seven months survival in the calf. *Trans Am Soc Artif Intern Organs* 1984;30:69–74.

Saito S, Nishinaka T, Westaby S. Hemodynamics of chronic nonpulsatile flow: implications for LVAD development. *Surg Clin North Am* 2004 Feb;84(1):61–74.

Samuels L. Biventricular mechanical replacement. *Surg Clin North Am* 2004 Feb;84(1):309–21.

Sanders GD, Hlatky MA, Owens DK. Cost-effectiveness of implantable cardioverter-defibrillators. *N Engl J Med* 2005 Oct 6;353(14):1471–80.

Saxena AK, Panhotra BR. Cardiovascular mortality and dialysis access-related infections: is there a link? *Am J Kidney Dis* 2005 Dec;46(6):1149–50.

Scholten A, Joosten S, Silny J. Unipolar cardiac pacemakers in electromagnetic fields of high voltage overhead lines. *J Med Eng Technol* 2005 Jul–Aug;29(4):170–5.

Seitelberger R, Bialy J, Gottardi R, Seebacher G, Moidl R, Mittelbock M, Simon P, Wolner E. Relation between size of prosthesis and valve gradient: comparison of two aortic bioprosthesis. *Eur J Cardiothorac Surg* 2004 Mar;25(3):358–63.

Shao J, Zydney AL. Effect of bleach reprocessing upon the clearance characteristics and surface charge of polysulfone hemodialyzers. *ASAIO J* 2004 May–Jun;50(3):246–52.

Shimada Y, Yaku H, Kawata M, Oka K, Shuntoh K, Okano T, Takahashi A, Fukumoto A, Hayashida K, Kitamura N. An operative case of inferior vena cava stenosis due to fibrosis around permanent pacemaker leads. *Pacing Clin Electrophysiol* 2002 Feb;25(2):223–5.

Silver MT. Primary prevention implantable cardioverter-defibrillators: economics and ethics. *Am Heart Hosp J* 2005 Summer;3(3):205–6.

Smith WA, Hete BF, Kiraly RJ, Fujimoto LK, Jacobs GB, Ishikawa M, Butler K, Nose Y. The E4T electric powered total artificial heart (TAH). *Artif Organs* 1988 Oct;12(5):402–9.

Song X, Throckmorton AL, Wood HG, Allaire PE, Olsen DB. Transient and quasi-steady computational fluid dynamics study of a left ventricular assist device. *ASAIO J* 2004 Sep–Oct;50(5):410–17.

Song X, Wood HG, Olsen D. Computational Fluid Dynamics (CFD) study of the 4th generation prototype of a continuous flow Ventricular Assist Device (VAD). *J Biomech Eng* 2004 Apr;126(2):180–7.

Stevenson LW, Rose EA. Left ventricular assist devices: bridges to transplantation, recovery, and destination for whom? *Circulation* 2003 Dec 23;108(25):3059–63.

Stragier A. Dialyzer reuse: the debate continues. *Nephrol News Issues* 2003 Nov;17(12):97.

Sudkamp M, Schmid M, Geissler HJ, Emmel M, Gillor A, Mehlhorn U, Hekmat K. VDD-pacemaker in children—a long-term therapy? *Thorac Cardiovasc Surg* 2005 Jun;53(3):158–61.

Sung K, Park PW, Park KH, Jun TG, Lee YT, Yang JH. Comparison of transprosthetic mean pressure gradients between Medtronic Hall and ATS valves in the aortic position. *Int J Cardiol* 2005 Mar 10;99(1):29–35.

Szathmary S, Hegyi E, Amoureux MC, Rajapakse N, Chicorka L, Szalai G, Reszegi K, Derbyshire Z, Paluh J, Dodson B, Grandics P. Characterization of the DialGuard device for endotoxin removal in hemodialysis. *Blood Purif* 2004;22(5):409–15. Epub 2004 Aug 13.

Throckmorton AL, Allaire PE, Gutgesell HP, Matherne GP, Olsen DB, Wood HG, Allaire JH, Patel SM. Pediatric circulatory support systems. *ASAIO J* 2002 May–Jun;48(3):216–21.

Toff WD, Camm AJ, Skehan JD; United Kingdom Pacing and Cardiovascular Events Trial Investigators. Single-chamber versus dual-chamber pacing for high-grade atrioventricular block. *N Engl J Med* 2005 Jul 14;353(2):145–55.

Tsukiya T, Taenaka Y, Tatsumi E, Takano H. Visualization study of the transient flow in the centrifugal blood pump impeller. *ASAIO J* 2002 Jul–Aug;48(4):431–6.

Unruh ML, Evans IV, Fink NE, Powe NR, Meyer KB; Choices for Healthy Outcomes in Caring for End-Stage Renal Disease (CHOICE) Study. Skipped treatments, markers of nutritional nonadherence, and survival among incident hemodialysis patients. *Am J Kidney Dis* 2005 Dec;46(6):1107–16.

Vahlhaus C. Heating of pacemaker leads during magnetic resonance imaging. *Eur Heart J* 2005 Jun;26(12):1243; author reply 1243–4. Epub 2005 May 4.

Ward RA, Ouseph R. Impact of bleach cleaning on the performance of dialyzers with polysulfone membranes processed for reuse using peracetic acid. *Artif Organs* 2003 Nov;27(11):1029–34.

Wernicke JT, Meier D, Mizuguchi K, Damm G, Aber G, Benkowski R, Nose Y, Noon GP, DeBakey ME. A fluid dynamic analysis using flow visualization of the Baylor/NASA implantable axial flow blood pump for design improvement. *Artif Organs* 1995 Feb;19(2):161–77.

Westaby S. Ventricular assist devices as destination therapy. *Surg Clin North Am* 2004 Feb;84(1):91–123.

Wheeldon DR. Mechanical circulatory support: state of the art and future perspectives. *Perfusion* 2003 Jul;18(4):233–43.

Wijnen E, Keuter XH, Planken NR, van der Sande FM, Tordoir JH, Leunissen KM, Kooman JP. The relation between vascular access flow and different types of vascular access with systemic hemodynamics in hemodialysis patients. *Artif Organs* 2005 Dec;29(12):960–4.

Woodruff J, Prudente LA. Update on implantable pacemakers. *J Cardiovasc Nurs* 2005 Jul–Aug;20(4):261–8; quiz 269–70.

Wu C, Hwang NH, Lin YK. Statistical characteristics of mechanical heart valve cavitation in accelerated testing. *J Heart Valve Dis* 2004 Jul;13(4):659–66.

Wu Y, Gregorio R, Renzulli A, Onorati F, De Feo M, Grunkemeier G, Cotrufo M. Mechanical heart valves: are two leaflets better than one? *J Thorac Cardiovasc Surg* 2004 Apr;127(4):1171–9.

Wu ZJ, Antaki JF, Burgreen GW, Butler KC, Thomas DC, Griffith BP. Fluid dynamic characterization of operating conditions for continuous flow blood pumps. *ASAIO J* 1999 Sep–Oct;45(5):442–9.

Yamane T, Miyamoto Y, Tajima K, Yamazaki K. A comparative study between flow visualization and computational fluid dynamic analysis for the sun medical centrifugal blood pump. *Artif Organs* 2004 May;28(5):458–66.

Yoganathan, AP, He Z, Jones SC. Fluid mechanics of heart valves. *Annu Rev Biomed Eng* 2004;6:331–62.

SIGNAL PROCESSING OF RANDOM PHYSIOLOGICAL SIGNALS

Charles S. Lessard, Ph.D.
Texas A&M University, College Station, USA

MORGAN & CLAYPOOL PUBLISHERS

ABSTRACT

This *lecture* book is intended to be an accessible and comprehensive introduction to random signal processing with an emphasis on the real-world applications of *biosignals*. Although the material has been written and developed primarily for advanced undergraduate biomedical engineering students it will also be of interest to engineers and interested biomedical professionals of any discipline seeking an introduction to the field. Within education, most biomedical engineering programs are aimed to provide the knowledge required of a graduate student while undergraduate programs are geared toward designing circuits and of evaluating only the cardiac signals. Very few programs teach the processes with which to evaluate brainwave, sleep, respiratory sounds, heart valve sounds, electromyograms, electro-oculograms, or random signals acquired from the body. The primary goal of this *lecture* book is to help the reader understand the time and frequency domain processes which may be used and to evaluate random physiological signals. A secondary goal is to learn the evaluation of actual mammalian data without spending most the time writing software programs. This publication utilizes "DADiSP", a digital signal processing software, from the DSP Development Corporation.

KEYWORDS

Signals, Processes, Time Domain, Frequency Domain, Random Data

Contents

CHAPTER 1

Biomedical Engineering Signal Analysis

1.1 INTRODUCTION

The purpose of this book is to present the most widely used techniques in signal and system analysis. Individuals should have sufficient working knowledge of mathematics through calculus and some physiology and be familiar with the elements of circuit theory (i.e., you can write both loop and node equations for passive and active circuits) to gain the most knowledge. Extensive programming ability is not necessary if an individual wishes to apply some of the signal-processing principles, and it is recommended that the individual link to DSP Corporation web site (http://www.dadisp.com/) and try the Student version of the digital signal-processing software package called "DADiSP." The material contained herein should serve as an introduction to the most widely used techniques in analog and discrete (digital) data and signal-processing analysis of physiological data.

1.2 GENERALIZED SYSTEMS ENGINEERING APPROACH

This chapter is concerned with methods of characterizing signals and systems. Systems can be studied from two main perspectives:

a) *Microscopic systems analysis approach*: The fine structure of a system is taken into account. This approach is an extremely difficult to analyze because of the complexity and large number of variables in the mathematical description of the system. Muscles are a good example. A model of a muscle must consider its components (fibrin, actin, myosin, and the level of action potential) in building up the comprehensive description of the system operation.

b) *Macroscopic system analysis method*: This method is the most common and most useful approach in system analysis. In this approach, the system is characterized in terms of subsystems and components (usually lumped parameter components), which is the analysis that we will be primarily concerned with.

Macroscopic analysis requires that the system be broken into a number of individual components. The various components are described in sufficient detail and in a manner so that the system operation can be predicted. The crux is the description of the component behavior, which is done in terms of a mathematical model.

For many components of engineering interest, the mathematical model is an expression of the response of the component (subsystem) to some type of stimulus or forcing function. In our mathematical models, the forcing function and the system response are variables, which we will categorize as signals. Signal or data (variables) could be displacement, voltage, price of stock, numbers in an inventory, etc.

As stated before, the purpose of an engineering analysis of a system is to determine the response of the system to some input signal or excitation. Response studies are used to:

1) establish performance specifications;
2) aide in selection of components;
3) uncover system deficiencies (i.e., instabilities);
4) explain unusual or unexplained behavior;
5) establish proper direction and ranges of variables for the experimental program. The final system design requires a combined analytical and experimental approach; and
6) analytical studies are invaluable in the interpretation of results.

Many systems cannot be analyzed as a whole because of the enormous complexity of the systems and the lack of a satisfactory mathematical model or models for the systemwide component; for example, large systems such as an aircraft, spacecraft, etc. Subsystems can be treated and studied independently: the body, brain, central nervous system, endocrine system, etc.

To study and analyze a system properly, the means by which energy is propagated through the system must be studied. Evaluation of energy within a system is done by specifying how varying qualities change as a function of time within the system. A varying quantity is referred to as a *signal.* Signals measure the excitation and responses of systems and are indispensable in describing the interaction among various components/subsystems. Complicated systems have multi-inputs and multioutputs, which do not necessarily have to be of the same number. Signals also carry information and coding (i.e., encoded messages), which has led to the field of signal theory in communication systems. Systems analysis is used to find the response to a specific input or range of inputs when

1) the system does not exist and is possible only as a mathematical model;
2) experimental evaluation of a system is more difficult and expensive than analytical studies (i.e., ejection from an airplane, automobile crash, forces on lower back); and
3) study of systems under conditions too dangerous for actual experimentation (i.e., a/c ejection, severe weather conditions).

System representation is performed by means of specifying relationships among the systems variables, which can be given in various forms, for example, graphs, tabular values, differential equations, difference equations, or combinations. We will be concerned primarily with the system representation in terms of ordinary linear differential equations with constant coefficients.

There are two main questions one needs to ask:

1. What is the appropriate model for a particular system?
2. How good a representation does the model provide?

Often how well a model represents the real system is measured through the "squared correlation coefficient" or its equivalent the "coherence function," which are often referred to as "The Measure of Goodness Fit." You will learn more about these subjects in the later chapters.

CHAPTER 2

System Classification

As mentioned in Chapter 1, a system as defined in engineering terms is "a collection of objects interacting with each other to accomplish some specific purpose." Thus, engineers tend to group or classify systems to achieve a better understanding of their excitation, response, and interactions among other system components. It should first be noted that the general conditions for any physical system can be reflected by the following general model equation (2.1):

$$A_n(t)\frac{d^n y}{dt^n} + A_{n-1}(t)\frac{d^{n-1} y}{dt^{n-1}} + \cdots + A_1(t)\frac{dy}{dt} + A_o(t)y$$
$$= B_m(t)\frac{d^m x}{dt^m} + B_{m-1}(t)\frac{d^{m-1} x}{dt^{m-1}} + \cdots + B_1(t)\frac{dx}{dt} + B_o(t)x \tag{2.1}$$

Thus, the necessary conditions for any physical system are (1) that for an excitation (input) the function $x\,(t) = f(\phi)$ exists for all $t < t_\circ$ and (2) that the corresponding response (output) function $y\,(t) = f(\phi)$ must also exist for all $t < t_\circ$. These conditions are important because natural physical system cannot anticipate an excitation and responds before the excitation is applied.

Classifying a signal typically involves answering several questions about the system, as shown in Table 2.1.

Let us begin with definitions of terms used to characterize systems.

2.1 ORDER

What is the order of a system? The order of a system is the highest order derivative. Determining the order of a system is important because it also characterizes the response

TABLE 2.1: How Are Systems Classified
When classifying a system, the following questions should be considered:
1. What is the order of the system?
2. Is it a causal or noncausal system?
3. Is it a linear or nonlinear system?
4. Is it a fixed or time-varying system?
5. Is it a lumped or distributed parameter system?
6. Is it a continuous or discrete time system?
7. Is it an instantaneous or a dynamic system?

of the system. For example, (2.2) may be used to represent a system as a differential equation and to classify the system as an "nth" order system.

$$a_n \frac{d^n y}{dt^n} + a_{n-1} \frac{d^{n-1} y}{dt^{n-1}} + \cdots + a_1 \frac{dy}{dt} + a_\circ (t)\, y = b_m \frac{d^m x}{dt^m} + b_{m-1} \frac{d^{m-1} x}{dt^{m-1}} + \cdots + b_1 \frac{dx}{dt} + b_\circ y \qquad (2.2)$$

2.2 CAUSAL VS. NONCAUSAL

What is a causal or noncausal system? A "Causal System" is defined as a physical nonanticipatory system whose present response does not depend on future values of the input, whereas a "Non-causal System" does not exist in the real world in any natural system. "Non-causal System" exists only when it is man-made by simulating the system with a time delay.

2.3 LINEAR VS. NONLINEAR

What is a linear or a nonlinear system? From previous classes you have learned that when the output vs. the input is a straight line, the system is considered to be linear. Do not use this definition. Let us extend the simple straight-line definition to a more general analytical or mathematical expression. By mathematical definition, "A linear system is a system that possesses the mathematical properties of associative, commutative, and distributive."

Other key facts noteworthy of consideration regarding limitations in determining linearity include the following.

1. *Superposition* applies for linear systems *only*, but superposition can be applied only to linear systems that possess the mathematical properties of associative, commutative, and distributive.
2. It should also be noted that in linear systems components do not change their characteristics as a function of the magnitude of excitation. For example, electical components R, L, and C do not change their values because of changes in the magnitude of voltage or current running through them, or in the time period of analysis.
3. Some components are linear within a range.

That a system is linear can be shown by proving that superposition applies. Note that superposition is expressed as input output, the following equations are applicable:

If $y_1(t)$ is the system response to $x_1(t)$, and
If $y_2(t)$ is the system response to $x_2(t)$, then

if the inputs are summed, the output of the system should yield the sum of the individual responses as shown in (2.3): Associative property,

$$y_1(t) + y_2(t) = x_1(t) + x_2(t) \tag{2.3}$$

Or, if the responses (outputs) of two weighted inputs are given in Eq. 2.4.

$$\begin{aligned} y_1\,(t) &= a x_1\,(t) \\ y_2\,(t) &= b x_2\,(t) \end{aligned} \tag{2.4}$$

Then, if the inputs are summed, the response (output) of the system should yield the sum of the individual responses (2.5) such that

$$y_1 + y_2 = ax_1 + bx_2 \quad \text{for all} \quad a, b, x_1(t), x_2(t) \tag{2.5}$$

A simple, but not an accurate way to define a linear system is as follows: "A system in which all derivatives of the excitation and the response are raised to the 1st power."

A nonlinear system is a system in which derivatives of excitation and response are greater than the 1st order (power). Nonlinear systems or components are often modeled as 2nd or higher order systems, that is, rectifiers, junction diodes, and nonlinear magnetic devices.

2.4 FIXED VS. TIME-VARYING

What is a fixed or time-varying system? A fixed system means that coefficients do *not* vary with time, whereas a time-varying system means that coefficients do vary with time.

Another way of thinking of this definition is that if the excitation, $x(t)$, results in a response, $y(t)$, then the excitation at some time later, $x(t-\tau)$, will result in a response, $y(t-\tau)$, for any excitation $x(t)$ and any delay τ.

As an example, in (2.1), it can be noted that the system is time varying, since the coefficients $a_i(t)$ and $b_i(t)$ are indicated as being functions of time; however, in Eq. 2.2, the coefficients a_i and b_i are not functions of time, hence the system is a "Fixed System."

Analysis of time-varying systems is very difficult; thus, this book will concentrate on differential equations with constant coefficients such that the system is said to be fixed, time-invariant, or stationary. One can add that for "Time Invariant Systems," the slope of the system response depends on the shape of the system excitation and not on the time when the excitation was applied.

2.5 LUMPED PARAMETER VS. DISTRIBUTED PARAMETER

What is a lumped or distributed parameter system? A lumped parameter system is defined as a system whose largest physical dimension is small compared to the wavelength of the highest significant frequency of interest. Equation (2.1) is representative of a lumped parameter system by virtue of being an ordinary differential equation. It implies that the physical size is of no concern, since the excitation propagates through the system instantaneously, which is valid only if the largest physical dimension of the system is small compared to the wavelength of the highest significant frequency considered.

A distributed parameter system is defined as a system whose dimensions are large (*NOT Small*) compared to the shortest wavelength of interest. Distributed-parameter systems are generally represented by partial differential equations. For example, waveguides, microwave tubes, and transmission lines (telephone and power lines) are all distributed parameter systems, because of their physical lengths; that is, Hoover Dam to Los Angeles are much larger than the highest frequency or shortest wavelength of interest.

Note: As the frequency increases (higher frequency), the wavelength decreases.

2.6 CONTINUOUS-TIME VS. DISCRETE TIME

What is a Continuous or Discrete time system? "Continuous-Time Systems" are those that can be represented by continuous data or differential equations, whereas "Discrete Time Systems" are represented by sampled digital data.

It should be noted that all natural physical systems are inherently continuous-time, since time is continuous. In addition, it should be noted that both Continuous- and Discrete time systems can also be

1. linear or nonlinear,
2. fixed or time varying, and
3. lumped or distributed.

2.7 INSTANTANEOUS VS. DYNAMIC

What is an Instantaneous or a Dynamic System? An Instantaneous system is defined as a system that has no memory, which means that the system is not dependent on any future or past value of excitation. The response at some point in time (t_1) of an Instantaneous system depends only on the excitation at time (t_1). For example, a fully resistive circuit has no memory elements, that is, stored charge in a capacitor, and is considered an instantaneous system. There is no need to express the system as a differential equation, only as constants.

If a system response depends on past value of excitation, then the system is classified as a Dynamic System, which has memory (meaning it stores energy). Circuits with capacitors and magnetic inductors components are Dynamic systems, since the capacitors initial state is based on past excitation. Similarly, magnetic components with residual magnetism store energy based on past excitation, thus making such systems "Dynamic." Dynamic system mathematical models are expressed as differential equations in either time or frequency domain.

In general, most problems are considered initially to be

1. causal,
2. linear,
3. fixed,
4. lumped-parameter,
5. continuous time, and
6. dynamic systems.

CHAPTER 3

Classification of Signals

In this chapter, an overview of various methods to describe signals in a quantitative manner is presented. Most medical devices in the 1950s and 60s presented their output in a strip chart for the physician to view and make some determination (diagnosis); however, with the advent of digital computers and advanced automated analytical processes, it has become necessary to classify and quantify signals.

So, the questions are: "Why is it necessary to quantify signals?" and "How is it useful?" Reasons for signal quantification are as follows:

1. By representing signals as mathematical models, the engineer/researcher is able to carry out system analyses under specified conditions
2. The engineer may be after the information that the signal carries.
   a. Can your thoughts be decoded from your brainwaves?
3. To specify the form of a signal that will be used in the design of systems to perform specific tasks.
   a. Examples:
      i. Petrofski's work in feeding electrical signals back into paralytic limbs in order to cause the patient to walk
      ii. Auditory or visual prostheses?
4. Finally, classification and quantification provide physical insight for the analyst or designer.
   a. If an engineer wants to design a "Star Trek Tricoder" for vital life signs, what signal characteristics and values does the engineer need to consider?

There are some useful approximations of a signal that can be used to quantify the signal such that the signal is represented in some mathematically precise format. The most general representation of a signal, which is a function of time, is the abstract symbol $x(t)$ or $y(t)$. These abstract symbols can be used with linear systems because the system characteristics are independent of the explicit forms of the signals. But also note that the abstract representation, $x(t)$, is ***NOT*** a quantitative representation, since the abstract notation does not specify the magnitude of the signal at any instant in time.

To specify the magnitude at any instant of time, the signal must be specified more explicitly by a function of time defined for all instants of time. Realistically, any quantitative representation is only an approximation. So that leaves the engineer with the next problem, which is that of deciding how good the approximation is in order to obtain meaningful results from its use.

3.1 HOW ARE SIGNALS CLASSIFIED

Let us begin to examine the various ways that signals are classified. In this section, only the most used classifications are presented. Engineers should ask the following questions (see Table 3.1).

TABLE 3.1: Signal Classification
1. Is the signal periodic or nonperiodic?
2. Is the signal random or nonrandom (deterministic)?
3. Is the signal an energy signal or a power signal?
4. What is the bandwidth of the signal?
5. What is the time duration of the signal?
6. What is time-bandwidth product?
7. What is the dimensionality of the signal?

3.1.1 Periodic vs. Nonperiodic Signals

Is the signal periodic or nonperiodic? A **periodic** signal is defined as a signal that repeats the sequences of values exactly after a fixed length of time (the period, T) as shown in (3.1), such that

$$x(t) = x(t + T) \quad \text{for all} \quad t \tag{3.1}$$

Keep in mind that a delay is written as a difference $(t - T)$ in (3.2):

$$y(t) = x(t - T) \tag{3.2}$$

The smallest positive T, which satisfies 3.1, is the period, which is the duration of one complete cycle. The fundamental frequency (f_0) of a periodic signal is defined as the reciprocal of the period (T) as in (3.3):

$$f_0 = \frac{1}{T} \tag{3.3}$$

A **nonperiodic** signal or an "aperiodic" (almost periodic) signal does not have a period (T). Typical nonperiodic signals include signals such as speech, electroencephalograms (brainwaves) measured by surface electrodes, electromyogram, and respiratory sounds.

From the standpoint of mathematical representation of signals, the periodic class has the greatest theoretical importance. In most cases, an explicit mathematical expression can be written for a periodic signal.

3.1.2 Random vs. Nonrandom Signals

Is the signal random or nonrandom (deterministic)? Another method of classifying signals is whether or not the signal is random. The author's definition of a **random** signal is a signal about which there is some degree of uncertainty before it actually occurs; that is, "I've always considered a random signal as a time varying variable or function whose magnitude varies erratically and in an unpredictable manner."

The formal definition of a **nonrandom** signal is a signal that has no uncertainty before it occurs. In other words, it is "**Deterministic**"; in most cases, an explicit math expression can be written.

3.1.3 Energy Signals vs. Power Signals

Is the signal an energy signal or a power signal? Since we are most familiar with electrical measurements, let us recall that most electrical signals are voltages or currents. The energy dissipated by a voltage in a resistor for a given period of time is given by (3.4).

$$E = \int_{t_1}^{t_2} \frac{v^2(t)}{R} dt \quad \text{watt seconds} \tag{3.4}$$

Similarly for current, the equation is given in (3.5):

$$E = \int_{t_1}^{t_2} Ri^2 dt \tag{3.5}$$

In either equation, the energy is proportional to the integral of the square of the signal. If the resistance (R) is made unity (R is set at 1 ohm: 1Ω), then (3.4) and (3.5) may be written in the more generalized form as shown in (3.6).

$$E = \int_{-\infty}^{+\infty} x^2(t) dt < \infty \tag{3.6}$$

Thus (3.4) is used to define an energy signal of infinite energy. An exponentially damped sinusoid is an example of an energy signal.

What is a power signal? **Average power** is also known as a "Power Signal." If the resistance is assumed to be 1 Ω, then the equation for a power signal is given by (3.7).

$$P = \frac{1}{t_2 - t_1} \int_{t_1}^{t_2} x^2(t) dt \tag{3.7}$$

From (3.7), one may conclude that the "Power" signal is the energy of a signal in a time interval divided by the time interval, or restated: Power is the average energy in an interval of time.

Average power (3.8) satisfies the following conditions: [Note the limits; it is bound.]

$$0 < \underset{T \to \infty}{\text{Lim}} \frac{1}{2T} \int_{-T}^{T} x^2(t)dt < \infty \tag{3.8}$$

A classic definition given by McGillem and Cooper is as follows:

> If the **power signal** has a non-zero value when the time interval becomes infinite, then the signal has a finite average power and is called a **Power Signal**. Then, an **energy signal** has zero average power, whereas a **power signal** has infinite energy.

3.2 SIGNAL CHARACTERIZATION (MEASURABLE PARAMETERS)

Thus far, we have simply provided discussions on signal representation. Let us turn to specification of the signals and the "Characterization" of the signals by a few significantly useful parameters, which include the following parameters:

1. Signal energy
2. Average power
3. Bandwidth, or
4. Time duration
5. Time–bandwidth product
6. Dimensionality of the signal

The first two, Energy and Power, were treated in the previous section. Thus, let us start with the question, "What is the time duration of a signal?" The "Time Duration" of a signal is defined in statistical terms as the normalized variance about its central moment where the 1st moment is the mean or its center of gravity and the variance is generally the second moment. The definition is stated in a statistical sense, which is especially useful for signals that do not have a definite starting or stopping time.

In the case of the Time Duration of a signal, the moments are normalized by dividing the moments by the energy associated with the signal [the generalized energy equation (3.9)]

$$E = \int_{-\infty}^{\infty} x^2 dt \tag{3.9}$$

Then, the normalized first moment is shown in (3.10):

$$t_0 = \frac{\int_{-\infty}^{\infty} t x^2(t) dt}{\int_{-\infty}^{\infty} x^2(t) dt} \tag{3.10}$$

And the normalized second-central moment of the square of the time function is given by (3.11):

$$T_e = \left[\frac{\int_{-\infty}^{\infty} (t - t_0)^2 \, x^2(t) dt}{\int_{-\infty}^{\infty} x^2(t) dt} \right]^{\frac{1}{2}} \tag{3.11}$$

where T_e defines the measure of the time duration, and t_0 is the normalized first moment or mean about the origin. Therefore, the time duration (T_e) is in reality the normalized **second** moment about the mean, t_0.

3.2.1 Bandwidth

What is the bandwidth of the signal? Most engineers are very familiar with the definition of bandwidth, but let us review the various definitions. Bandwidth is intended to indicate a range of frequencies within which most of the energy lies. Engineers use the half power or 3 dB point, if a frequency response plot is available (Fig. 3.1).

However, the bandwidth of a general signal is defined in terms of the derivative of the time function as shown in (3.12).

$$W_e = \frac{1}{2\pi} \left[\frac{\int_{-\infty}^{\infty} \left(\frac{dx(t)}{dt} \right)^2 dt}{\int_{-\infty}^{\infty} x^2(t) dt} \right]^{\frac{1}{2}} \text{Hz} \tag{3.12}$$

Caution: This definition has problems with discontinuities in the signals.

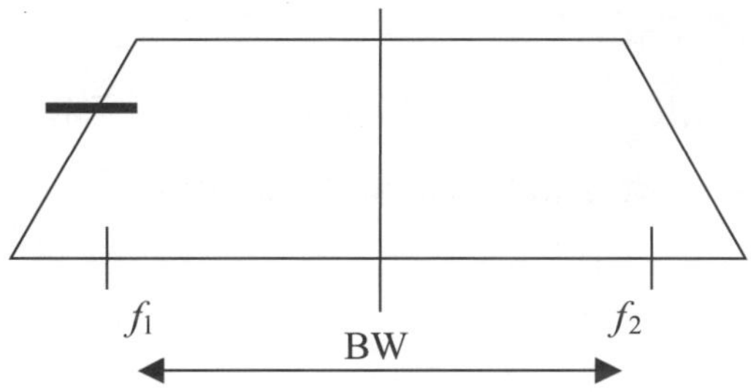

FIGURE 3.1: Frequency response plot. The 3 dB or 0.707 magnitude point shown in bold horizontal line determines the low- and high-frequency cutoff points

3.2.2 Time–Bandwidth Product

What is time–bandwidth product? Another concept in characterizing signals is time–bandwidth product, which is the product $(T_e W_e)$ of the signal time duration and the signal bandwidth as given in (3.13).

$$TW = (T_e)(W_e) = \frac{\int_{-\infty}^{\infty} (t - t_0)^2 x^2(t)dt}{\int_{-\infty}^{\infty} x^2(t)dt} \frac{1}{2\pi} \left[\frac{\int_{-\infty}^{\infty} \left(\frac{dx(t)}{dt} \right)^2 dt}{\int_{-\infty}^{\infty} x^2(t)dt} \right]^{\frac{1}{2}} \tag{3.13}$$

No signal can have a time–bandwidth product smaller than $\frac{1}{4\pi}$. Small values of the time–bandwidth product are associated with very simple signals. Large values imply complex structure and large information content. The time–bandwidth product is a parameter for judging the usefulness of a given signal in conveying information.

3.2.3 Dimensionality

What is the dimensionality of a signal? Signal dimensionality is a concept closely related to time–bandwidth product. Dimensionality of a signal is the smallest number of basis functions needed to achieve a reasonable approximation with a finite number of coefficients. Signal dimensionality is defined as twice the time–bandwidth product plus one. The equation for the dimensionality of a signal is given in (3.14).

$$D \cong 2WT + 1 \tag{3.14}$$

where W is the bandwidth of the signal, and T the time duration of the signal.

3.3 REFERENCE

McGillem, C.D. and G.R. Cooper, "Continuous and Discrete Signal and System Analysis", Holt, Rinehart, and Winston, Inc., 1974.

CHAPTER 4

Basis Functions and Signal Representation

4.1 INTRODUCTION TO BASIS FUNCTIONS

To obtain a quantitative description of a signal, it is necessary to represent the signal in terms of explicit time functions. Mathematical convenience dictates that the signal can be represented as a linear combination of a set of elementary time functions called **Basis Functions**. Equation (4.1) is the general equation representing a signal, $x(t)$, in terms of a set of basis functions designated as $\Phi_0(t)$, $\Phi_1(t)$, $\Phi_2(t), \ldots, \Phi_N(t)$.

$$x(t) = \sum_{n=0}^{N} a_n \Phi_n(t) \tag{4.1}$$

4.2 DESIRABLE PROPERTIES OF BASIS FUNCTIONS

One property that is desirable for a set of basis functions is the *finality of coefficients*, which allows the determination of a coefficient without needing to know any other coefficients. This means that more terms can be added without making changes or recalculating the earlier or previous coefficient. To achieve "finality of coefficients," the basis functions must be orthogonal over the time interval of interest.

The condition of orthogonality requires that the integral of two functions satisfies two conditions. The first condition is when the two basis functions are not equal, then the resulting integral must be zero as shown in (4.2).

$$\int_{t_1}^{t_2} \Phi_n(t)\Phi_k(t)dt = 0; \qquad k \neq n \tag{4.2}$$

The second condition is when the two basis functions are equal, then the resulting integral will equal some value "Lambda," λ_k, as shown in (4.3).

$$\int_{t_1}^{t_2} \Phi_n(t)\Phi_k(t)dt = \lambda_k; \; k = n \quad \text{for all } k \text{ and } n \tag{4.3}$$

If both sides of (4.3) are divided by λ_k and lambda λ_k is made unity ($\lambda_k = 1$) for all k, then the basis function is called **orthonormal**. Note that the limits of integration can be defined as finite interval or an infinite or semi-infinite interval.

4.3 EVALUATION OF COEFFICIENTS

Recall the general expression of a signal represented in terms of a basis function, as the weighted summation of the elementary functions:

$$x(t) = \sum_{n=0}^{N} a_n \Phi_n(t) \tag{4.4}$$

Hence, to evaluate the coefficients a_n of the basis function, the following steps are necessary:

(1) Multiply both sides by $\Phi_j(t)$ as in (4.5).

$$\Phi_j(t)x(t) = \Phi(t)\sum_{n=0}^{N} a_n \Phi_n(t) \tag{4.5}$$

(2) Integrate both sides of the equation over the specified interval $t_2 - t_1$ (4.6).

$$\int_{t_1}^{t_2} \Phi_j(t)x(t)dt = \int_{t_1}^{t_2} \Phi_j(t)\left[\sum_{n=0}^{N} a_n \Phi_n(t)\right] dt = \sum_{n=0}^{N} a_n \int_{t_1}^{t_2} \Phi_j(t)\Phi_n(t)dt \tag{4.6}$$

Applying the condition of orthogonality, the right side of the equation becomes (4.7).

$$\int_{t_1}^{t_2} \Phi_j(t)x(t)dt = a_j\lambda_j \tag{4.7}$$

Note that when the indexes of the basis function are equal, $j = n$, the resulting integral will be some value, lambda (4.8);

$$\int_{t_1}^{t_2} \Phi_j(t)\Phi_n(t)dt = \lambda_j \tag{4.8}$$

otherwise, if indexes of the basis function are not equal, $j \neq n$, the integral will equal zero. The equation may be rewritten to solve for the coefficients as shown in the following equation:

$$\frac{1}{\lambda_j}\int_{t_1}^{t_2} \Phi_j(t)x(t)dt = a_j \tag{4.9}$$

Recall the general equation for an energy or average power signal as shown in the following equation:

$$E = \int_{t_1}^{t_2} x^2(t)dt \tag{4.10}$$

By substituting the general basis function representation of a signal (4.11) into (4.10);

$$x(t) = \sum_{n=0}^{N} a_n\Phi_n(t) \tag{4.11}$$

the energy equation (4.10), may be expressed as (4.12):

$$E = \int_{t_1}^{t_2} x^2(t)dt = \int_{t_1}^{t_2} x(t)\sum_{n=0}^{N} a_n\Phi_n(t)dt = \sum_{n=0}^{N} a_n \int_{t_1}^{t_2} \Phi_n(t)x(t)dt \tag{4.12}$$

From the orthogonality condition, (4.12) is rewritten in terms of the coefficient and lambda as given in (4.13),

$$E = \sum_{n=0}^{N} a_n(a_n\lambda_n) \tag{4.13}$$

or as given in (4.14).

$$E = \int_{t_1}^{t_2} x^2(t)dt = \sum_{n=0}^{N} a_n^2\lambda_n \tag{4.14}$$

Since $a_n^2\lambda_n$ is the energy in the nth basis function, each term of the summation is the energy associated with its index, nth component of the representation. Thus, the TOTAL energy of a signal is the sum of the energies of its individual orthogonal coefficients, which is often referred to as "Parseval's theorem."

Many functions are, or can be, made orthogonal over an interval, but it does not mean that the function may be a desirable function to use as a basis function. As engineers, we tend to use the trigonometric functions, sine and cosine, in many analytical applications. Why are the sinusoids so popular? Three important facts about sinusoidal expressions stand out:

(1) Sinusoidal functions are very useful because they remain sinusoidal after various mathematical operations (i.e., integration, derivative). Also, exponents of sinusoidal functions can be expressed as exponents by Euler's identity ($\cos\theta + j\sin\theta = e^{j\theta}$).

(2) The sum or difference of two sinusoids of the *same frequency* remains a sinusoid.

(3) This property combined with the superposition properties of linear systems implies that representing a signal as a sum of sinusoids may be a very convenient technique, which is used for periodic signals.

4.4 SIGNAL REPRESENTATIONS

4.4.1 Fourier Transforms

The Trigonometric Fourier Transform Function. As indicated in the previous section, it is convenient to use basis functions that are invariant with respect to the mathematical operations, that is, integration, derivative, or summation. The "Sinusoidal Basis Function" has these properties and is commonly used to represent periodic, complex signals. It is possible to use sines and cosines as basis functions as given by (4.15), but it is more convenient to use complex exponentials and to write the "Exponential Fourier Transform" as (4.16).

$$x(t) = a_0 + \sum_{n=1}^{\infty} [a_n \cos(n\omega_0 t) + b_n \sin(n\omega_0 t)] \tag{4.15}$$

$$x(t) = c_n e^{-jn\omega_0 t} \tag{4.16}$$

You should be familiar with the relationship given in (4.17), and note that the summation may be positive or negative.

$$e^{\pm jnw_0 t} = \cos(nw_0 t) \pm j \sin(nw_0 t) \tag{4.17}$$

Consider a function of time for which you would like to obtain the Fourier series representation. The function is periodic over the interval, $T = t_2 - t_1$. For the series to converge to the true value, the Dirichlet conditions requires that the function

1. be single-valued within the interval, $t_2 - t_1$;
2. have a finite number of maximums and minimums in a finite time;
3. satisfy the inequalities of (4.18)

$$\int_{t_1}^{t_2} |x(t)|\, dt < \infty; \text{ and} \tag{4.18}$$

4. be a finite number of discontinuities.

For the exponential Fourier Series, the basis function is defined as given by (4.19).

$$e^{\pm jnw_0 t} = \Phi_n(t), \text{ where } n = \pm(0, 1, 2, \ldots, \infty), \text{ and} \tag{4.19}$$

where: $\omega_0 \frac{2\pi}{T}$ (the fundamental frequency is in radians).
To prove that the functions are orthogonal over the interval start with (4.20).

$$\begin{aligned} \int_{t_1}^{t_1+T} e^{jn\omega_0 t} * e^{-jn\omega_0 t} dt &= 0 \quad \text{when} \quad n \neq k \\ &= T \quad \text{when} \quad n = k \end{aligned} \tag{4.20}$$

The coefficients for the series can be expressed as (4.21).

$$a_n = \frac{1}{T} \int_{t_1}^{t_1+T} x(t) e^{-jnw_0 t} dt \tag{4.21}$$

The coefficients are usually complex and can be expressed as $a_{-n} = \alpha_n{}^*$ (* means conjugate). The signal $x(t)$ is expressed as (4.22):

$$x(t) = \sum_{n=-\infty}^{\infty} \alpha_n e^{jnw_0 t} \tag{4.22}$$

The accuracy of an approximate representation using a finite number of terms is obtained from the energy ratio given by (4.23) and (4.24).

$$\eta_m = \frac{\text{error energy}}{\text{signal energy}} = 1 - \frac{1}{E} \sum_{n=0}^{M} \lambda_n a_n^2 \tag{4.23}$$

where

$$E = \int_{t_1}^{t_2} x^2(t) dt \tag{4.24}$$

For the Fourier series, $\lambda_n = T$, since $\alpha_{-n} = \alpha_n{}^*$. Then, $|\,\alpha_n\,|^2 = |\,\alpha_{-n}\,|^2$.

Then from $-M$ to M, the fractional error is given by (4.25):

$$\eta_m = 1 - \frac{1}{E}\sum_{n=0}^{M} T|\alpha_n|^2$$
$$\eta_m = 1 - \frac{1}{E}\left[\alpha_0^2 + 2\sum_{n=0}^{M} T|\alpha_n|^2\right] \tag{4.25}$$

Example: Consider the periodic sequence, $x(t) = A$ for $0 < t < t_a$, and $x(t) = 0$ for $t_a < t < T$: as in (4.26) and (4.27).

$$\alpha_n = \frac{1}{T}\int_0^{t_a} Ae^{-jnw_0 t}dt + 0 \text{ (for } t_a < t < T) \tag{4.26}$$

$$\alpha_n = \frac{A}{T}\frac{-1}{jnw_0}e^{-jnw_0 t}\Big|_0^{t_a}$$
$$\alpha_n = \frac{A}{T}\frac{-1}{jnw_0}e^{-jnw_0 t_a} - \frac{A}{T}\left(\frac{-1}{jnw_0}\right)(1) \tag{4.27}$$
$$\alpha_n = \frac{A}{jnw_0 T}\left(1 - e^{-jnw_0 t}\right)$$

By rewriting the terms in the bracket, you get (4.28):

$$e^{-j\frac{x}{2}}\left(e^{j\frac{x}{2}} - e^{-j\frac{x}{2}}\right), \text{ where } x = nw_0 t \tag{4.28}$$

and α_n becomes (4.29):

$$\alpha_n = \left[\frac{e^{j\frac{nw_0 t_a}{2}} - e^{-j\frac{nw_0 t_a}{2}}}{jnw_0}\right]\left(e^{-j\frac{nw_0 t_a}{2}}\right) \tag{4.29}$$

The [ ] (square bracket) term is a trig identity [(4.30) and (4.31)] with circular functions where

$$\sin(x) = \frac{e^{jx} - e^{-jx}}{2j} \tag{4.30}$$

$$\text{With } x = \frac{nw_0 t_a}{2} \tag{4.31}$$

So, to begin solving this problem, you first need to multiply the denominator by 2/2, as shown by (4.32), (4.33), and (4.34).

$$\alpha_n = \frac{A}{T}\frac{2}{2}\frac{\sin\left(nw_0\frac{t_a}{2}\right)}{nw_0/2}e^{-j\frac{nw_0 t_a}{2}} \tag{4.32}$$

Then, multiply by t_a/t_a.

$$\alpha_n = \frac{A}{T}\frac{t_a}{t_a}\frac{\sin\left(nw_0\frac{t_a}{2}\right)}{\frac{nw_0}{2}}e^{-j\frac{nw_0 t_a}{2}} \tag{4.33}$$

$$\alpha_n = \frac{At_a}{T}\left[\frac{\sin\left(nw_0\frac{t_a}{2}\right)}{\frac{nw_0 t_a}{2}}\right]e^{-j\frac{nw_0 t_a}{2}} \tag{4.34}$$

Since $w_0 = 2\pi f_0$ and $f_0 = 1/T$, then $w_0 = (2\pi)/T$. Thus, alpha can be rewritten as (4.35).

$$\alpha_n = \frac{At_a}{T}\left[\frac{\sin(\pi n t_a/T)}{n\pi t_a/T}\right]e^{-j\frac{2n\pi t_a}{2T}} \tag{4.35}$$

The complete series is written as (4.36) or equations in (4.37).

$$x(t) = \sum_{n=-\infty}^{\infty} \alpha_n e^{jnw_0 t} \quad \text{or} \tag{4.36}$$

$$x(t) = \sum_{n=-\infty}^{\infty}\left\{\frac{At_a}{T}\left[\frac{\sin(n\pi t_a/T)}{\frac{n\pi t_a}{T}}\right]e^{-j\frac{2n\pi t_a}{2T}}e^{j\frac{n2\pi t}{T}}\right\}$$

$$x(t) = \sum_{n=-\infty}^{\infty}\left\{\frac{At_a}{T}\left[\frac{\sin(n\pi t_a/T)}{n\pi t_a/T}\right]e^{-j\frac{2n\pi\left(t-\frac{t_a}{2}\right)}{T}}\right\} \tag{4.37}$$

Keep in mind that the signal is being represented in terms of sinusoids having frequencies that are multiples of the fundamental frequency $1/T$. The coefficients, α_n, give the magnitude and phase of these sinusoids, which constitute an approximate frequency-domain description of the explicit time-domain signal. The Fourier series is by far the most commonly used orthogonal representation.

4.4.2 Legendre Functions

Other signal representation includes the Legendre functions, which are an orthonormal set of basis functions in the time interval from -1 to 1. The general basis function is given by (4.38) and (4.39).

$$\Phi_n(t) = \sqrt{\frac{2n+1}{2}} P_n(t) \quad \text{for} \quad -1 \le t \le 1. \tag{4.38}$$

where $P_n(t)$ is the Legendre polynomials, given by (4.39) and (4.40).

$$P_n(t) = \frac{1}{2^n n!} \frac{d^n}{dt^n} (t^2 - 1)^n \quad \text{for} \quad n = 0, 1, 2, \ldots \text{ for } -1 \le t \le 1 \tag{4.39}$$

$$\begin{aligned}
\Phi_0(t) &= \frac{1}{\sqrt{2}} \\
\Phi_1(t) &= \sqrt{\frac{3}{2}} t \\
\Phi_2(t) &= \sqrt{\frac{5}{2}} \left(\frac{3}{2} t^2 - \frac{1}{2} \right) \\
\Phi_3(t) &= \sqrt{\frac{7}{2}} \left(\frac{5}{2} t^3 - \frac{3}{2} t \right)
\end{aligned} \tag{4.40}$$

These basis functions may be convenient when the signals have a predominant linear or quadratic term.

4.4.3 Laguerre Functions

When the time interval of the representation is from 0 to ∞, the Laguerre functions form a complete orthonormal set defined by (4.41):

$$\Phi_n(t) = \frac{1}{n!} e^{-\frac{t}{2}} \ln(t) \quad \text{for} \quad 0 \le t < \infty \tag{4.41}$$

ln is not natural logarithm, but rather as defined by (4.42):

$$\ln(t) = e^t \frac{d^n}{dt^n} (t^n e^{-t}) \tag{4.42}$$

4.4.4 Hermite Functions

Hermite functions (4.43) are orthonormal over the range from $-\infty$ to ∞.

$$\Phi_n(t) = (2^n n! \sqrt{\pi})^{-1/2} e^{-\frac{t^2}{2}} H_n(t)$$
$$H_n(t) = (-1)^n e^{t^2} \frac{d^n}{dt^n}(e^{-t^2}) \tag{4.43}$$

Cardinal functions are given by (4.44).

$$\Phi_n(t) = \frac{\sin \pi(2wt - n)}{\pi(2wt - n)} \quad \text{for } n = 0, \pm 1, \pm 2, \ldots \tag{4.44}$$

Another property of orthogonal functions has to do with the accuracy of the representation when not all of the terms can be used for an exact representation. In most cases, N is infinity for exact representation, but the coefficient, a_n, becomes smaller as n increases.

The approximation or estimation of $x(t)$ is expressed as $x'(t)$ as in (4.45).

$$x'(t) = \sum_{n=0}^{M} a'_n \Phi_n(t), \text{ where } M \text{ is finite.} \tag{4.45}$$

A measure of the closeness or goodness of approximation is the integral squared error (4.46).

$$I = \int_{t_1}^{t_2} [x(t) - x'(t)]^2 dt \quad \text{and} \quad I = 0 \text{ when } x'(t) = x(t) \tag{4.46}$$

The smaller the value of I, the better the approximation. The integral squared error is applicable *only* when $x(t)$ is an energy signal or when $x(t)$ is periodic. When $x(t)$ is an energy signal, the limits of integration may range from $-\infty$ to $+\infty$. When $x(t)$ is periodic, then $t_2 - t_1$ is equal to the period T.

To find the optimum values of the estimator for the basis coefficient, we use the integral squared error or sum of squared error (SSE) as shown in (4.47).

$$I = \int_{t_1}^{t_2} \left[x(t) - \hat{x}(t)\right]^2 dt \tag{4.47}$$

Next substitute (4.48) into (4.47) to obtain the least squared error (I), which is given by (4.49).

$$\hat{x}(t) = \sum_{n=0}^{M} \hat{a}_n \phi_n(t) \tag{4.48}$$

$$I = \int_{t_1}^{t_2} \left[x(t) - \sum_{n=0}^{M} \hat{a}_n \phi_n(t) \right]^2 dt \tag{4.49}$$

The next step is to find the set of estimators coefficients $\hat{a}_n$ that minimize the value of the integral (least squared error) in (4.47). One method is to differentiate the least squared error (I) with respect to each of the coefficients and set each derivative equal to 0. There are other alternative procedures that may be used, but will not be discussed in this chapter.

CHAPTER 5

Data Acquisition Process

At this point, I would like to change pace and have a discussion of general steps or procedures in data acquisition and processing. The required operations may be divided into 5 primary categories (Fig. 5.1):

1. Data collection,
2. Data recording/transmission,
3. Data preparation,
4. Data qualification, and
5. Data analysis.

5.1 DATA COLLECTION

The primary element in data collection is the instrumentation transducer. In general, a transducer is any device which translates power from one form to another. I will not get into this but in general a model might be as shown in Fig. 5.2.

In some cases it is possible to perform all desired data processing directly on the transducer signals in real time. In most cases, this is not practical. After preamplification some initial filtering and postamplification, most signals are recorded on magnetic media. At the present time, some form of storage of the amplified and preconditioned transducer signals are still required.

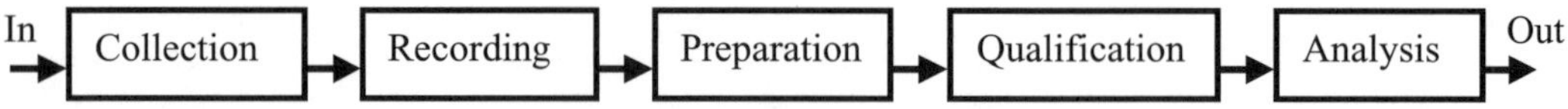

FIGURE 5.1: Data acquisition and analysis process

5.2 DATA RECORDING/TRANSMISSION

As in the past, magnetic storage systems are the most convenient and desirable type of data storage. In the 1950s through the 1980s, magnetic tape recordings were the only way to store large quantities of data. If a magnetic tape recorder was used, the most widely used method was referred to as "Frequency Modification" (FM) for analog type recordings. Beyond the limitation inherent in the magnetization-reproduce and modulation-demodulation operations, there are a number of miscellaneous problems.

Among the more important of these problems are the errors associated with variations in the speed at which the tape passes over the record and/or reproduce heads. These errors were called "Time Based Errors," with the most common being "Flutter," which is the variation of tape velocity from the normal. For the frequency modulation (FM) recording systems, the time based (frequency) is used to represent the amplitude variations of the signal. The error in good high-quality FM recorders is about 0.25%.

Currently, few laboratories still use magnetic tape recordings. The large "Gigabite" disk memories and optical disc with compression techniques are the most common approach to storage of data; however, the data must be in digital format, and thus must be converted from an analog format to a digital format. The advantage of storing the raw data is that different signal-processing methods may be used on the same data and the results compared.

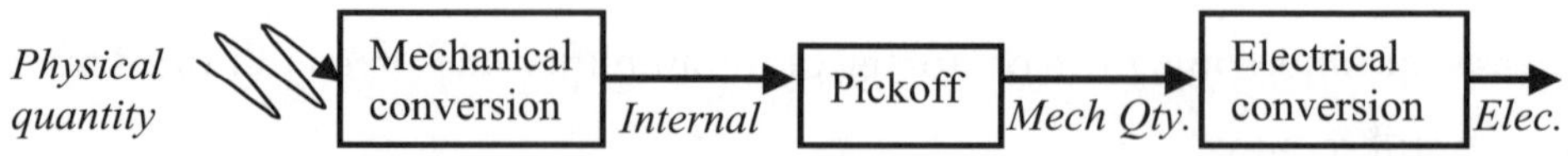

FIGURE 5.2: General model of a mechanical transducer for data collection, that is, a blood pressure transducer

5.3 DATA PREPARATION

The third step is data preparation of the *raw data* supplied as voltage time histories of some physical phenomenon. Preparation involves three distinct processes as shown in Fig. 5.3:

1. editing;
2. conversion; and
3. preprocessing.

Data editing refers to preanalysis procedures that are designed to detect and eliminate superfluous and/or degraded data signals. The undesirable parts of the signal may be from acquisition or recording problems that result in excessive noise, movement artifacts, signal dropout, etc. Editing may be through visual inspection of graphic recordings. Normally, experts in the field are invited to perform the inspection and identify those sections of the data that should not be analyzed. Some laboratories use a Hewlett-Packard Real-time spectrum analyzer and preview the data. In short, "Editing is critical in digital processing" and it is best performed by humans, and not by machines.

Data preparation beyond editing usually includes conversions of the physical phenomena into engineering units (calibration) and perhaps digitization. Digitization consists of the two separate and distinct operations in converting from analog to digital signals:

1. *Sampling interval*, which is the transition from continuous time to discrete time, and
2. *Sample quantization*: the transition from a continuous changing amplitude to a discrete amplitude representation.

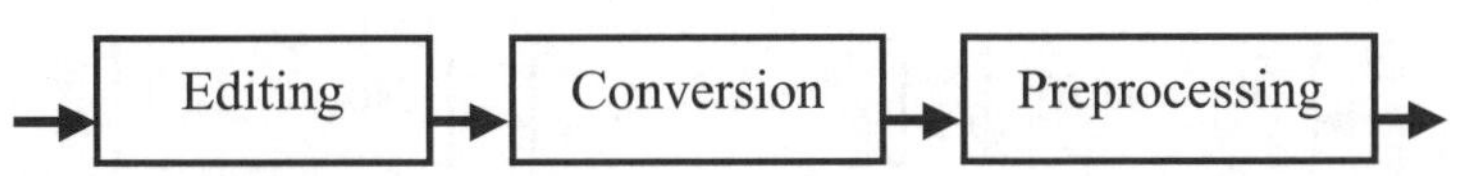

FIGURE 5.3: Data preparation

Considerations that must be taken into account when digitizing an analog signal will be covered in two chapters: one on Sampling Theory and the second on Analog-to-Digital Conversion.

Prepossessing may include formatting of the digital data (Binary, ASCII, etc., format), filtering to clean up the digital signal, removing any DC offset, obtaining the derivative of the signal, etc. But before getting into digital conversion, let us continue with the data-acquisition process by examining the fourth step of "Random Data Qualification."

5.4 RANDOM DATA QUALIFICATION

Qualification of data is an important and essential part in the statistical analysis of random data. Quantification involves three distinct processes as shown in Fig. 5.4:

1. stationarity,
2. periodicity, and
3. normality.

It is well known in statistics that researchers should not use "parametric statistical methods" unless the data are shown to be "Gaussian Distributed" and "Independent." If the data are not Gaussian Distributed and Independent, parametric statistics cannot be used and the researcher must then revert to "nonparametric statistical methods."

Likewise, necessary conditions for Power Spectral Estimation require that the random data be tested for "Stationarity" and "Independence," which may be tested parametrically by showing the mean and autocorrelation of the random data to be time invariant or by using the nonparametric "RUNS" Test.

After showing the data to be stationary, researchers may examine the data for "Periodicities" (cyclic variations with time). Periodicity may be analyzed via the

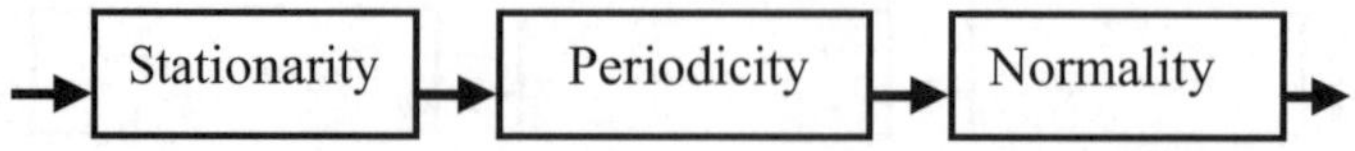

FIGURE 5.4: Data qualification

FIGURE 5.5: Random data analysis

"Correlation Function" using either Time or Frequency Domain approach. Periodicity may also be analyzed via the "Power Spectral Analysis." The next step regarding random data may be to examine the distribution of the data ("Is the data Normality distributed?") or to determine all the basic statistical properties (Moments) of the signal (data).

5.5 RANDOM DATA ANALYSIS

The fifth and final step is data analysis of the data supplied as time histories of some physical phenomenon. Data analysis involves three distinct and independent processes dependent on the experimental question and the data set as shown in Fig. 5.5:

1. analysis of individual records,
2. analysis of multiple records, and
3. linear relationships.

If the analysis is to be accomplished on an individual record, that is, one channel of electrocardiogram recording, or one channel of electroencephalogram recording, etc., then either "Autocorrelation Function Analysis" (Time-domain Analysis) or "Autospectral Analysis" (Frequency-domain Analysis) would be preformed on the random data.

If the analysis is to be accomplished on multiple records, that is, more than one channel of recording, then either "Cross-correlation Function Analysis" (Time-domain Analysis) or "Cross-spectral Analysis" (Frequency-domain Analysis) would be preformed on the random data. Multiple Records would be used to determine the "Transfer Functions" of systems when the input stimulus and the output responses are random signals.

In a relational study, cause and effect, the primary aim would be to determine the relationship of a stimulus to the response. The most popular method for determining linear relationships is to use the least squares regression analysis and to determine the correlation coefficient; however, the least squares regression analysis is difficult to apply in many random data cases. Therefore, an alternative method is to determine the model or transfer function via auto- and cross-spectra. In addition, for any model or transfer function, it is necessary to determine how well the transfer functions represent the system process; hence, the investigator must perform a "Coherence Function Analysis" to determine the "Goodness Fit."

CHAPTER 6

Sampling Theory and Analog-to-Digital Conversion

6.1 BASIC CONCEPTS

Use of a digital computer for processing an analog signal implies that the signal must be adequately represented in digital format (a sequence of numbers). It is therefore necessary to convert the analog signal into a suitable digital format.

By definition, an **analog signal** is a continuous physical phenomenon or independent variable that varies with time. Transducers convert physical phenomena to electrical voltages or currents that follow the variations (changes) of the physical phenomena with time. One might say that an electrical signal is an *Analog Signal* if it is a continuous voltage or current that represents of a physical variable. Analog signals are represented as a deterministic signal, $x(t)$, for use in a differential equation. Or, one may simply state that an analog signal is any signal that is not a digital signal, but then you must define a digital signal. There are three classes of analog signals.

a) Continuous in both time and amplitude (Fig. 6.1).

b) Continuous in time (Fig. 6.2) but taking on only discrete amplitude values, that is, analog signals used to approximate digital bit streams in serial communication, which are used primarily for time codes or markers.

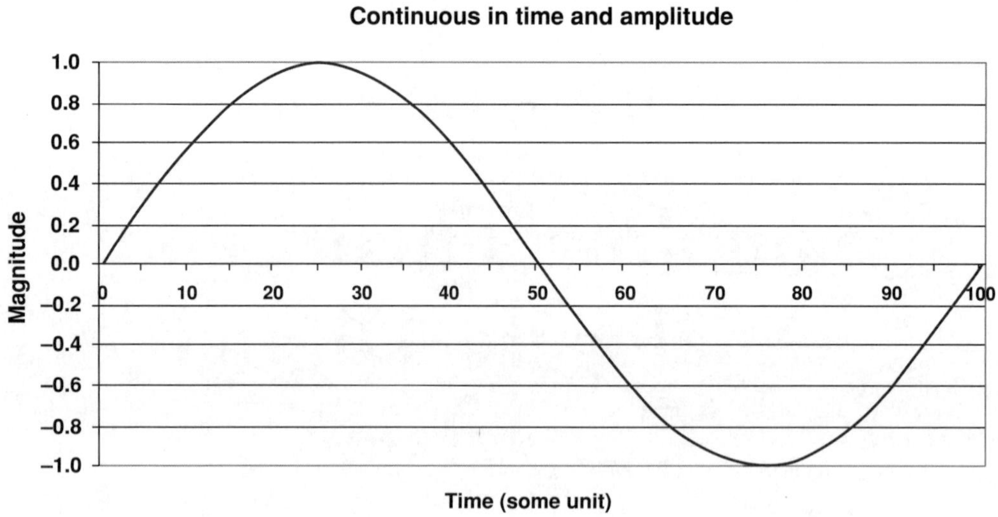

FIGURE 6.1: Analog signal that is continuous in both time and magnitude

c) Continuous in amplitude (Fig. 6.3) but defined only at discrete points in time, for example, sampled analog signals.

By definition, a digital signal is an ordered sequence of numbers, each of which is represented by a finite sequence of bits; that is, of finite length. It is important to note that a digital signal is defined only at discrete points in time and takes on only one of a

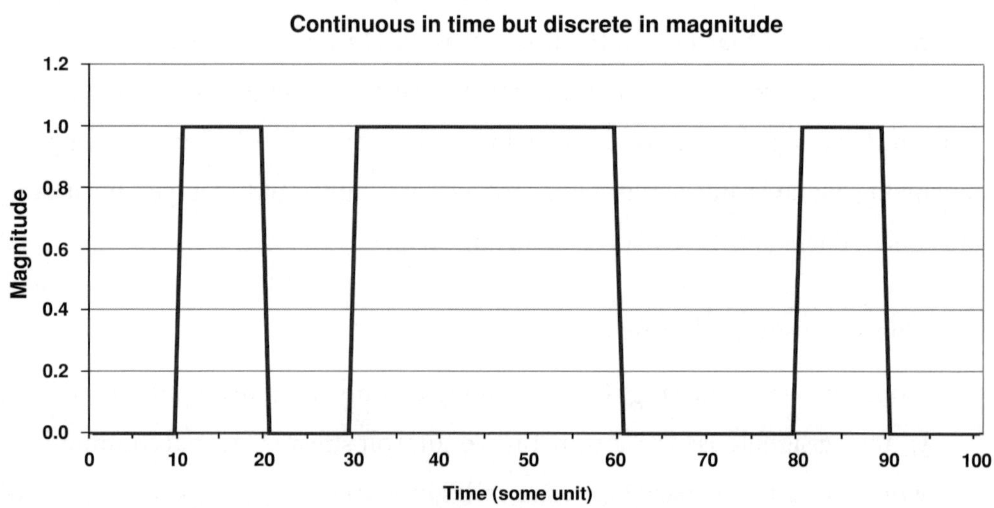

FIGURE 6.2: Analog signal that is continuous in both time and discrete in magnitude

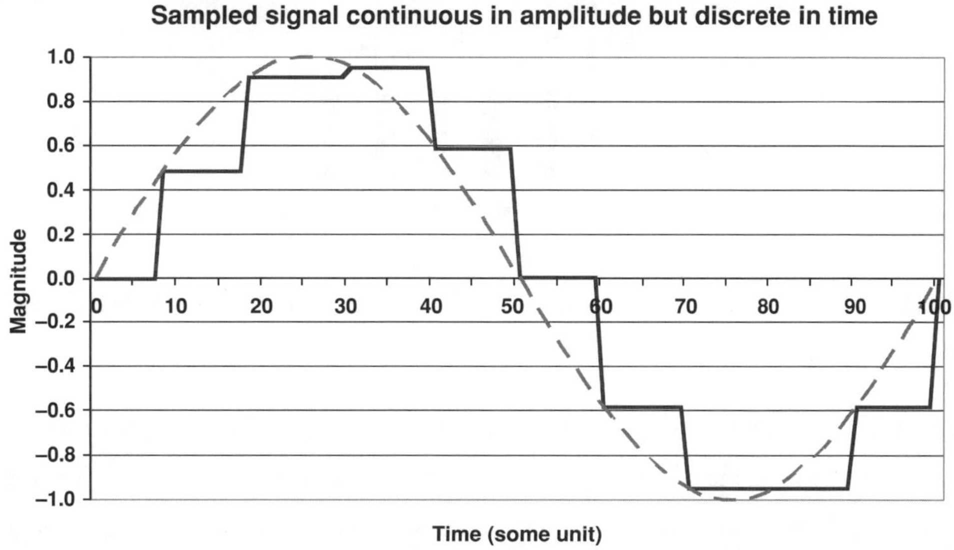

FIGURE 6.3: An analog signal that has been sampled such that the sampled signal is "continuous in amplitude but defined only at discrete points in time." The dashed line is the original sinusoidal signal

finite set of discrete values at each of these points. A digital signal can represent samples of an analog signal at discrete points in time.

There are two distinct operations in converting from analog to digital signals:

1. *Sampling*, which is the transition from continuous-time to discrete-time
2. *Sample quantization*, the transition from a continuous changing amplitude to a discrete amplitude representation

Reversing the process from digital to analog involves a similar two-step procedure in reverse.

6.2 SAMPLING THEORY

Before getting into the topic on analog-to-digital conversion, it is necessary to have some knowledge and understanding of the theory behind the sampling of data. The mechanism for obtaining a sample from an analog signal can be modeled in two ways. Ideally, an instantaneous value of the signal is obtained at the sample time. This model

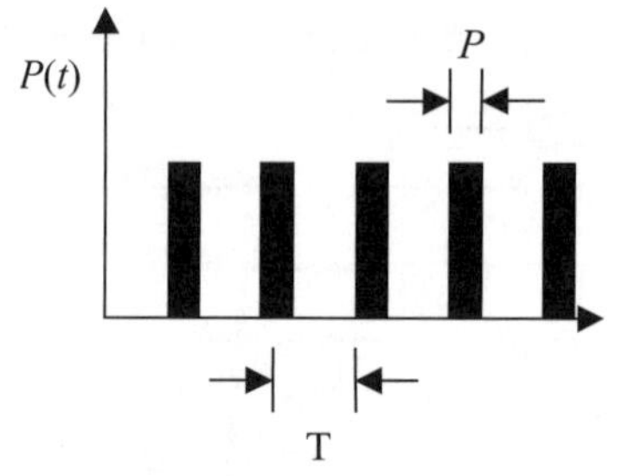

FIGURE 6.4: The realistic Delta function model. Note the uniform magnitude and finite pulse width

is known as a "delta function." The model is attractive in analytical expressions, but the delta function (model) is really impossible to generate in the real world with real circuit components. The Realist model is considered as modulation of the original signal by a uniform train of pulses, with a pulse finite width of "*P* " and interval of the sample period "*T*," as shown in Fig. 6.4.

Each pulse represents the instant of time that the value of an analog signal is multiplied by the delta function. It may be easier to think of the process as grabbing a value from the analog signal at each pulse such that the resulting collection of values is no longer continuous in time but only exists at discrete points in time.

Sampling Theorem: The sampling theory is generally associated with the works of Nyquist and Shannon. The Sampling Theorem for band-limited signals of finite energy can be interpreted in two ways:

1. Every signal of finite energy and Bandwidth W Hz may be completely recovered from the knowledge of its samples taken at the rate of 2 W per second (called the Nyquist rate). The recovery is stable in the sense that a small error in reading sample values produces only a corresponding small error in the recovered signal.
2. Every square-summable sequence of numbers may be transmitted at the rate of 2 W per second over an ideal channel of bandwidth W Hz. By being represented as the samples of a constructed band limited signal of finite length.

What does it all mean? It has been proven by Landau that

(1) stable sampling can*not* be performed at a rate lower than the Nyquist rate; and

(2) data can*not* be transmitted as samples at a rate higher than the Nyquist rate.

To put it in simpler terms, Lessard's paraphrase of the sampling theorem is: "Sampling rate must be at least twice the highest frequency of interest in the signal."

The Fourier transform of the periodic analog signal, $x(t)$, is given by (6.1):

$$\tilde{f}(x(t)) = X(j\omega) = \int_{-\infty}^{\infty} x(t)e^{-j\omega t}dt \tag{6.1}$$

When the analog signal is sampled, you have the resulting *sampled* signal (6.2):

$$x_s(t) = x(t)p(t) \tag{6.2}$$

where $p(t)$ is the sampling function with the period T.

The Fourier Series of the sampling function is as shown in (6.3):

$$p(t) = \sum_{n=-\infty}^{\infty} c_n e^{jw_s nt} \tag{6.3}$$

where $w_s = 2\pi/T$ (sampling rate in radians/second) and c_n is the Complex Fourier coefficient.

The Fourier coefficients are evaluated by (6.4):

$$c_n = \frac{1}{T}\int_{-T/2}^{T/2} p(t)e^{-j\omega nt}dt \tag{6.4}$$

The result is given by (6.5):

$$c_n = \frac{P}{T}\left(\frac{\sin\left(n\omega_s {}^{P}\!/_{2}\right)}{n\omega_s {}^{P}\!/_{2}}\right) \tag{6.5}$$

For the Sampling Function, where P is the pulse width of $p(t)$, the Fourier Transform of the Sampled function (6.6) is written in the integral format as (6.7),

$$x_s(t) = x(t)p(t) = \sum_{n=-\infty}^{\infty} x(t)c_n e^{jw_s nt} \tag{6.6}$$

$$\tilde{f}(x_s(t)) = \int_{-\infty}^{\infty} \left[\sum_{n=-\infty}^{\infty} x(t)c_n e^{jw_s nt} \right] e^{-j\omega t} dt \tag{6.7}$$

which results in the expression given by (6.8):

$$\sum_{n=-\infty}^{\infty} c_n X(\omega - n\,\omega_s) \tag{6.8}$$

where for the ideal delta function, $c_n = 1/T$ and $t = nT$.

The Fourier Transform of the sampled function is often called "Discrete Fourier Transform of a Time Series." It should be noted that the Discrete Fourier Transform DFT is a transform in its own right as is the Fourier integral transform, and that the DFT has mathematical properties analogous to the Fourier Integral Transform. The usefulness of the DFT is in obtaining the Power Spectrum analysis or filter simulation on digital computers. The Fast Fourier Transform, which will be covered, is a highly efficient procedure of a time series. Restrictions on the period T of the pulse train $p(t)$ to avoid ambiguity in the sampled version. The Nyquist Sampling Theorem must be followed.

Let us examine the 1st step in converting from an analog signal to a digital signal. Consider an analog signal $v(t) = V \sin(\omega t)$. The sampled values will give the exact value of $v(t)$ only at the sampled time; thus, a discrete-time sample of $v(t)$. If the period or time between samples is constant, T (the sampled period), the discrete approximation is expressed as $v(nt)$, where n is an integer index. Sampling with a constant period is referred to as "uniform sampling," which is the most widely used sampling method because of the ease of implementing and modeling. As is known, the sample period, T, effects how accurately the original signal is represented by the sampled approximation. The ideal sampled signal $X_s(t)$ is zero except at time $t = nT$ and the signal is represented

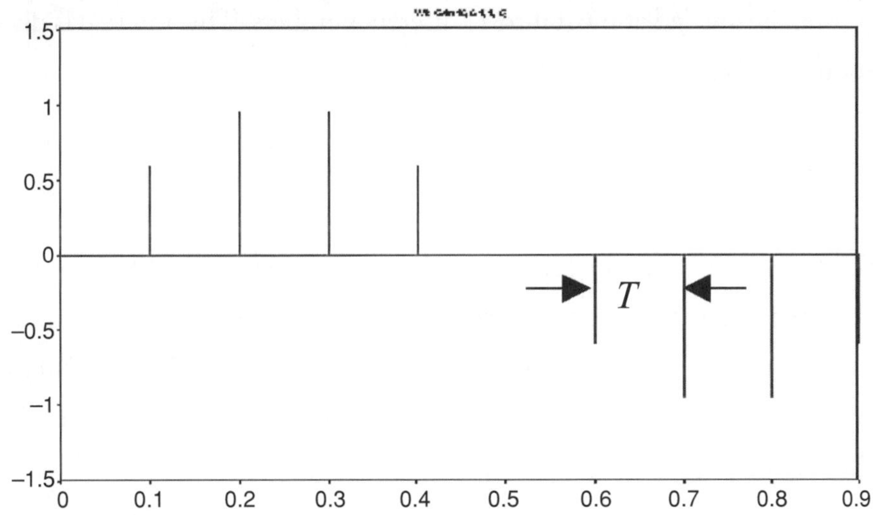

FIGURE 6.5: Digital signal sampled at a rate of $1/T$, where T is the sampling interval

as a sequence of numbers $x(n)$ representing sampled values. Figure 6.5 shows a sampled sinusoid with a sampling interval of 1/10 the period of the sinusoid.

$$v(t) = \sin(t) \qquad T = 2\pi/10$$

6.3 QUANTIZATION

Now, let us consider the quantization operation since the magnitude of each data sample must be expressed by some fixed number of digits, since only a fixed set of levels is available for approximating the infinite number of levels in continuous data. The accuracy of the approximating process is a function of the number of available levels. Since most digitizers produce binary outputs compatible with computers, the number of levels may be described by the number of binary digits (bits) produced.

6.3.1 Resolution or Quantization Step

In digital systems, sampled values must be represented in some binary number code that is stored in finite-length register. A binary sequence of length n can represent at most 2^n different numbers: that is, a 4-bit register can represent 2^4 or 16 numbers from 0 to 15 inclusively. If the amplitude of the analog signal is to be between $-v$ and $+v$ volts, the interval can be divided at most into 2^n discrete numbers. This interval is divided into quantization step size of

$$\frac{2(v)}{2^n} = E_0 \tag{6.9}$$

The size of the quantization step (E_0) or aperture introduces an amplitude approximation which limits the potential accuracy of the subsequent processing. For example, a 4-bit A/D converter with an input range of -5 to $+5$ volts had a resolution (E_0) of 10 volts divided by 16 levels, results in a quantizing step with a voltage resolution of 0.625 volts as given by (6.10):

$$\frac{10v}{2^4} = \frac{10}{16} = 0.625 \text{ volts} \tag{6.10}$$

A change in the least significant bit will cause a step from one level to the next. The most common types of sample quantization are as follows:

1) *Rounding*: The quantization level nearest the actual sample values is used to approximate the sample;
2) *Truncation*: The sample value is approximated by the highest quantization level that is *not* greater than the actual value;

and

3) *Signed Magnitude Truncation*: Positive samples are truncated but negative values are approximated by the nearest quantization level that is not less than the actual sample values.

The digital signal $X(nT)$ can often be expressed as two components: $X(nT) = X_0(nT) + e(nT)$, where $X_0(nT)$ can be thought of as the actual signal value at $t = nT$ and $e(nT)$ is an added error signal called "Quantization noise." The error signal $e(nT)$ is modeled as a uniformly distributed random sequence with the exact nature of the distribution dependent on the type of quantization involved. The variance for Truncation and rounding is the same as shown in (6.11):

$$\sigma_e^2 = \frac{E_0^2}{12} \tag{6.11}$$

where E_0 is the quantization step size.

6.3.2 Signal-to-Noise Ratio of an A/D Converter

The signal-to-noise ratio introduced by the quantization is defined as the ratio of peak-to-peak signal to RMS noise in decibels (6.12).

$$S/N = 20\log\left(2^n\sqrt{12}\right)\text{dB} \cong 6.02n + 10 \tag{6.12}$$

Hence one should know that the signal-to-noise ratio increases by approximately 6 dB for each added bit in the converter.

6.3.3 Dynamic Range of an A/D Converter

The dynamic range of the converter is defined as the ratio of the full-scale voltage to quantization step size as given by (6.13):

$$\text{Dynamic range} \; = 20 \; \log 2^n = 6.02n\,\text{dB} \tag{6.13}$$

The required dynamic range of an analog-to-digital converter (ADC) is an important attribute of a physical problem if the amplitude of the signal is to be retained for processing. One would think that increasing the number of bits (n), which decreases the quantization size, would lead to any arbitrary degree of accuracy. But recall that analog signals are limited in accuracy by the thermal noise of the analog components. Real ADC converters are analog devices, which exhibit other types of instrumentation errors, that is,

a) Offset errors

b) Gain error or scale factor

c) *Nonlinearity*: Linear offset and gain errors may be adjustable with a potentiometer; however, most nonlinearities tend to be uncorrectable.

Digital Codes: The next problem is to code the numbers from the ADC to a format suitable for further computation. So what are the most common formats used in digital processing of fixed points? If one thinks about it, 2^n different numbers can be represented by an n-bit digital word. The position of the binary point is assumed fixed. Bits to the left of the point represent the integer part. Bits to the right of the point represent the fractional part.

There are three common fixed-point formats for representing negative numbers.

1. *Sign magnitude*: The left most bit (most significant bit) represents the sign of the number, and the remaining bits ($n - 1$) represent the magnitude of the number. A well-known problem is that in a 4-bit sign magnitude, both the codes 1000 and 0000 represent the value of 0.
2. *One's compliment*: Positive numbers are represented in sign magnitude format. The negative of a number is obtained by complimenting all the bits of the number, the negative of 0101 is 1010. But again, there are two representations for 0, which are 0000 and its compliment 1111.
3. *Two's compliment* : The representation depends on the sign of the number.
   a) Positive numbers are represented in sign-magnitude format.
   b) Negative numbers are represented in two's compliment.

To obtain the two's compliment requires two operations; first, take the one's complements, and then add 1 to the least significant (right most) bit. For example, The negative of 0101 = 1011. Zero (0) is then 0000 (only one representation), with the largest positive number being 0111 for $n = 4$, while the largest negative number is 1000.

6.4 DIGITAL FORMAT

6.4.1 Floating Point Format

Floating point representation of a number consists of two fixed-point numbers representing a mantissa and an exponent. The floating-point number, n, is the product of mantissa, m, and raising base (2) to exponent a, as shown in (6.14).

$$n = m2^a \tag{6.14}$$

Sign of a floating-point number is represented by the sign bit of the mantissa while the sign bit of the exponent indicates a negative for numbers with magnitudes less than 0.5. The mantissa is normalized to lie within the range, as shown in (6.15).

$$\frac{1}{2} \leq m < 1 \tag{6.15}$$

6.4.2 Block Floating Point Format

Another format is the block floating point format where a single exponent is used for a block of fixed points. Most ADC produce fixed point numbers. Some ADC contain analog prescaling that allows sample values to be output in floating point or block floating point. Some ADC encode as differential encoding. The output is not the absolute value of the sample but rather the difference between successive sample values can be represented with fewer bits.

6.5 SIGNAL RECONSTRUCTION—DIGITAL-TO-ANALOG CONVERSION (DAC)

Two steps, which are the reverse of Analog-to-Digital Conversion (ADC), are required to reconstruct the analog signal from digitized data. The first step is to transform from discrete amplitude representation to continuous value. Second, transform the discrete time sequence to a continuous-time sample.

In most cases, a sample and hold circuit is used to hold the sample value constant until the next sample value. The results would look like the solid trace in Fig. 6.3. Usually, the zero-order hold circuit is followed by a low pass filter to remove the high-frequency components from sharp amplitude charges.

CHAPTER 7

Stationarity and Ergodic Random Processes

7.1 INTRODUCTION

In Chapter 3, signals were classified as either deterministic or random. Random signals can be further subdivided and classified as stationary and nonstationary. Properties necessary to a stationary process were provided for first-order, second-order, wide-sense, and strictly stationary. Two different methods of determining the functions necessary for a random process to be classified as a "Stationary Process," which include ensemble and extension in time, are discussed in this chapter. In addition, the necessary conditions for a random process to be an "Ergodic Process" are also presented. Nonparametric tests and examples are also included to provide the reader with the necessary information and tools to identify or classify random processes as being a "Stationary Process" and/or an "Ergodic Process."

In qualifying data, one must first determine whether the signal is deterministic or random. If a signal can be predicted exactly for a particular time span, it is said to be deterministic. Examples of deterministic signals are given in (7.1), (7.2), and (7.3).

$$x(t) = 10 \sin 2\pi t \qquad \text{Sine wave} \tag{7.1}$$

$$x(t) = 1, t > 0, \text{ and} \qquad \text{Unit step} \tag{7.2}$$
$$x(t) = 0, t \geq 0$$

$$x(t) = 1 - e^{-t}, t \geq 0, \text{ and} \qquad \text{Exponential Response} \tag{7.3}$$
$$x(t) = 0, t < 0$$

By definition, a random signal is an unpredictable signal; therefore, there is always some element of chance associated with it. Examples of random signals are given in (7.4), (7.5), and (7.6).

$$x(t) = 10\ \sin\,(2\pi t + \theta) \tag{7.4}$$

where θ is a random variable uniformly distributed between 0 and 2π.

$$x(t) = A\ \sin\,(2\pi t + \theta) \tag{7.5}$$

where θ and A are independent random variables with known distributions.

$$x(t) = A \tag{7.6}$$

where A is a noise-like signal with no particular deterministic structure.

Random signals are further divided into those signals that are classified as "Stationary" and those that are classified as "Nonstationary." Only stationary random processes will be discussed, since available standard statistical analysis techniques are applicable only to stationary random processes.

A random process is "Stationary" if its statistical properties (mean and autocorrelation) are "Time Invariant." The two methods used in obtaining the statistical properties of any random process include the following:

1. *Ensemble method* : Where the assumption is that a large number of systems equipped with identical instruments are available and that a set of instantaneous values from multiple channels are obtained at the same time.
2. *Extension-in-time method (Ergodic Testing)* : Reduces the number of observations on a system to a single long record taken over an extended period of time.

7.2 ENSEMBLE METHOD TO TEST FOR STATIONARITY

Parameters derived from the Ensemble method will be expressed as $x(t)$. As seen in Fig. 7.1, parameters derived from the Ensemble method, such as the ensemble mean,

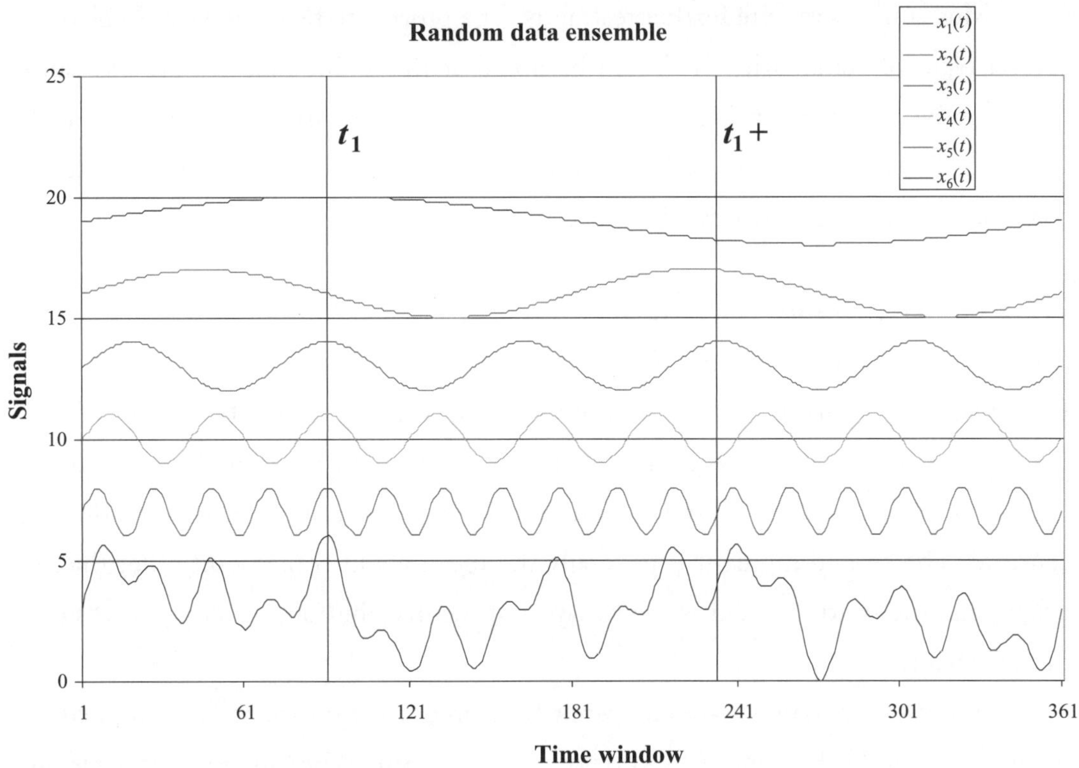

FIGURE 7.1: Random data ensemble. Values at an instant of time are taken across channels

require the selection of a specific time (t_1). Each function's value at t would be summed and divided by N (where N is the number of functions) to determine the ensemble average, where the subscript of x denotes a separate channel or function as in (7.7).

$$\overline{x}\,(t_1) = \frac{[x_1\,(t_1) + x_2\,(t_1) + x_3\,(t_1) + \cdots\cdots + x_N\,(t_1)]}{N} \tag{7.7}$$

Accordingly, a random process is stationary if the ensemble's averages are independent of time (referred to as *time invariant*) at which the group averages are computed, and if time averages are independent of functions making up this group. Thus, from the ensemble point of view, a stationary random process is one in which the statistics derived from observations on the ensemble members at any two distinct instants of time are the same (Fig. 7.1).

Most authors require further restraints for a process to be stationary. Some define several types of stationarity. It should be noted at this point that if a process can be proven to be stationary in even the weakest sense, stationarity of the random process can be assumed and analysis techniques requiring stationarity can then be conducted. Various types of stationarity are described in the subsequent section.

7.2.1 Statistics and Stationarity

7.2.1.1 First-Order Stationary Random Process

In statistics, the general method is to obtain a measure of a distribution by calculating its moments, that is, the 1st moment about the origin is often referred to as the "Mean" or "Average Value" about the origin, which is a measure of central tendency or centroid. The second moment is generally the mean about some point other than the origin, and the third moment is the variance of the distribution, which is a measure of disbursement.

If first-order statistics are independent of time (time invariant), it is implied that the moments of the ensemble values do not depend on time. The function associated with the random variable describing the values taken on by the different ensemble members at time t_1 is identical to the function associated with the random variable at any other time t_2.

First-order statistics for mean, mean square value, and variance are shown in (7.8), (7.9), and (7.10), respectively.

$$\overline{x}(t) = \int_{-\infty}^{\infty} x f_x(t)dt = \bar{x} \qquad \text{Mean Value} \qquad (7.8)$$

$$x^2(t) = \int_{-\infty}^{\infty} x^2 f_x(t)dt = x^2 \qquad \text{Mean Squared Value} \qquad (7.9)$$

$$T^2 x(t) = \int_{-\infty}^{\infty} [x - \bar{x}(t)]^2 f_x(t)dt = \sigma^2 \qquad \text{Variance} \qquad (7.10)$$

7.2.1.2 Second-Order Stationary Random Process

If all second-order statistics associated with random variables $X(t_1)$ and $X(t_2)$ are dependent at most on the time differences defined as $\tau = |t_1 - t_2|$ and not on the two values,

$t_1$and t_2, the implications are that the joint function is obtained for any two random variables that describe the process at separate times. If a second-order random process is stationary, the first-order is also stationary. Second-order statistics are described by (7.11) and (7.12).

$$R_{xx}(\tau) = \int x(t)x(t+\tau)d\tau \qquad \text{Autocorelation function} \qquad (7.11)$$

$$L_{xx} = [x(t) - \bar{x}][[x(t-\tau) - \overline{x}] \qquad \text{Covariance} \qquad (7.12)$$

7.2.1.3 Wide-Sense Stationary

A random process is stationary in the wide sense if the mean from the first-order statistics and the autocorrelation from the second-order statistics are time invariant. Therefore, a process considered to be of second-order stationarity is also stationary in the wide sense. However, wide-sense stationarity does not imply second-order stationarity since covariance is not a statistic included in wide-sense stationarity.

7.2.2 Ensemble Method

7.2.2.1 Weakly Stationary

By definition, an ensemble is said to be a Stationary Random process as *Weakly Stationary* or *Stationary in the Wide Sense*, when the following two conditions are met:

1. **The mean,** $\mu_x(t)_1$, or the first moment of the random processes at all times, t, and
2. **The autocorrelation function,** R_{xx} $(t_1,\ t_1 + \lambda)$, or joint moment between the values of the random process at two different times do not vary as time t_1 varies.

The conditions are shown in (7.13) and (7.14).

$$\mu_x(t_1) = \mu_x = \lim_{N\to\infty} \frac{1}{N} \sum_{k=1}^{N} x_k(t_1) \qquad (7.13)$$

$$R_{xx}\,(t_1, t_1 + \lambda) = R_{xx}(\lambda) = \lim_{N\to\infty} \frac{1}{N} \sum_{k=1}^{N} x_k(t_1)x_k(t_1 + \lambda) \qquad (7.14)$$

7.2.2.2 Strictly Stationary

If one could collect all the higher moments and all the joint moments, then one would have a complete family of probability distribution functions describing the process. In the case where all possible moments and joint moments are time invariant, the random process is *Strongly Stationary* or *Strictly Stationary*. If a random process is strictly stationary, it must be stationary for any lower case.

7.3 EXTENSION-IN-TIME METHOD

The second method to test for stationarity, which deals with those random processes derived from the "Extension-in-Time Method." The extension-in-time method requires dividing a single long signal into segments of equal length of time, then the mean squared value (magnitude) of each segment is obtained. If the signal is a continuous analog signal, the average magnitude for the segment is given by (7.15).

$$\overline{X_1}(t) = \int_0^{t_1} x(t)dt \tag{7.15}$$

If the signal were a discrete sampled signal, then the segment average value would be obtained by dividing the sum of the data points in a segment by the total number of data points in the segment. From the extension-in-time point of view, a stationary random process is one in which the statistics measured from observations made on one system over an extended period of time are independent of the time in which observed time interval is located.

7.3.1 Ergodic Random Process

The previous section discussed how the properties of a random process can be determined by computing ensemble averages at specific instances of time. Often, it is possible to describe the properties of a stationary random process by computing time averages over specific sample functions in the ensemble. However, for the case of the extension-in-time method, the random process is referred to as an *Ergodic* process and the test is for *Ergodicity*.

Some processes that are stationary may also be categorized as Ergodic, but not "all" stationary processes are Ergodic. Yet, "all" Ergodic processes are considered weakly stationary. Orders of ergodicity can also be established. First-order ergodicity requires that all first-order statistics have equal ensemble and time series values. The same is true for second-order and weak-sense ergodicity. For most analyses techniques, weak-sense ergodicity is sufficient.

It should be noted that only Stationary Random Processes can be classified as either being *Ergodic* or being *Nonergodic*. In addition, if the signal (random process) is Ergodic, then the signal is also considered to be *Stationary in the Weak Sense*. It is also important to note that in accordance to Bendat and Piersol, that for many applications the verification of *Weak Stationarity* justifies the assumption of *Strong Stationarity*.

For example, one can obtain the mean value $\mu_x(k)$ and the autocorrelation function $R_x(\lambda, k)$ of the kth sample function with (7.16) and (7.17).

$$\mu_x(k) = \lim_{T\to\infty} \frac{1}{T} \int_0^T x_k(t)dt \tag{7.16}$$

$$R_x(\lambda, k) = \lim_{T\to\infty} \frac{1}{T} \int_0^T x_k(t)x_k(t+\lambda)dt \tag{7.17}$$

For the random process $\{x(t)\}$ to be stationary, the time averaged mean value $\mu_x(k)$, and the autocorrelation function $R_x(\lambda, k)$ of the segments should NOT differ (statistically) when computed over different sample functions; in that case, the random process is considered to be Ergodic.

7.4 REVIEW OF BASIC PARAMETRIC STATISTICS AND MOMENT-GENERATING FUNCTIONS

7.4.1 Moment-Generating Function

Given a random variable for continuous distribution as in (7.18),

$$M(t) \equiv \langle e^{tx} \rangle = \int e^{tx} P(x)dx \tag{7.18}$$

or for discrete distributions as given by (7.19).

$$M(t) \equiv \langle e^{tx} \rangle = \sum e^{tx} P(x) \tag{7.19}$$

where $M(t) \equiv \langle e^{tx} \rangle$ is the moment-generating function. Expansion of (7.18) yields (7.20), the general moment-generating function.

$$\begin{aligned} M(t) &= \int_{-\infty}^{\infty} (1 + tx + \frac{1}{2!}t^2x^2 + \frac{1}{3!}t^3x^3 + \cdots)P(x)dx \\ &= 1 + tm_1 + \frac{1}{2!}t^2 m_2 + \frac{1}{3!}t^3 m_3 + \cdots + \frac{1}{r!}t^r m_r \end{aligned} \tag{7.20}$$

where m_r is the rth moment about zero.

For independent variables x and y, the moment-generating function may be used to generate joint moments, as shown in (7.21).

$$M_{x+y}(t) = \left\langle e^{t(x+y)} \right\rangle = \langle e^{tx} \rangle \langle e^{ty} \rangle = M_x(t) M_y(t) \tag{7.21}$$

If $M(t)$ is differentiable at zero, the rth moments about the origin are given by the following equations (7.22):

$$\begin{aligned} M\,(t) &= \langle e^{tx} \rangle, \text{ which is } M\,(0) = 1 \\ M'(t) &= \langle xe^{tx} \rangle, \text{ which is } M\,'(0) = \langle x \rangle \\ M\,''(t) &= \langle x^2 e^{tx} \rangle, \text{ which is } M\,''(0) = \langle x^2 \rangle \\ M\,'''(t) &= \langle x^3 e^{tx} \rangle, \text{ which is } M\,'''(0) = \langle x^3 \rangle \\ M\,^r(t) &= \langle x^r e^{tx} \rangle, \text{ which is } M\,^r(0) = \langle x^r \rangle \end{aligned} \tag{7.22}$$

Therefore, the mean and variance of a distribution are given by equations in (7.23).

$$\begin{aligned} \mu &\equiv \langle x \rangle = M'(0) \\ \sigma^2 &\equiv \langle x^2 \rangle - \langle x \rangle^2 = M''(0) - \left[M'(0)\right]^2. \end{aligned} \tag{7.23}$$

7.5 SUMMARY

The moments may be simply computed using the moment-generating function. The nth raw or first moment, that is, moment about zero; the origin (μ'_n) of a distribution $P(x)$

is defined as $\mu\prime_n = \langle x^n \rangle$; where for Continuous Distributions (7.24):

$$\langle f(x) \rangle = \int f(x) P(x) dx \tag{7.24}$$

and for Discrete Distributions (7.25)

$$\langle f(x) \rangle = \sum f(x) P(x) \tag{7.25}$$

μ'_n, the mean, is usually simply denoted as $\mu = \mu_1$.

If the moment is taken about a point a rather the origin, the moment is termed the second moment and is generated by (7.26).

$$\mu_n(a) = \langle (x - a)^n \rangle = \sum (x - a)^n P(n) \tag{7.26}$$

The moments are most commonly taken about the mean (a point a), which are also referred to as "Central Moments" and are denoted as μ_n. The Central or second moment about the origin is defined by (7.27), where $n = 2$.

$$\mu_n \equiv \langle (x - \mu)^n \rangle = \int (x - \mu)^n P(x) dx \tag{7.27}$$

It should be noted that the third moment ($n = 3$) about the origin may also be called the second moment about the mean, $\mu_2 = \sigma^2$, which is equal to the variance or disbursement of the distribution; and the square root of the variance is called the standard deviation.

7.6 SUGGESTED READING

Papoulis, A. *Probability, Random Variables, and Stochastic Processes,* 2nd ed. New York: McGraw-Hill, pp. 145–149, 1984.

Press, W. H., Flannery, B. P., Teukolsky, S. A., and Vetterling, W. T. "Moments of a Distribution: Mean, Variance, Skewness, and So Forth." §14.1 in *Numerical Recipes in FORTRAN: The Art of Scientific Computing,* 2nd ed. Cambridge, England: Cambridge University Press, pp. 604–609, 1992.

Kenney, J. F. and Keeping, E. S. "Moment-Generating and Characteristic Functions," "Some Examples of Moment-Generating Functions," and "Uniqueness Theorem for Characteristic Functions." §4.6–4.8 in *Mathematics of Statistics,* Pt. 2, 2nd ed. Princeton, NJ: Van Nostrand, pp. 72–77, 1951.

Weisstein, E. W. *Moment Generating Functions*. Wolfram Research, Inc., CRC Press LLC, 1999–2003.

Bendat, J. S., and Piersol, A. G. *Random Data: Analysis and Measurement Procedures*, 2nd ed. New York: Wiley-Interscience, John Wiley & Sons, 1986.

Bendat, J. S., and Piersol, A. G. *Engineering Applications of Correlation and Spectral Analysis*, 2nd ed. New York: Wiley-Interscience, John Wiley & Sons, 1993.

CHAPTER 8

Nonparametric Statistic and the Runs Test for Stationarity

8.1 INTRODUCTION

8.1.1 Conditions for a Random Process to Be Considered as Stationary

Recall that Chapter 7 contained a discussion on how the properties of a random process can be determined to be stationary by computing the ensemble averages at specific instants of time. A stationary random process was considered to be weakly stationary or stationary in the wide sense, when the following conditions were met:

(a) the mean, $\mu_x(t_1)$, termed the first moment of the random processes at all times t_i and

(b) the autocorrelation function, $R_x\,(t_1, t_1 + \tau)$, or joint moment between the values of the random process at two different times does not vary as time t varies.

That is, both the ensemble mean $\mu_x(t1) = \lim\limits_{N\to\infty} \frac{1}{N}\sum_{k=1}^{N} x_k(t1)$ and the autocorrelation function $R_x(t_1, t_1 + \tau) = R_x(\lambda) = \lim\limits_{N\to\infty} \frac{1}{N}\sum_{k=1}^{N} x_k(t_1)x(t_1 = \tau)$ are time invariant.

If all the higher moments and all the joint moments could be collected, one would have a complete family of probability distribution functions describing the processes. For the case where all possible moments and joint moments are time invariant, the random process is strongly stationary or stationary in the strict sense.

Unfortunately, it *is not* often possible to describe the properties of a stationary random process by computing time averages over specific sample functions in the ensemble, since many times an ensemble cannot be collected, but rather a single long recording of a signal is acquired. In this case, the long recording is divided into smaller segments and the data are tested for the Ergodic properties.

For example, the mean value $\mu_x(k)$ and the autocorrelation function $R_x(\lambda, k)$ of the kth sample function can be obtained with (8.1) and (8.2).

$$\mu_x(k) = \lim_{T\to\infty} \frac{1}{T} \int_0^T x_k(t)dt \tag{8.1}$$

$$R_x(\lambda, k) = \lim_{T\to\infty} \frac{1}{T} \int_0^T x_k(t)\, x_k(t+\tau)dt \tag{8.2}$$

The random process $\{x(t)\}$ is ***Ergodic*** and Stationary if the mean value $\mu_x(k)$ and the autocorrelation function $R_x(\tau, k)$ of the short-time segments **do not** differ when computed over different sample functions. It should be noted that only stationary random processes can be **Ergodic**. Another way of thinking is "all Ergodic Random Processes are Stationary." For many applications, verification of weak stationarity justifies an assumption of strong stationary (Bendat and Piersol).

8.1.2 Alternative Methods to Test for Stationarity

So far, the methods described require integral functions and parametric statistical testing or moments describing the properties of the random process. A disadvantage in the use of parametric statistics is that the random data must be tested for independence and normality. By normality of the distribution, we mean that the data must be "Gaussian distributed." By far the most common method used is to test whether the data are Ergodic. In general, three assumptions are made, when random data are tested for Ergodicity. The basic assumptions are as follows.

1. That any given sample record will properly reflect the nonstationary character of the random process.

2. That any given sample record is very long compared to the lowest frequency component in the data, excluding a nonstationary mean. What is meant by this assumption is that the sample record has to be long enough to permit nonstationary trends to be differentiated from the random fluctuations of the time history.
3. That any nonstationarity of interest will be revealed by time trends in the mean square value of the data.

With these assumptions, the stationarity of random data can be tested by investigating a single long record, $x(t)$, by the following procedure:

(1) Divide the sample record into N equal time intervals where the data in each interval may be considered independent.

(2) Compute a mean square value (or mean value and variance separately) for each interval and align these samples values in a time sequence as follows: $\overline{X}_1^2, \overline{X}_2^2, \overline{X}_3^2, \ldots \overline{X}_N^2$.

(3) Test the sequence of mean square values for the presence of underlying trends or variations other than those due to expected sampling variations. The final test of the sample values for nonstationary trends may be accomplished in many ways. A *nonparametric* approach, which *does not* require knowledge of the sampling distributions of data parameter, is desirable [3 and 4]. One such test is the Runs Test. There are other nonparametric tests that may be used. These tests will be presented in the section on review of nonparametric statistics.

8.1.2.1 Runs Test

Because of the requirements in testing with parametric statistics, that is, independence and normality of the distribution, nonparametric statistics are often used. The advantage of the nonparametric test is that *a priori* knowledge regarding the distribution is not necessary.

Runs test considers a sequence of N observed values of a random variable x in which it was decided to use the mean square value ($\overline{X}_i^2$), where each observation is classified into one of two mutually exclusive categories, which may be identified simply as $+$ or $-$. For example,

$$x_i \geq \overline{X} \text{ is} + \quad \text{and} \quad x_i < \overline{X} \text{ is} -$$

Runs test for independence of data may be extended to two simultaneous sequences of data (x_i,y_i). For example, if two simultaneous sequence of two sets of measures x_i and y_i, $i = 1, 2, 3 \ldots, N$, then

$$x_i \geq y_i(+) \quad \text{or} \quad x_i < y_i(-).$$

Back to the single variable data set, let us hypothesize that the sequence of sample mean square values $(\overline{x}_1^2, \ldots, \overline{x}_N^2)$ is each independent sample values of a random variable with mean square value of ψ_x^2, where $R_x(0) = \psi_x^2$ is the autocorrelation function with zero time displacement $\lambda = 0$, as shown in (8.3).

$$\psi_x^2 = \lim_{T\to\infty} \frac{1}{T} \int_0^T x^2(t)dt \quad \text{or} \quad R_x(0) = \psi_x^2 \tag{8.3}$$

If the hypothesis is true, the variations in the sequence of sample values will be random and display no trend. Hence, the number of runs in the sequence, relative to say the median value, will be as expected for a sequence of independent random observations of the random variables as in Table 8.2. If the number of runs is significantly different from the expected number given in the table, then the hypothesis of stationarity would be rejected; otherwise, the hypothesis is accepted. The acceptance region for the hypothesis is

$$r_{\frac{N}{2}}; 1 - \alpha/2 < r \leq r_{\frac{N}{2}}; \alpha/2$$

Note that this testing procedure is independent of frequency bandwidth of the data or the averaging time used. The Runs test is *not* limited to a sequence of mean square values, since it works equally well on mean values, RMS values, standard deviations, mean absolute values, or any other parameter estimate. Furthermore, it is not necessary for the data under investigation to be free of periodicities, but the fundamental period must be short compared to the averaging time used to compute sample values.

The number of runs that occur in a sequence of observations gives an indication as to whether or not results are independent random observations of the same random variables. Specifically, if a sequence of N observations is independent observations of the same random variable, that is, the probability of a (+) or (−) result does not change from one observation to the next, then the sampling distribution of the number of runs in the sequence is a random variable r with a mean value as shown in (8.4) and the variance is given by (8.5).

$$\mu_r = \frac{2N_1N_2}{N} + 1 \tag{8.4}$$

where N_1 is the number of (+) observations and N_2 is the number of (−) observations.

$$\sigma_r^2 = \frac{2N_1N_2\,(2N_1N_2 - N\,)}{N^2\,(N_1 - 1\,)} \tag{8.5}$$

For the case where $N_1 = N_2 = N/2$ then the mean and variance quations become (8.6 and 8.7.)

$$\mu_r = \frac{N}{2} + 1 \tag{8.6}$$

and

$$\sigma_r^2 = \frac{N(N-2)}{4\,(N-1)} \tag{8.7}$$

There distribution function is given by Table 8.2.

$$r_\eta \text{ is,} \quad \text{where } \eta = \frac{N}{2} = N_1 = N_2, \quad r_\eta; 1 - \frac{\alpha}{2} \quad \text{and} \quad r_n; \frac{\alpha}{2}$$

In any case, a sequence of plus and minus observations is obtained from which the number of “RUNS” is determined. A “RUN” is defined as a sequence of identical observations that is followed and preceded by a different observation or no observation at all. There are tables for the distribution function of runs. Runs test can be used to evaluate data involving the question of testing a single sequence of observations for independence.

TABLE 8.1: Mean Squared Values of the Segments $N = 20$. Significance Level: $\alpha = 0.05$

$(\overline{x}_1^2)$	5.5	$(\overline{x}_6^2)$	5.7	$(\overline{x}_{11}^2)$	6.8	$(\overline{x}_{16}^2)$	5.4
$(\overline{x}_2^2)$	5.1	(7)	5.0	(12)	6.6	(17)	6.8
(3)	5.7	(8)	6.5	(13)	4.9	(18)	5.8
(4)	5.2	(9)	5.4	(14)	5.4	(19)	6.9
(5)	4.8	(10)	5.8	(15)	5.9	(20)	5.5

Let us look at an example where the random data are from a sample record of 480 seconds in duration. The sample record is divided into 20 segments of equal lengths or 24 seconds of durations with the mean squared values shown in Table 8.1.

8.1.2.2 Median = 5.6

Hence, mean squared values greater than the median value will be designated a (+) and those values less than the median will be designated as (−).

$$x_i \geq 5.6 = + \qquad x_i < 5.6 = -$$

Result

$$\underbrace{--}_{1}\ \underbrace{\pm}_{2}\ \underbrace{--}_{3}\ \underbrace{\pm}_{4}\ \underbrace{-}_{5}\ \underbrace{\pm}_{6}\ \underbrace{-}_{7}\ \underbrace{\pm\pm\pm}_{8}\ \underbrace{--}_{9}\ \underbrace{\pm}_{10}\ \underbrace{-}_{11}\ \underbrace{\pm\pm\pm}_{12}\ \underbrace{-}_{13}$$

$r = 13$ runs are represented by the sequence of 20 observations.

$$\text{for} \quad \alpha = 0.05 \quad \text{or} \quad \frac{\alpha}{2} = 0.025$$

From Runs Distribution, (Table 8.2),

$$r_{10}; 1 - \frac{\alpha}{2} = r_{10} : 0.975 = 6$$
$$r_{10}; \frac{\alpha}{2} = r_{10} : 0.025 = 15$$

Then,

$$r_{10} : 0.975 = 6 < r = 13 < r_{10} : 0.025 = 15$$

TABLE 8.2: Percentage Points of Run Distribution Values of $r_{n;\alpha}$ Such That Prob $[r_n > r_n] = \alpha$, Where $n = N_1 = N_2 = N/2$.

	α					
$n = N/2$	0.99	0.975	0.95	0.05	0.025	0.01
5	2	2	3	8	9	9
6	2	3	3	10	10	11
7	3	3	4	11	12	12
8	4	4	5	12	13	13
9	4	5	6	13	14	15
10	5	6	6	15	15	16
11	6	7	7	16	16	17
12	7	7	8	17	18	18
13	7	8	9	18	19	20
14	8	9	10	19	20	21
15	9	10	11	20	21	22
16	10	11	11	22	22	23
18	11	12	13	24	25	26
20	13	14	15	26	27	28
25	17	18	19	32	33	34
30	21	22	24	37	39	40
35	25	27	28	43	44	46
40	30	31	33	48	50	51
45	34	36	37	54	55	57

(*cont.*)

TABLE 8.2: (*Continued*)

	α					
$n = N/2$	0.99	0.975	0.95	0.05	0.025	0.01
50	38	40	42	59	61	63
55	43	45	46	65	66	68
60	47	49	51	70	72	74
65	52	54	56	75	77	79
70	56	58	60	81	83	85
75	61	63	65	86	88	90
80	65	68	70	91	93	96
85	70	72	74	97	99	101
90	74	77	79	102	104	107
95	79	82	84	107	109	112
100	84	86	88	113	115	117

The hypothesis is accepted, since $r = 13$ falls within the range between 6 and 15; hence, the data are independent and ergodic; thus stationary. There is no reason to question independence of observations since there is no evidence of an underlying trend.

8.1.2.3 Summary of Ergodic Test Via the Runs Test

The statistical properties of a random process can be determined by computing ensemble averages at specific instances of time or from segments of a single long record. If the time averaged mean value $\mu_x(k)$ and the autocorrelation function $R_x(\lambda,k)$ are time invariant when computed over different sample functions, then the random process is Ergodic and Stationary. An easier approach in testing a single random signal for stationarity is to use the Nonparametric Runs Test.

8.2 REVIEW OF NONPARAMETRIC STATISTICS USED FOR TESTING STATIONARITY

8.2.1 Kolmogorov–Smirnov One-Sample Test

The Kolmogorov–Smirnov one-sample test is concerned with the degree of agreement between the distribution of a set of observed scores and some specified theoretical distribution [5 and 6]. The test method requires calculation of the maximum deviation, D, with (8.8).

$$D = \text{max}imum\left|F_0(X) - SN(X)\right| \tag{8.8}$$

where $F_0(X)$ equals the theoretical cumulative distribution under the null hypothesis, H_0. Then for any value of X, the value $F_0(X)$ is the proportion of cases expected to have scores equal to or less than X. $SN(X)$ is the observed cumulative frequency distribution of a random sample with N observations. D, maximum deviation, is the largest value of the difference between $F_0(X)$ and $SN(X)$.

If the maximum deviation, D, is greater than or equal to the table value, then the null hypothesis is rejected, and it is concluded that the data have trends and is, therefore, nonstationary.

An example of the Kolmogorov–Smirnov test when testing for stationarity of average rainfall data is shown in Table 8.3.

TABLE 8.3: Example of Kolmogorov-Smirnov Test

	x_1	x_2	x_3	x_4	x_5
f	2	3	3	1	1
$F_0(X)$	1/5	2/5	3/5	4/5	5/5
$SN(X)$	2/10	5/10	8/10	9/10	10/10
$F_0(X) - SN(X)$	0	1/10	2/10	1/10	0

Note: Calculated test value: $D = 0.2$; Table value: $D_{10,0.05} = 0.410$; and $\Delta x = 0.5$; such that, $x_1 = 0.5$, $x_2 = 1.0$, $x_3 = 1.5$, $x_4 = 2.0$, and $x_5 = 2.5$.

The test failed to reject H_0, since the calculated test value of D is less than the table value ($0.2 < 0.410$). It is concluded that the data are stationary.

8.2.2 One-Sample Runs Test

For the One-Sample Runs test, the sample record must be illustrated by symbols; generally, by taking the median of the sample data and assigning all values above the median a symbol, n_1, and all values below the median another symbol, n_2. Once the values are represented by symbols, the number of runs can be determined. A run is defined as a succession of identical symbols.

For small samples, n_1 and $n_2 < 20$, the number of runs, r, is then compared to the value from the Runs table. If r falls within the critical values, then the null hypothesis is accepted, which infers failure to reject the null hypothesis, H_0.

If n_1 or n_2 is larger than 20, the table value is not used. For large samples, an approximation of the sampling distribution of r is the normal distribution. Therefore, a Z score is computed to test H_0, as shown in (8.9):

$$Z = \frac{r - \mu r}{\sigma r} = \frac{r - \left(\dfrac{2n_1n_2}{n_1 + n_2} + 1\right)}{\sqrt{\dfrac{2n_1n_2(2n_1n_2 - n_1 - n_2)}{(n_1 + n_2)^2(n_1 + n_2 - 1)}}} \tag{8.9}$$

The calculated Z value from (8.9) is then compared to values in the Normal Distribution Table, which shows the one-tailed probability. For a two-tailed test, double the probability value (p) given in the table. If the p associated with the observed value of r is less than or equal to the significant level (α), then the null hypothesis H_0 is rejected and it is concluded that the data are nonstationarity.

8.2.3 Testing for Ergodicity

8.2.3.1 Wilcoxon-Matched Pairs Signed-Ranks Test

The Wilcoxon-matched pairs test is a powerful test because it considers relative magnitude and direction of differences. The process steps are as follows:

TABLE 8.4: Example of Wilcoxon-Matched Pairs Signed-Ranked Test

INTERVAL	DIFFERENCE	RANK OF DIFFERENCE	RANK WITH LESS FREQUENT SIGN
1	2	3	
2	−1	−1.5	1.5
3	3	4	
4	0		
5	−1	−1.5	1.5
6	5	5	
7	6	6	
	$N = 6$		$T = 3.0$

1. Take signed difference for each matched pair.
2. Rank differences without respect to sign; if there is a tie, assign the average of the tied ranks.
3. Affix "+" or "−" to rank according to the difference.
4. Determine T, where T is the smaller of the sums of like ranked signs.
5. Determine N, where N is the number of differences having signs (zeroes do not count).
6. If $N < 25$, then the critical value is obtained from the "Table of Critical Values of T in the Wilcoxon-Matched Pairs Signed-Ranks Test." If the observed T is less than or equal to the table value, then the null hypothesis, H_0, is not accepted (rejected).
7. If $N > 25$, then the Z value must be calculated from (8.10) and the table of Critical Values of T used.

$$Z = \frac{T - \dfrac{N(N+1)}{4}}{\sqrt{\dfrac{N(N+1)(2N+1)}{24}}} \tag{8.10}$$

To test for ergoticity of a process, we must determine the ensemble average and time average for each interval. The ensemble average and time average are then paired for each interval and their differences computed. The results of an example are given in Table 8.4.

The value from the table at $T_{6,0.05}$ is 0. Since the calculated value of 3.0 is not less than or equal to zero (3.0 is not <0.0), the null hypothesis, H_0, is accepted (failure to reject H_0) and it is concluded that the ensemble average and time average do not differ; thus, the process is Ergodic.

For more detailed information on nonparametric statistical procedures, see Hollander and Wolfe [1] and Siegel [2].

8.3 REFERENCES

1. Hollander, M., and Wolfe, D. A. *Nonparametric Statistical Methods*. New York: John Wiley and Sons, 1973.
2. Siegel, S. *Nonparametric Statistics*. New York: McGraw-Hill Book Company, 1956.
3. Au, T., Shane, R. M., and Hoel, L. A. *Fundamentals of Systems Engineering: Probabilistic Models*. Reading, MA: Addison-Wesley Publishing Company, 1972.
4. Brown, R. G. *Introduction to Random Signal Analysis and Kalman Filtering*. New York: John Wiley and Sons, 1983.
5. O'Flynn, M. *Probabilities, Random Variables, and Random Processes*. New York: Harper & Row, Publishers, 1982.
6. Panter, P. F. *Modulation, Noise, and Spectral Analysis*. New York: McGraw-Hill Book Company, 1965.

CHAPTER 9

Correlation Functions

In the general theory of harmonic analysis, an expression of considerable importance and interest is the correlation function given in (9.1).

$$\frac{1}{T}\int_{-T_{1/2}}^{T_{1/2}} f_1(t)f_2(t+\tau)dt \tag{9.1}$$

For the case of periodic functions where $f_1(t)$ and $f_2(t)$ have the same fundamental angular frequency ω_1 and where τ is a continuous time of displacement in the range $(-\infty\,,\infty)$, independent of t (the ongoing time).

One important property of the correlation expression is the fact that its Fourier Transform is given by (9.2), where $f_1(t)$ has the complex spectrum $F_1(n)$, and $f_2(t)$ has $F_2(n)$.

$$\overline{F_1}(n)F_2(n) \tag{9.2}$$

The bar ($\overline{F_1}(n)$) indicates the complex conjugate of the quantity (function) over which it is placed. To establish this fact, the Fourier expansion of $f_1(t)$ and $f_2(t)$ can be expressed as (9.3) and (9.4).

$$f_1(t) = \sum_{n=-\infty}^{\infty} F_1(n)e^{jn\omega_1 t} \tag{9.3}$$

$$f_2(t) = \sum_{n=-\infty}^{\infty} F_2(n)e^{jn\omega_1 t} \tag{9.4}$$

where the complex spectrums are given by (9.5) and (9.6), respectively.

$$F_1(n) = \frac{1}{T_1}\int_{-T_{1/2}}^{T_{1/2}} f_1(t)e^{-jn\omega 1t}dt \tag{9.5}$$

and

$$F_2(n) = \frac{1}{T_1} \int_{-T_{1/2}}^{T_{1/2}} f_2(t) e^{-jn\omega 1t} dt \tag{9.6}$$

However, it should be noted from (9.1) that the original equation is not for $f_2(t)$ but rather for $f_2(t+\tau)$. Thus, it was found that caution must be taken in the manner that the correlation function is expressed:

$$\frac{1}{T} \int_{-T_{1/2}}^{T_{1/2}} f_1(t) f_2(t+\tau) dt = \frac{1}{T} \int_{-T_{1/2}}^{T_{1/2}} f_1(t) dt \sum_{n=-\infty}^{\infty} F_2(n) e^{jn\omega_1(t+\tau)} \tag{9.7}$$

The manner in which the right-hand side of the equation is written indicates that summation with respect to n comes first and is followed by integration with respect to t, as shown in (9.7).

At this point, if the order of summation and integration is inversed, the result would be given by (9.8):

$$\frac{1}{T_1} \int_{-T_{1/2}}^{T_{1/2}} f_1(t) f_2(t+\tau) dt = \sum_{n=-\infty}^{\infty} F_2(n) e^{jn\omega_1\tau} \frac{1}{T_1} \int_{-T_{1/2}}^{T_{1/2}} f_1(t) e^{jn\omega_1 t} dt \tag{9.8}$$

Note that in (9.8), the exponent in the second term (the integral) has a positive, $+j$, rather than a negative, $-j$, as shown in the integral ($\int dt$) of (9.5), where the last integral is recognized as the complex conjugate of $F_1(n)$. Hence, the result yields (9.10),

$$\frac{1}{T_1} \int_{-T_{1/2}}^{T_{1/2}} f_1(t) f_2(t+\tau) dt = \sum_{n=-\infty}^{\infty} [\overline{F_1}(n) F_2(n)] e^{jn\omega_1\tau} \tag{9.10}$$

which results in the general form of the inverse Fourier Transform (9.11) for a periodic function of fundamental angular frequency ω_1 and complex spectrum $\overline{F_1}(n)F_2(n)$ of (9.10).

$$f(t) = \sum_{n=-\infty}^{\infty} F(n) e^{jn\varpi 1t} \tag{9.11}$$

However, note that (9.10) is in terms of delays, τ, instead of time, t, as in (9.11). Then by applying the definition of the Fourier Transform of the periodic function $f(t)$ as in (9.12),

$$F(n) = \frac{1}{T_1} \int_{-T_{1/2}}^{T_{1/2}} f(t) e^{-jn\omega_1 t} dt \quad \text{for} \quad n = 0, \pm 1, \pm 2, \pm \ldots \tag{9.12}$$

and substituting the convolution function for $f(t)$ as in (9.13):

$$\frac{1}{T} \int_{-T_{1/2}}^{T_{1/2}} f_1(t) f_2(t+\tau) dt = f(t) \tag{9.13}$$

the complex spectrum $\overline{F_1}(n)F_2(n)$ becomes (9.14):

$$\overline{F_1}(n)F_2(n) = \frac{1}{T_1} \int_{-T_{1/2}}^{T_{1/2}} e^{-jn\omega_1 \tau} d\tau \int_{-T_{1/2}}^{T_{1/2}} f_1(t) f_2(t+\tau) dt \tag{9.14}$$

Again, the right integration is preformed first with respect to t and the resulting function of τ is then multiplied by $e^{-jn\omega_1 \tau}$. Then, the product is integrated with respect to τ in order to obtain a function of n.

Thus the equations in (9.15) are Fourier Transforms of each other.

$$\frac{1}{T_1} \int_{-T_{1/2}}^{T_{1/2}} f_1(t) f_2(t+\tau) dt \quad \text{and} \quad \overline{F_1}(n)F_2(n) \tag{9.15}$$

This relationship is called the "Correlation Theorem" for periodic functions.

9.1 THE CORRELATION PROCESS

The correlation process involves three important operations:

(1) The periodic function $f_2(t)$ is given as delays or time displacement, τ.

(2) The displacement function is multiplied by the other periodic function of the same fundamental frequency.

(3) The product is averaged by integration over a complete period.

These steps are repeated for every value τ in the interval $(-\infty, \infty)$ so that a function is generated. In summary, the combination of the three operations—displacement, multiplication, and integration—are termed "Correlation," whereas folding, displacement, multiplication, and integration are referred to as "Convolution," a totally different process. Correlation is not limited to periodic functions, as it can be applied to aperiodic and random functions.

9.1.1 Autocorrelation Function

Basic statistical properties of importance in describing single stationary random processes include autocorrelation functions and autospectral density functions. In the correlation theorem for periodic functions, it was found that under the special conditions, the two functions may be identical, $f_1(t) = f_2(t)$. In this case, the correlation equation becomes (9.16) and is referred to as the "Autocorrelation Function."

$$\frac{1}{T_1}\int_{-T_{1/2}}^{T_{1/2}} f_1(t) f_1(t+\tau)dt = \sum_{n=-\infty}^{\infty} \overline{|F_1(n)|}^2 e^{jn\omega_1\tau} \tag{9.16}$$

For the case in which $\tau = 0$, it means that the squared value of the function $f_1(t)$ is equal to the sum of the squares of the absolute values over the entire range of harmonics in the spectrum of the function. Because the expression on the left divides the integrated results, the autocorrelation at zero delay is referred to as the *Mean Squared Value* of the function. The autocorrelation function is often subscripted as φ_{11} in the time domain or as Φ_{11} in the frequency domain. The reciprocal relations expressed in (9.17) and (9.18) form the basis for the "Autocorrelation Theorem" for periodic functions.

Expressed in words, the theorem states that the autocorrelation function and the power spectrum of a periodic function are "Fourier" transforms of each other. Equation (9.17) is the time domain expression, stating that the autocorrelation is equal to the "Inverse Fourier" transform of the power spectrum.

$$\varphi_{11}(\tau) = \sum_{n=-\infty}^{\infty} \Phi_{11}(n) e^{jn\omega_1\tau} \tag{9.17}$$

Equation (9.18) is the frequency domain expression, stating that the "Fourier" transform of the autocorrelation function is equal to the power spectrum of the function.

$$\Phi_{11}(n) = \int_{-\frac{T_1}{2}}^{\frac{T_1}{2}} \varphi_{11} e^{-jn\omega_1\tau} d\tau \tag{9.18}$$

9.2 PROPERTIES OF THE AUTOCORRELATION FUNCTION

One of the properties of an autocorrelation function is that although it retains all harmonics of the given function, it discards all their phase angles. In other words, all periodic functions having the same harmonic amplitudes but differing in their initial phase angles have the same autocorrelation function or that the power spectrum of a periodic function is *independent* of the phase angles of the harmonics.

Another important property is that the autocorrelation function is an *Even Function* of τ. An even periodic function satisfies the expression given by (9.20).

$$\varphi_{11}(-\tau) = \varphi_{11}(\tau) \tag{9.20}$$

The proof is as follows: writing (9.16) as a function of a negative delay, $-\tau$, gives (9.21).

$$\varphi_{11}(-\tau) = \frac{1}{T_1} \int_{-T_{1/2}}^{T_{1/2}} f_1(t) f_1(t-\tau) dt \tag{9.21}$$

Using the change of variable, since $x = (t - \tau)$ or $x = (t + \tau)$, then the limits become $T_{1/2} - \tau$; $-T_{1/2} - \tau$, and $f_1(t - \tau)$ becomes $f_1(x + \tau - \tau) = f_1(x)$. Likewise, $f_1(t)$ becomes $f_1(t + \tau)$. Also since $\varphi_{11}(\tau)$ is a periodic function of period T_1, integration over the interval, $-T_{1/2} - \tau$to $T_{1/2} - \tau$, is the same as $T_{1/2}$ to $-T_{1/2}$. The result is 9.22:

$$\varphi_{11}(-\tau) = \frac{1}{T_1} \int_{-T_{1/2}}^{T_{1/2}} f_1(t) f_1(t+\tau) dt \quad \text{or that} \quad \varphi_{11}(-\tau) = \varphi_{11}(\tau) \tag{9.22}$$

9.3 STEPS IN THE AUTOCORRELATION PROCESS

The autocorrelation process requires the five steps listed as follows:

1. Change of variable from time (t) to delay (τ)
2. Negative translation along the horizontal x-axis (N is the total number of delays)

3. Positive incremental translation along the horizontal x-axis (one delay)
4. Multiplication of the two functions, $f_1(\tau)$ and $f_2(t-\tau)$
5. Integration of area

Steps 3 through 5 are repeated until $2N$ delays are completed

The process might be synopsized as determining the area under the curves (product) as the folded function is slid along the horizontal axis. Note the direction of translation; the conventional notation is as follows:

a. Movement to the right for positive (+) time
b. Movement to the left for negative (−) time

At this point, let us go through an example to visualize the autocorrelation process. The function f_1 is shown in Fig. 9.1 as two repeated graphs. To simplify the example, the change of variable will not be preformed or shown.

The second step is to translate or move one of the functions in the negative direction without reversing the function. Figure 9.2 shows what the graph of the two functions would look like. Note that $x = 0$ is in the middle of the axis.

The next step is that of moving the translated function in a positive direction by 2 delays (from 0 to +2) along the x-axis as shown in Fig. 9.3. Note that the overlapping area is the area of both functions between 0 and +2 on the x-axis.

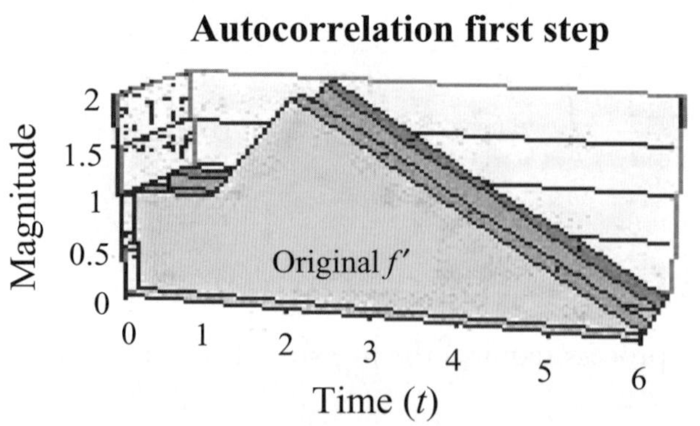

FIGURE 9.1: Starting the autocorrelation process. The function f_1 is shown as two separate 3-D objects

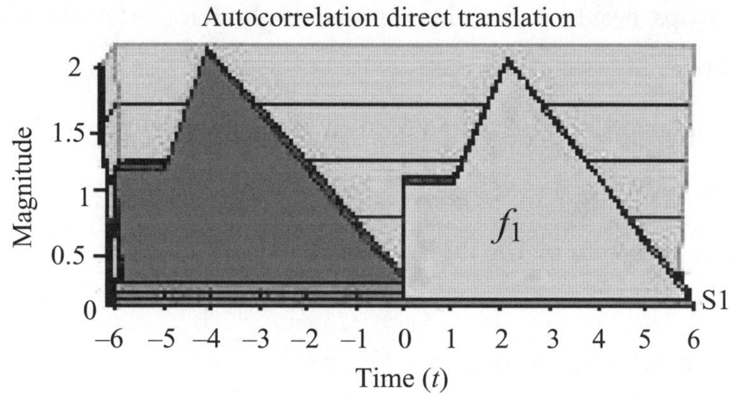

FIGURE 9.2: Step 2: Translation. Negative translation of function along the x-axis

9.4 NUMERICAL CORRELATION: DIRECT CALCULATION

9.4.1 Strips of Paper Method

One of the numerical methods is to use two strips of paper in the following steps:

1. Tabulate values of the function at equally spaced intervals on two strips of paper
2. Reverse tabulate one function; $f(-\lambda)$, which is equivalent to folding a function

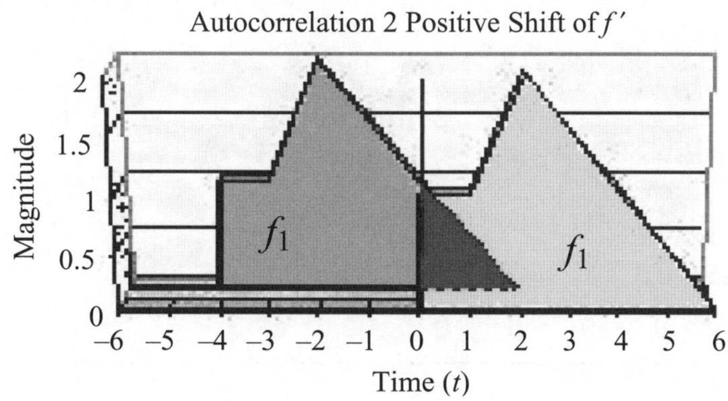

FIGURE 9.3: Positive translation. The second function is moved in a positive direction by 2 delays along the x-axis. Note the overlapping area of both functions between 0 and +2 on the x-axis

3. Place strips beside each other and multiply across the strips and write the products
4. Slide one strip, and repeat step 3, until there are no more products
5. Sum the rows
6. Multiply each sum by the interval width

9.4.2 Polynomial Multiplication Method

Another method is the polynomial multiplication method (Table 9.1). For this method, the values of the function are written twice in two rows; one function is placed above the other (reverse tabulated the second time the function is written) as in Table 9.1. It should be noted that the functions are aligned as in normal multiplication; from right to left. Reverse tabulation of the second function is important; otherwise, an

TABLE 9.1: Autocorrelation by Multiplication Method Correct Answer

				t		0	1	2	3	4	5	6
		f_1	D	T		1	1	2	1.5	1	0.5	0
		f_1	R	T		0	0.5	1	1.5	2	1	1
						1	1	2	1.5	1	0.5	0
					1	1	2	1.5	1	0.5	0	
				2	2	4	3	2	1	0		
			1.5	1.5	3	2.3	1.5	0.8	0			
		1	1	2	1.5	1	0.5	0				
	0.5	0.5	1	0.8	0.5	0.3	0					
0	0	0	0	0	0	0						
0	0.5	1.5	3.5	6.3	8	9.5	8	6.3	3.5	1.5	0.5	0

Note: Original function (f_1); Time t ; DT is Direct Transcription; and RT is Reverse Transcription.

TABLE 9.2: Autocorrelation by Multiplication Method Incorrect Answer

			t			*0*	*1*	*2*	*3*	*4*	*5*	*6*
		f_1	*D*	*T*		1	1	2	1.5	1	0.5	0
		f_1	*D*	*T*		1	1	2	1.5	1	0.5	0
						0	0	0	0	0	0	0
					0.5	0.5	1	0.8	0.5	0.3	0	
				1	1	2	1.5	1	0.5	0		
			1.5	1.5	3	2.3	1.5	0.8	0			
		2	2	4	3	2	1	0				
	1	1	2	1.5	1	0.5	0					
1	1	2	1.5	1	0.5	0						
1	2	5	7	9	9	7.3	5	2.5	1	0.3	0	0

incorrect answer is obtained as shown in the calculations of Table 9.2. The number of terms in the results should equal twice the number of terms in the function. The final results of the correlation operation are obtained by summing the column and multiplying by the interval width. The graph of the correct calculation is shown in Fig. 9.4; note that the autocorrelation function is an even function (symmetrical about the y-axis).

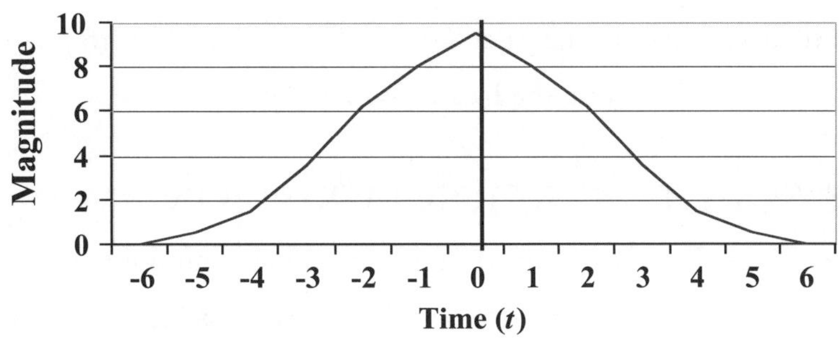

FIGURE 9.4: Correct calculation of autocorrelation. The graph shows the result of Table 9.1 calculations. Note that the autocorrelation function is an even function

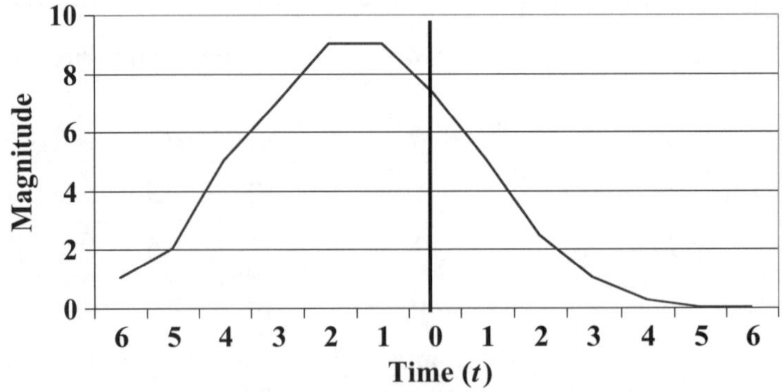

FIGURE 9.5: Incorrect autocorrelation results. Note the lack of symmetry in the graph of the incorrect calculation results of Table 9.2

For the Direct Numerical Method via polynomial multiplication, use the following steps:

1. directly tabulate values of the function in the second row with the increment of time above the function as in Table 9.1;
2. reverse tabulate the same function in the third row;
3. do normal polynomial multiplication (from right to left);
4. write the products in columns following normal multiplication procedures;
5. when there are no more products, sum the columns; and
6. multiply each sum by the interval width.

It is important to "Reverse tabulate" the second function; otherwise, the incorrect answer is obtained as shown in the calculations of Table 9.2. The graph of the incorrect results is shown in Fig. 9.5. Note the lack of symmetry.

9.5 CROSS-CORRELATION FUNCTION

For pairs of random data from two different stationary random processes, the joint statistics of importance include the cross-correlation and cross-spectral density functions. The cross-correlation theorem for periodic functions $f_1(t)$ and $f_2(t)$ applies to two different periodic functions, but with the stipulation that the functions must have the same fundamental frequency. If the condition of the same fundamental frequency is met, the cross-correlation function is defined as given by (9.23) for continuous data or by (9.24) for discrete data,

$$\varphi_{12}(\tau) = \frac{1}{T_1} \int_{-T_{1/2}}^{T_{1/2}} f_1(\tau) f_2(t+\tau) dt \tag{9.23}$$

$$r_{xy}(\tau) = \frac{1}{N} \sum_{i=1}^{n} x_i(t)\, y_t(t+\tau) = \overline{x(t)y(t+\tau)} \tag{9.24}$$

and cross-power spectrum of the functions as (9.25).

$$\Phi_{12}(n) = \overline{F_1(n)} F_2(n) \tag{9.25}$$

The subscripts 12 indicate that the cross-correlation involves $f_1(t)$ and $f_2(t)$ with the second numeral (2) referring to the fact that $f_2(t)$ has the displacement τ.

It is important to keep the numerals in the double subscripts in proper order; because, if we interchange the order of the subscripts, then by definition the resulting cross-correlation function is given by (9.26).

$$\varphi_{21}(\tau) = \frac{1}{T_1} \int_{-T_{1/2}}^{T_{1/2}} f_2(\tau) f_1(t+\tau) dt \tag{9.26}$$

Some text books will use the notation, R_{xy}, which is referred to as the correlation of x with y. The correlation of y with x is then expressed as (9.27). Note that the first subscript in either equation is the function without $(t+\tau)$. The preferred equation for the cross-correlation expression is (9.26).

$$R_{yx} = \int_{0}^{\infty} x(t+\tau) y(t) d\tau \tag{9.27}$$

The subscripts for the cross-spectrum are given in (9.28); note that the conjugated function is the first subscript.

$$\Phi_{21}(n) = \overline{F_2(n)} F_1(n) \tag{9.28}$$

The importance of keeping the numerals in the double subscript in the proper order

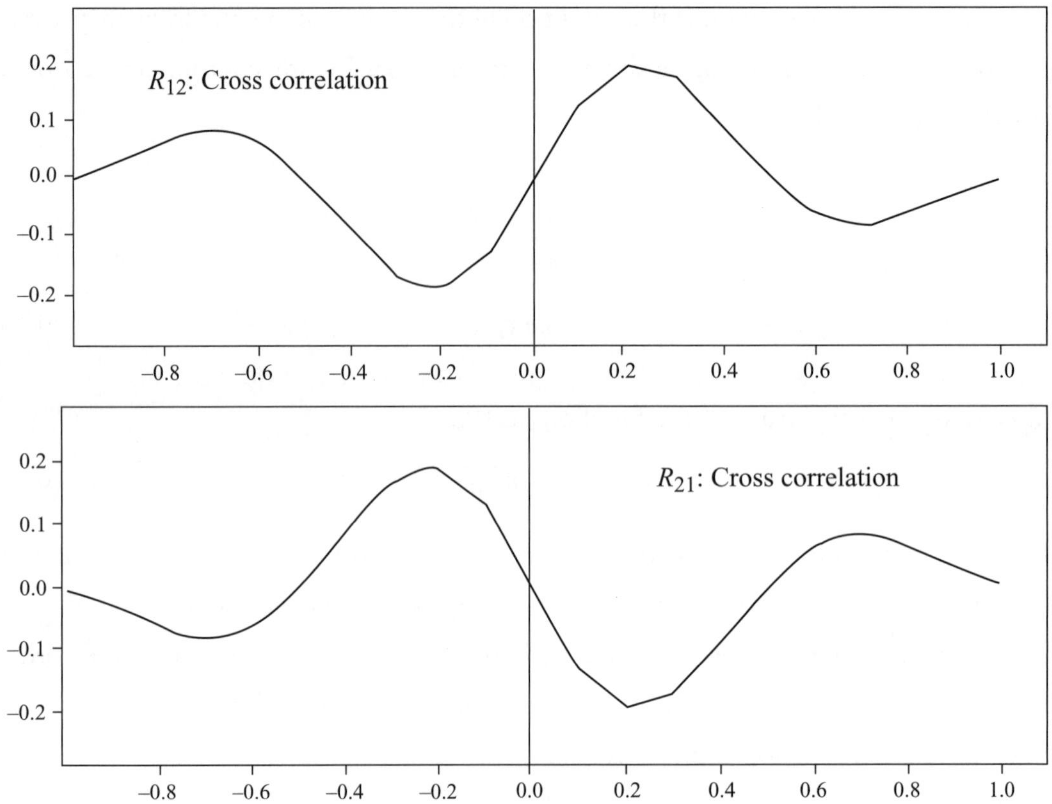

FIGURE 9.6: Cross-correlation images. The graphs show the comparison of the cross-correlation function $\Phi_{12}(n)[R_{12}]$ to its complex conjugate image $\Phi_{21}(n)[R_{21}]$

can not be overstressed. If the subscripts are interchanged, the results will not be what is expected. This interchange can be made by changing $+\tau$ in the definition to $-\tau$ and letting $x = t - \tau$, leading to the expression (9.29). The fact is that $\varphi_{12}(-\tau)$ is the mirror image of $\varphi_{21}(\tau)$ with respect to the vertical axis.

$$\varphi_{12}(-\tau) = \varphi_{21}(\tau) \tag{9.29}$$

The representation of the property in the frequency domain is given by (9.30).

$$\Phi_{12}(n) = \overline{\Phi}_{21}\,(n) \tag{9.30}$$

In summary, $\Phi_{12}(n)$ and $\Phi_{21}(n)$ are related to each other as complex conjugate quantities. Figure 9.6 shows the graphs comparing the cross-correlation function $\Phi_{12}(n)[R_{12}]$ to its complex conjugate image $\Phi_{21}(n)[R_{21}]$.

9.5.1 Properties of the Cross-Correlation Function

The cross-correlation function possesses several properties which are valuable in applications to signal and data processing:

1. $R_{xy}(\tau) = R_{xy}(-\tau)$ The cross-correlation is not an "Even Function."
2. $\left|R_{xy}(\tau)\right| \leq R_{xy}(0)R_{yx}(0) = \overline{x^2} * \overline{y^2}$(the product of the RMS values of x and y)
3. If x and y are independent variables, then
   $R_{xy}(0) = R_{yx}(0) = \overline{x} * \overline{y}$ (the product of the mean values of x and y)
4. If x and y are both periodic with period $= T_0$, then R_{xy} and R_{yx} are also periodic with period $= T_0$.
5. The cross-power spectrum of x and y is computed as the Fourier Transform of the cross-correlation function.

9.5.2 Applications of the Cross-Correlation Function

1. An indication of the predictive power of x to y (or vice versa) is given by the magnitude of the correlation function. Delays of the effects of x on y are indicated by the value of t at which R_{xy} reaches a maximum.
2. The RMS and mean values of x or y can be estimated through the use of properties 2 and 3 given above, provided that the values of one function are known.
3. Periodicities in x and y can be detected in the cross-correlation function (Property 4).
4. The cross-power spectrum, given by the Fourier transform of the cross-correlation, is used in the calculation of the coherence of the functions.

CHAPTER 10

Convolution

The convolution integral dates back to 1833. It is one of the methods used to determine the output response of a system to a specific input if the system transfer function is known. Convolution applies the superposition principal. Conceptually,

1. the input signal is represented as a continuum of impulses;
2. the response of the system to a single impulse is obtained;
3. the response of the system to each of the elementary impulses, representing the input, is computed; and then
4. the total input response is obtained by superposition.

Often signals are represented in terms of elementary linear basis functions, but not as a continuum of impulses. Therefore let us examine the representation of an arbitrary time function by a continuum of impulses. Let us begin with the approximation of a function $f(t)$ in the interval $-T$ to $+T$ by rectangles (rectangular pulses) whose height is equal to the function at the center of the rectangular pulse, and the width is ΔT (Fig. 10.1).

The equation describing the impulse representation of the continuous signal is given by (10.1).

$$f(t) = \int_{-T}^{+T} f(\tau)\delta(t-\tau)\,\partial\tau \quad \text{from} \quad -T < t < T \tag{10.1}$$

Since the convolution is used in the time domain analysis of a filter response, let us

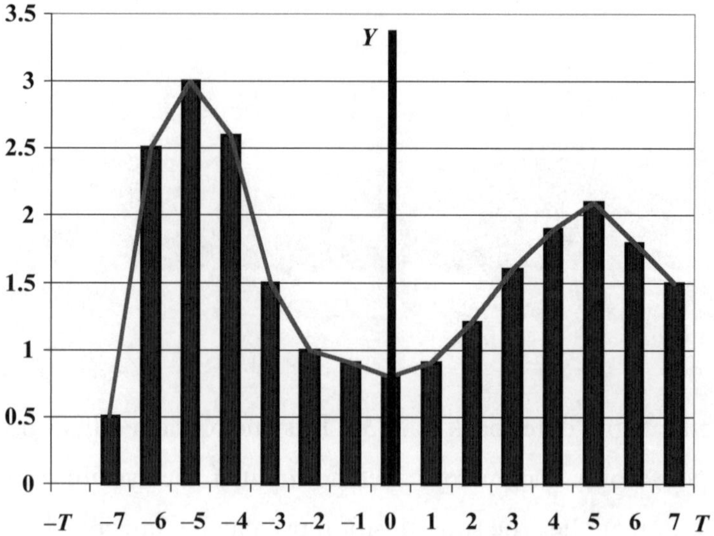

FIGURE 10.1: A signal represented in terms of a continuum of impulses in the interval, $-T$ to $+T$

examine how the convolution integral is used to obtain the system impulse response of a simple RC circuit or low pass filter as shown in Fig. 10.2.

If a single unit pulse of width ΔT and amplitude $1/\Delta T$ is applied, the output voltage increases exponentially toward the value $1/\Delta T$; when the pulse terminates at ΔT, the output voltage decays toward zero as shown in Fig. 10.3.

If the impulse is taken to its limit as $\Delta t \rightarrow 0$, then the impulse approaches a theoretical impulse; the *delta function*, δ, and the response to the impulse

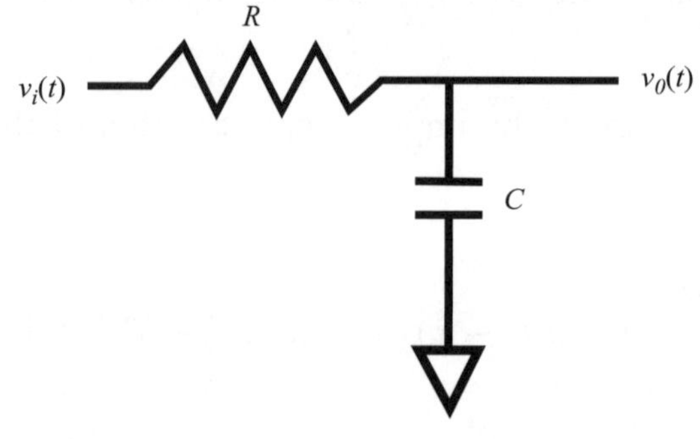

FIGURE 10.2: Low-pass RC filter circuit

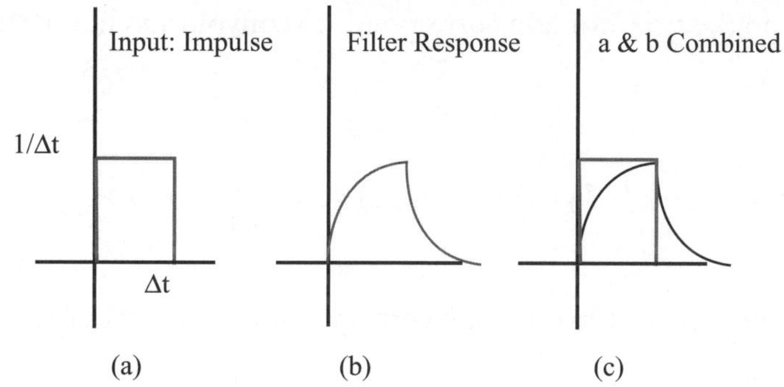

FIGURE 10.3: Low-pass filter response. (a) It shows a single unity pulse of Δt duration and $1/\Delta t$ magnitude. (b) It shows the output response of the low-pass filter shown in Fig. 10.2 to the input pulse, (a). (c) It shows the combined input and output signals

becomes (10.2).

$$h(t) = \frac{1}{RC} e^{-\frac{t}{RC}} \tag{10.2}$$

What Fig. 10.4 implies is that at the instant the impulse occurs a current flows in the circuit and is supposed to charge the capacitor to a voltage of 1/RC, instantaneously. However, the instantaneous charging that theoretically occurs in the instant that the impulse is applied is not physically possible because of the principle of conservation of momentum. As a note of information, the LaPlace transformation of the delta function, δ, is unity (1); therefore, the frequency domain is a better and easier method by which to analyze the response to an impulse function.

It is important to understand the time/frequency relationships that exist when functions are transformed into another domain. For example, the product (or *Modulation*)

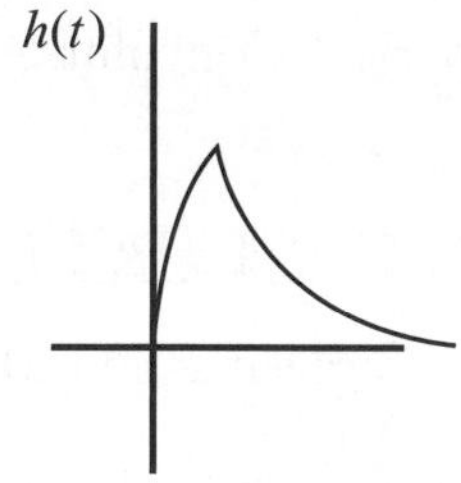

FIGURE 10.4: Low-pass filter response to a delta function as $\Delta t \rightarrow 0$

of two functions in the time domain corresponds to convolution in the frequency domain as shown in (10.3).

$$f_1(t)\, f_2(t) = \frac{1}{2\pi}\int_{-\infty}^{\infty} F_1(\zeta)\, F_2(\omega - \zeta)\, \partial\zeta = \frac{1}{2\pi} F_1(\omega) \times F_2(\omega) \tag{10.3}$$

Whereas convolution in the time domain corresponds to product in the frequency domain as shown by (10.4.)

$$f_1(t) \times f_2(t) = \int_{-\infty}^{\infty} f_1(\lambda)\, f_2(t-\lambda)\, \partial\lambda \Leftrightarrow F_1(\omega)\, F_2(\omega) \tag{10.4}$$

Convolution is commutative for time-invariant systems are given by (10.5) and (10.6).

$$y(t) = \int_{-\infty}^{\infty} h(\lambda) x(t-\lambda)\, \partial\lambda \tag{10.5}$$

$$y(t) = \int_{-\infty}^{\infty} x(\lambda) h(t-\lambda)\, \partial\lambda \tag{10.6}$$

Often, the convolution operation is denoted by an asterisk as shown in (10.7).

$$f_1(t) \times f_2(t) = \int_{-\infty}^{\infty} f_1(\lambda)\, f_2(t-\lambda)\, \partial\lambda \tag{10.7}$$

The integration limits vary with the particular characteristics of the functions being convolved. Next, let us examine and discuss the manner of "Evaluation and Interpretation of Convolution Integrals and Numerical Convolution."

10.1 CONVOLUTION EVALUATION

Formally the convolution operation is given as in (10.8).

$$f_3 = f_1 \times f_2 = \int_{-\infty}^{\infty} f_1(\lambda)\, f_2(t-\lambda)\, \partial\lambda \tag{10.8}$$

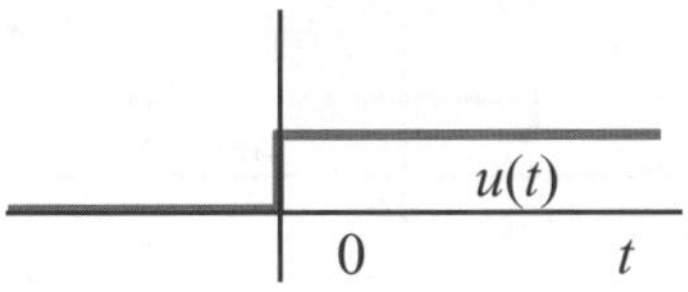

FIGURE 10.5: Unit step function, $u(t)$. The step function is similar to throwing a switch

The value f_3 for any particular instant of time t is the area under the product of $f_1(\lambda)$ and $f_2(t - \lambda)$. To use the convolution technique efficiently, one should be able to visualize the two functions. If convolution is a new topic to the reader, then a quick review of wave form synthesis is essential for visualization. Let us begin with a *step function*, $u(t)$, which is initiated at $t = 0$ (Fig. 10.5).

Figure 10.6 shows the unit step function translated, $u(t - a)$, along the time axis, $+t$, by some length, $+a$, whereas Fig. 10.7 shows the unit step function translated backward along the time axis, $-t$, by some length, $-a$.

In addition, the unit step function may be rotated about the y-axis, which is referred to as *folding* about some axis. A folded (flipped) unit step function about the y-axis is written as, $u(-t)$, and is shown in Fig. 10.8.

Figure 10.9 shows the result of positive translation of a folded unit step function. Note that this function is written as, $u(a - t)$.

In addition to rotation about the y-axis, the unit step function may also be folded about the x-axis as shown in Fig. 10.10; note the negative sign in front of the unit step function, $-u(t)$.

Figure 10.11 shows the translation of a unit step function folded about the x-axis displaced by some time value, a. This step function is written as $-u(t - a)$; note the negative signs.

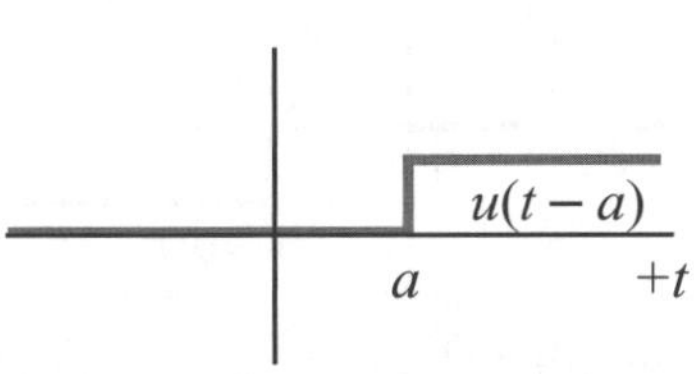

FIGURE 10.6: Positive translation. The unit step function of Fig. 10.5 is translated along the time axis, $+t$, by some length, $+a$

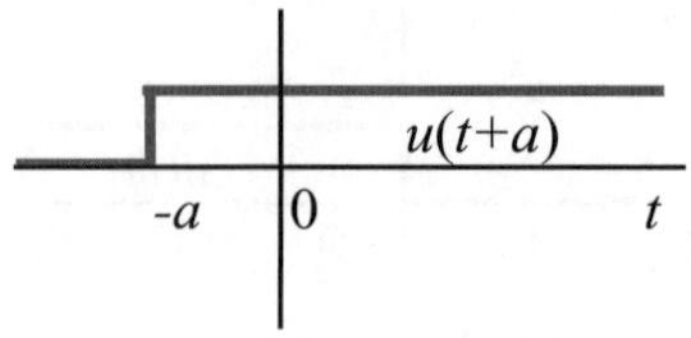

FIGURE 10.7: Negative translation. The unit step function, $u(t+a)$, translated backward along the time-axis, $-t$, by some length, $-a$

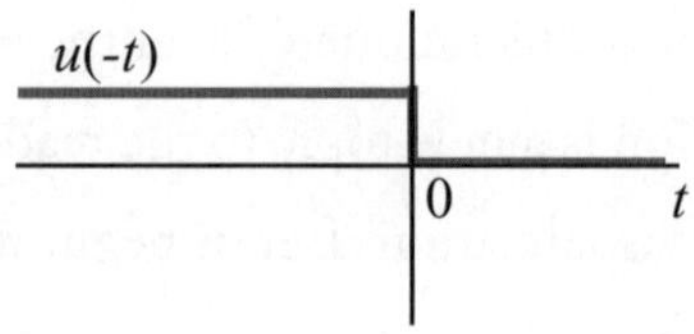

FIGURE 10.8: Folding. A unit step function rotated about the y-axis, $u(-t)$

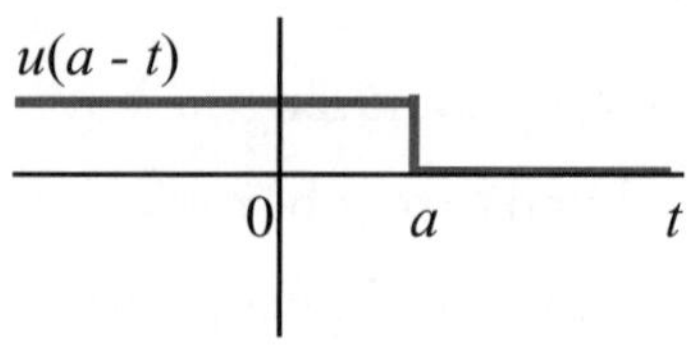

FIGURE 10.9: Positive translation of folded step function. Shown is the result of a positive translation in time (length, $+a$) of a folded unit step function, $u(a-t)$

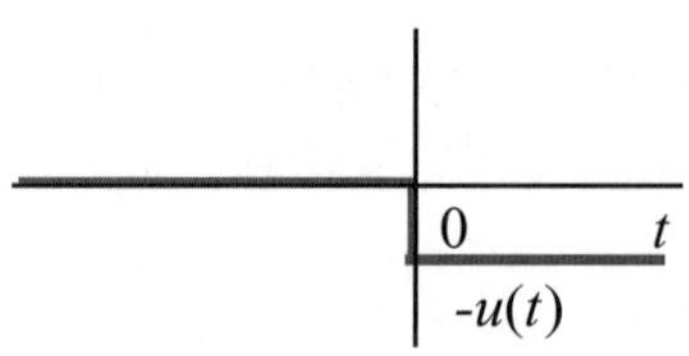

FIGURE 10.10: Folding about the x-axis. A unit step function is shown folded about the x-axis, $-u(t)$

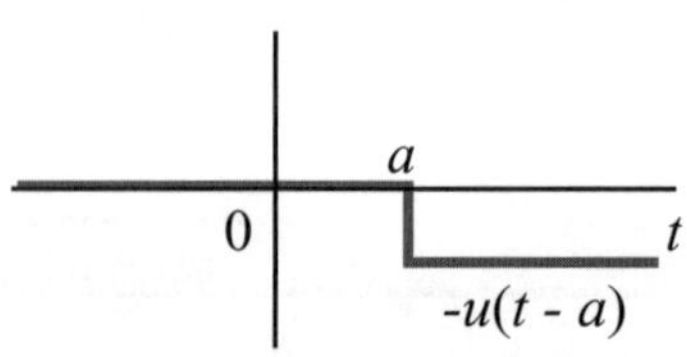

FIGURE 10.11: Folding and translation about the x-axis. A unit step function folded about the x-axis, t, then translated or displaced by some time value, $+a$

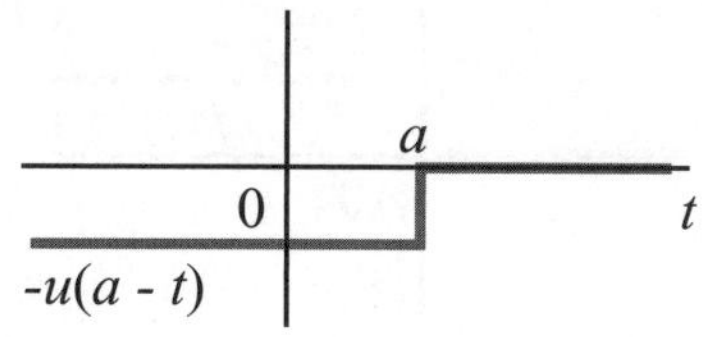

FIGURE 10.12: Combining the operations of folding and translating about both axes. The step function folded about both axes and displaced by $+a$ gives the unit step function expressed as $-u(a-t)$

Combining the operations of folding about the y-axis, then folding about the x-axis, and translating the step function by a positive displacement of $+a$ gives the results shown in Fig. 10.12. The unit step function is then expressed as $-u(a-t)$.

10.1.1 Real Translation Theorem

One can summarize the operation of translation along an axis by a well-known theorem which states that any function $f(t)$, delayed in its beginning to some time $t=a$, can be represented as a time-shifted function as in (10.9).

$$f(t-a)\,\mu(t-a) \tag{10.9}$$

where $\mu(t-a)=\left[\frac{1,\ -t\geq a}{0,\ -t<a}\right.$

Proof of the time shift theorem is based in LaPlace Transform of the two functions, which is often referred to as the *Real Translation Theorem* or simply the *Shifting Theorem* as shown in the resulting (10.10).

$$L\left[\mu\left(t-a\right)\right]=\frac{e^{-as}}{s} \tag{10.10}$$

where, e^{-as} shifts the transform from $t=0$ to $t=\infty$.

Let us return to the visualization of convolution. The function $f(\tau)$ should not be a problem since it is similar to the function $f(t)$, except for the change of independent variable from units of time t to units of delay τ. It is important to keep in mind that once the variable is changed from time t to delay τ that the unit of delay τ becomes the independent variable and that time t in the function $f(t-\tau)$ is the same as displacement a in the function $f(a-t)$.

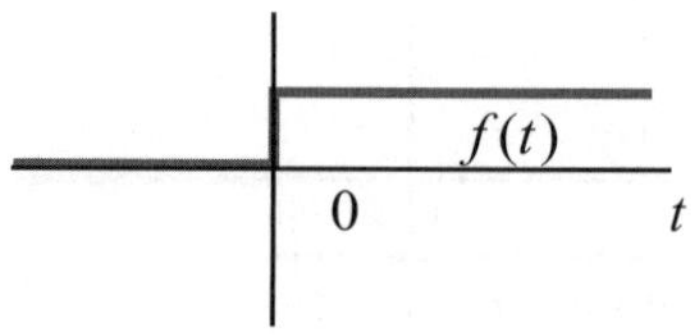

FIGURE 10.13: Simple unity step function

Before discussing the complete process of convolution, the reader should know that calculation of the convolution requires the following steps, which were previously described.

1. Change of variable from time, t, to delay, τ
2. Folding about the vertical axis
3. Translation

Let us look at an example, using the three steps to visualize and examine the process with a simple function, $f(t)$, as shown in Figs. 10.13–10.16.

The first step in the convolution process is the change of independent variable from a function of time, $f(t)$, to a function of delay, $f(\tau)$, as in Fig. 10.14.

The second step is to fold the step function about the y-axis, $f(-\lambda)$, as in Fig. 10.15.

The third step is to translate the step function along the x-axis some distance, t, as shown in Fig. 10.16. Note that the folded and translated step function is written as $f(t-\tau)$.

Let us examine another example; assume the function $f(t) = Ae^{-at}$. Figure 10.17 shows the function and the first step, which is the change of variable.

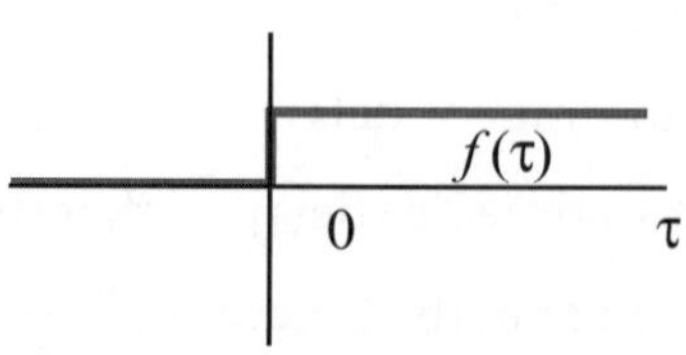

FIGURE 10.14: First step. Change the independent variable from a function of time, $f(t)$, to a function of delay, $f(\tau)$

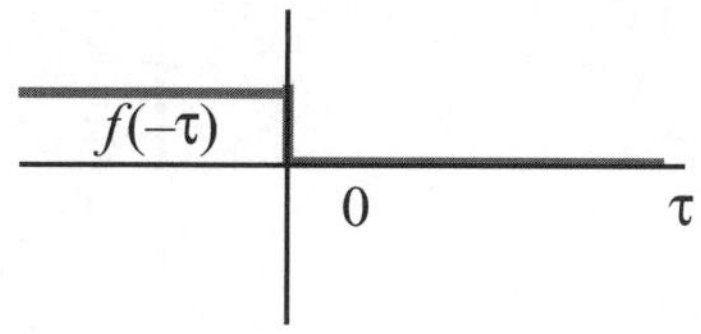

FIGURE 10.15: Second step. The step function is folded about the y-axis, $f(-\lambda)$

Figure 10.18 shows the folding and translation steps. Note that the translation is to the right. The function $f(t - \tau)$ can be thought of as $f\{-(\tau - t)\}$, then $f(-\tau')$ where τ' is $(\tau - t)$, which results in movement to the right (positive direction).

In the next example, illustrated in Fig. 10.19, the original does not begin at $t = 0$, but is delayed by some magnitude a. The amount of displacement of the function is measured from the position of $f(-\tau)$ which is $f(t - \tau)$ at $t = 0$. Note that the amount of delay is carried in the folding about the y-axis, which implies that the zero values must be accounted for in calculation of the convolution integral.

10.1.2 Steps in the Complete Convolution Process

The entire process of convolution requires the five steps listed as follows:

1. Change of variable
2. Folding about the vertical y-axis
3. Translation along the horizontal x-axis
4. Multiplication of the two functions, $f_1(\tau)$ and $f_2(t - \tau)$
5. Integration of areas

The process might be synopsized as determining the area under the curves (product) as the folded function is slid along the horizontal axis. Note the direction of translation; the conventional notation is as follows:

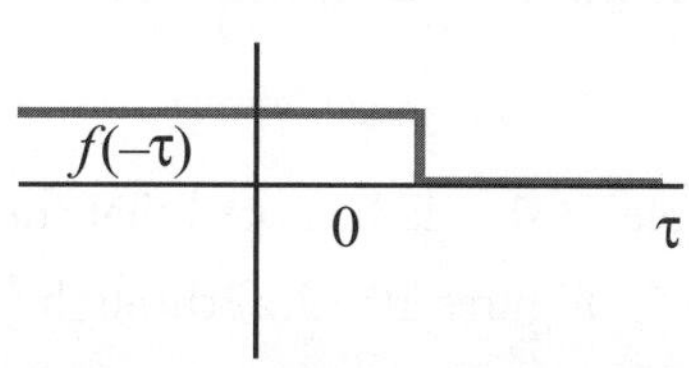

FIGURE 10.16: Third step. Translation of the function along the x-axis some distance, t

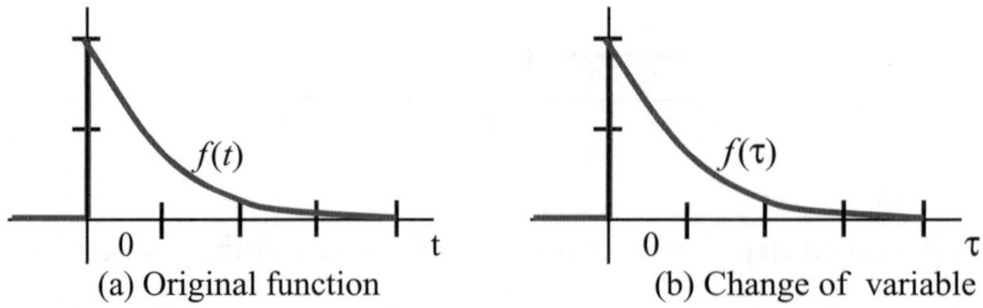

FIGURE 10.17: First step. (a) It shows the original function $f(t) = Ae^{-at}$. (b) It shows the change of variable for the function

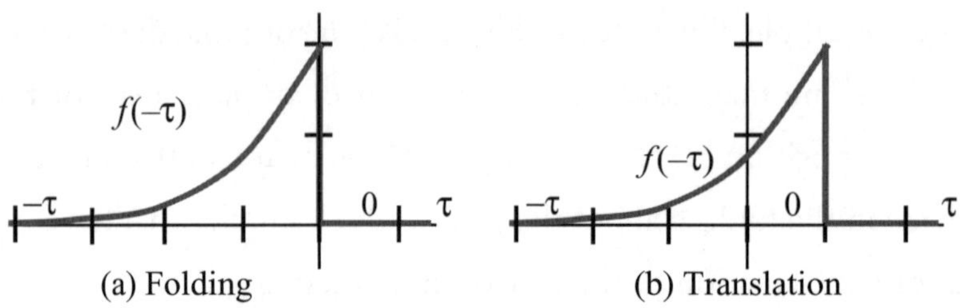

FIGURE 10.18: Next two steps. (a) It shows folding of the function about the y-axis and (b) it shows the translation step

a. Movement to the right for positive (+) time

b. Movement to the left for negative (−) time

Let us visualize an example, in which two functions of unity magnitude are to be convolved. Assume the two functions are given by (10.11) and (10.12).

$$f_1(t) = 1\mu(t) \tag{10.11}$$

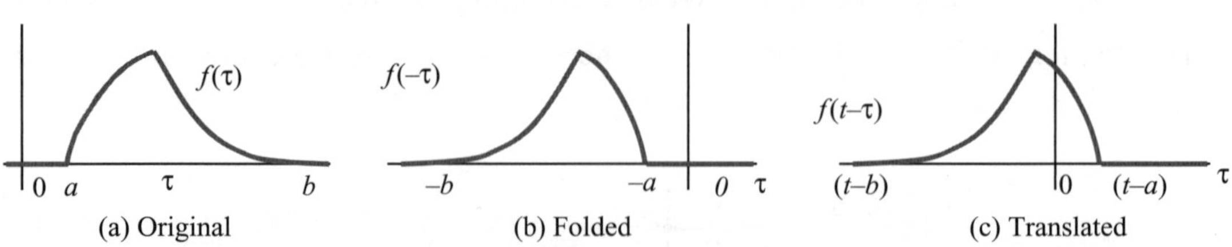

FIGURE 10.19: Another Example. It should be noted that the original does not begin at $t = 0$, but is delayed by some magnitude a. Figures 10.19(a) through (c) show that the amount of delay is carried through the Folding and Translation steps

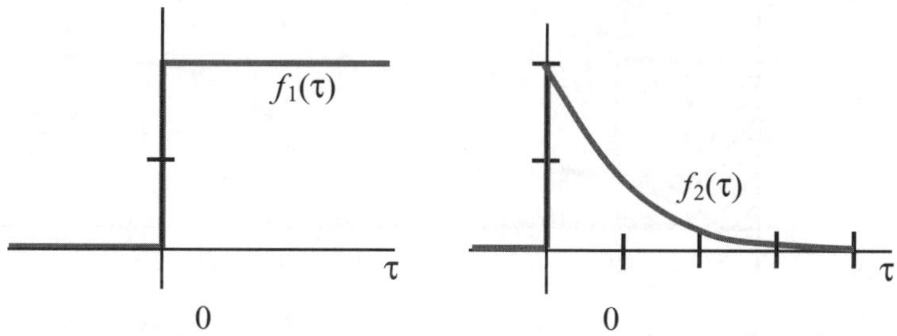

FIGURE 10.20: First step. Graphs of the two equations are shown after changing the independent variable

and

$$f_2(t) = e^{-at} \tag{10.12}$$

After changing the independent variable from time t to delay τ, the expressions are written as functions of τ, which are given in (10.13) and (10.14).

$$f_1(\tau) = \mu(\tau) \tag{10.13}$$

and

$$f_2(\tau) = e^{-a\tau} \tag{10.14}$$

Graphs of the two equations are shown in Fig. 10.20.

The next step is to fold one of the two functions about the y-axis. Either variable may be folded, since the results will be the same. In this case the step function was selected as the function to fold. Thus, $f_1(\tau) = \mu(\tau)$ becomes $f_1(-\tau) = \mu(-\tau)$ as shown in Fig. 10.21. Note that the second function, $f_2(\tau) = e^{-a\tau}$, remains the same.

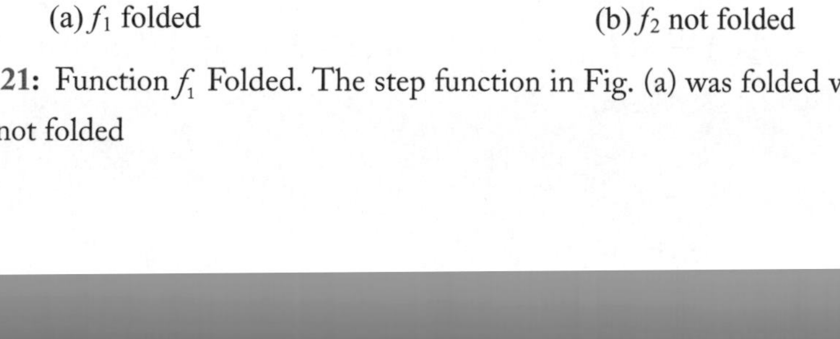

FIGURE 10.21: Function f_1 Folded. The step function in Fig. (a) was folded while the second function was not folded

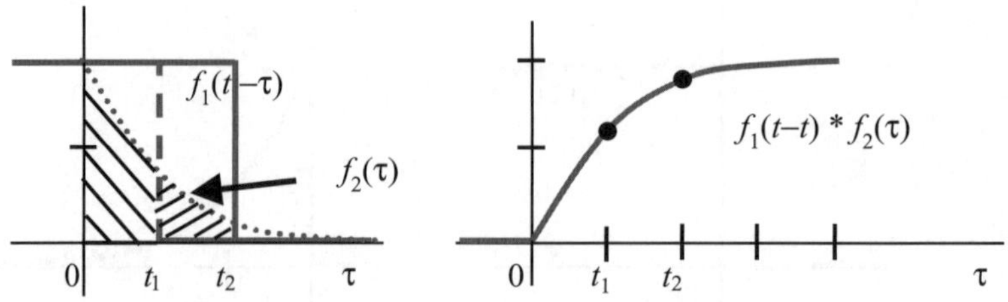

(a) Translation, product, and integration (b) Resulting convolved function

FIGURE 10.22: Last three steps. The final steps of translation, product, and integration in the convolution process are shown in (a) with the resulting function shown in (b)

The final steps of translation, product, and integration in the convolution process are shown in Fig. 10.22(a) with the final convolution function being shown in Fig. 22(b). The reader should keep in mind that the integrated product is the same as the area under the product graph. In the first translation (t_1), the integrated quotient is shown as the area with the cross hatches slanting downward to the right. The new area of the integrated quotient for the second translation (t_2) is shown with the cross hatches slanting upward to the right. Note in the results, shown in Fig. 10.22(b), that the result at t_2 is the total area (both cross-hatched areas). The action of the functions f_1 and f_2 on each other is thought of as a smoothing or weighting operation.

Let us examine the convolution of a pulse and another function: assume the functions are defined as follows and are described by (10.15) and (10.16).

1. $f_1(t)$ is a rectangular pulse of height 3 and width 2
2. $f_2(t)$ is an exponential decay height 2 at $t = 0$

$$f_1(t) = 3[\mu(t) - \mu(t-2)] \tag{10.15}$$

$$f_2(t) = 2e^{-2t}\mu(t) \tag{10.16}$$

where $f_2(t) = 0$ for $t < 0$.

Evaluation of the convolution integral is given by (10.17).

$$f_3 = f_1(t) \times f_2(t) = 6\int_{-\infty}^{\infty} [\mu(t-\tau) - \mu(t-2-\tau)]\, e^{-2\tau}\mu(\tau)\, \partial\tau \tag{10.17}$$

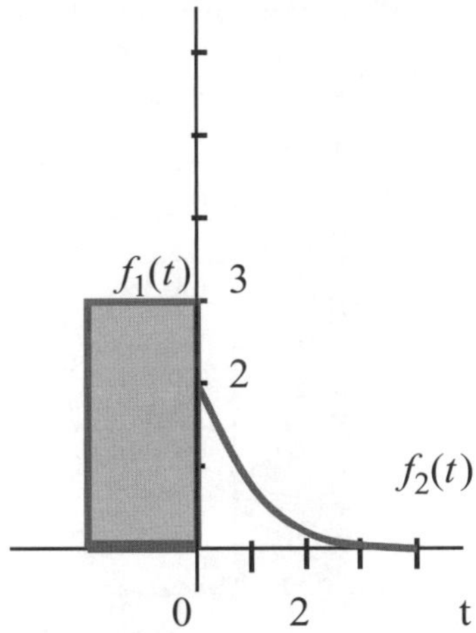

FIGURE 10.23: Second step: Folding. The figure shows the two functions after folding the pulse. It should be noted that there is no overlap of the functions when $t \leq 0$

Figure 10.23 shows the two functions after folding the pulse; note that at the start, when $t \leq 0$, there is no overlap and the resulting value of the convolution integral (f_3) at $t = 0$ is zero, ($f_3 = 0$).

Following the folding step, the exponential function, f_2, is multiplied by the pulse function, f_1, with a magnitude of 3, then the area between $0 < t < 1$ is integrated. The result of the integration is the first value of the convolution function, f_3, which is the shaded area in Fig. 10.24.

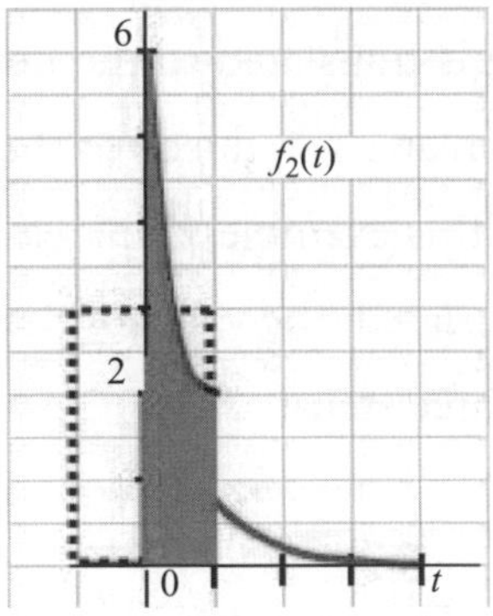

FIGURE 10.24: Result of integration. The first value of the convolution function, f_3, is shown by the shaded area in the figure

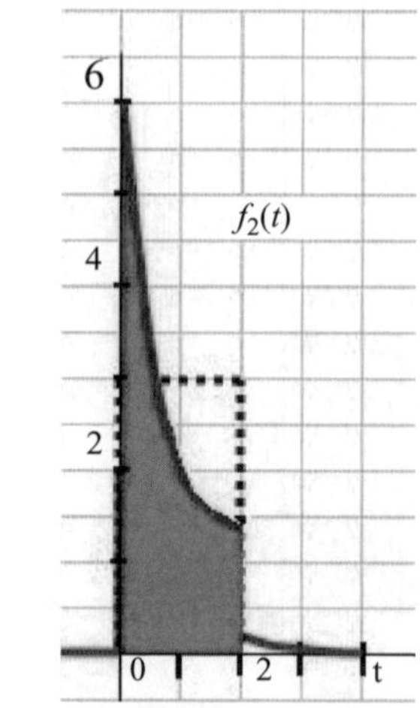

FIGURE 10.25: Result from the second translation. The second value after integration of the convolution function, f_3, with the limits between $0 < t < 2$, is shown by the shaded area in the figure

The next step is to repeat the translation step by moving the pulse function one increment to the right; hence, the limits of integration are changed to the interval of $0 < t < 2$, as given by (10.18).

$$f_3(t) = 6 \int_0^t e^{-2\tau} \partial\tau \tag{10.18}$$

The result of the second translation and including the multiplication and integration steps is shown in Fig. 10.25. Note that the width of the area is equal to the width of the pulse function. As the pulse is translated to the next interval between $0 < t < 3$, it should be noted that the common area is reduced to the interval between $1 < t < 3$ (the width of the pulse). Hence, for this example, the area is smaller than the previous one. As the pulse is continuously translated, the value of the integrated area approaches zero.

The process repeats the three steps:

1. Translation along the horizontal x-axis
2. Multiplication of the two functions, $f_1(\tau)$ and $f_2(t-\tau)$
3. Integration of area bound by the limits of integration

In the example, for the next translations the limits are reset to the interval between $2 < t < \infty$ as shown in (10.19).

$$f_3(t) = 6 \int_{t-2}^{t} e^{-2\tau} \partial\tau \tag{10.19}$$

10.1.3 Convolution as a Summation

In evaluation of the convolution integral by digital computer, it is necessary to replace the integral (10.20) with a finite summation.

$$v_2(t) = \int_{0}^{t} v_1(\tau)\, h(t-\tau)\, \partial\tau \tag{10.20}$$

To accomplish this conversion, the infinitesimal $\partial\tau$ is replaced with a time interval of finite width T and τ is replaced with nT. The convolution integral is written as the summation for $KT \leq t < (K+1)T$ as shown in (10.21).

$$v_2(t) = T \sum_{n=0}^{K} v_1(nT)\, h(t-nT) \tag{10.21}$$

where K is one of the values of the integer n.

Equation (10.22) gives the convolution of (10.21) in expanded form.

$$\begin{aligned} v_2(t) = {} & T[v_1(0)\, h(t) + v_1(T) h(t-T) \\ & + v_1(2T) h(t-2T) + v_1 3T) h(t-3T) + \cdots] \end{aligned} \tag{10.22}$$

Each term in (10.22) is the impulse response of magnitude of $v_1(nT)$ shifted along the t-axis and later is multiplied by T. Summation of the terms in the equation approximates the function. The accuracy of the approximation will depend on the magnitude of T. The smaller the value of T is, the better the approximation.

10.2 NUMERICAL CONVOLUTION

Numerical evaluation of the convolution integral is straightforward. When either function is of finite duration, there will be a finite number of terms in each summation.

However, when the functions are of infinite duration, the computations become lengthy as the overlapped area becomes very large. There are alternative methods that can be applied for simple problems or when a computer is available.

10.2.1 Direct Calculation

10.2.1.1 Strips of Paper Method

One of the numerical methods is to use two strips of paper and to follow the following steps:

1. tabulate values of the two functions at equally spaced intervals on two strips of paper;
2. reverse tabulate one function; $f(-\lambda)$, which is equivalent to folding a function;
3. place strips beside each other and multiply across the strips and write the products;
4. slide one strip, and repeat step three, until there are no more products;
5. sum the rows; and
6. multiply each sum by the interval width.

10.2.1.2 Polynomial Multiplication Method

Another method is the polynomial multiplication method (Table 10.1). For this method, the values of the two functions are placed in two rows one function above the other (without any reverse tabulation) as in Table 10.1. It should be noted that the functions are aligned to the left and not to the right as in normal multiplication. The alignment and multiplication from left to right serves as equivalence to folding of a function about the y-axis. The number of terms in the results should equal the sum of the number of terms in the two functions. The final results of the convolution operation are obtained by summing the columns and multiplying the resulting summations by the interval width.

At this point, it is worth discussing an area of possible confusion: that of establishing the proper relationship of the time variable to the sample values employed in the

TABLE 10.1: Convolution by Multiplication Method

$t =$	1	2	3	4	5	6	
$f_1 =$	0.5	1.5	2.5	3	3	3	
$f_2 =$	9.5	8.5	7.5	6.5	5.5		
at $n = 19.5 \times f_1(n = 1..6)$	4.75	14.25	23.75	28.5	28.5	28.5	
at $n = 18.5 \times f_1(n = 1..6)$		4.25	12.75	21.25	24.5	24.5	24.5
at $n = 17.5 \times f_1(n = 1..6)$			3.75	11.3	...	...	...
at $n = 16.5 \times f_1(n = 1..6)$				3.25	...	...	...
at $n = 15.5 \times f_1(n = 1..6)$					2.75	...	...
$f_1 \times f_2 = (\Sigma \times T)$	4.75	18.5	40.25	64.3	...	...	...

calculations. The question arises as to where to take the values from the time varying function within a fixed interval. Specifically, the question may be rephrased as "Where should the interval be located?" For example, a function, $f(t) = -1\lambda + 8$, can be sampled with some interval of time, T, in the following manner: Let $T = 2$ and the value be selected at midinterval, as in Fig. 10.26. The general procedure is to take the average value of the function $f(t)$ in the interval, kT, where k is the sample number.

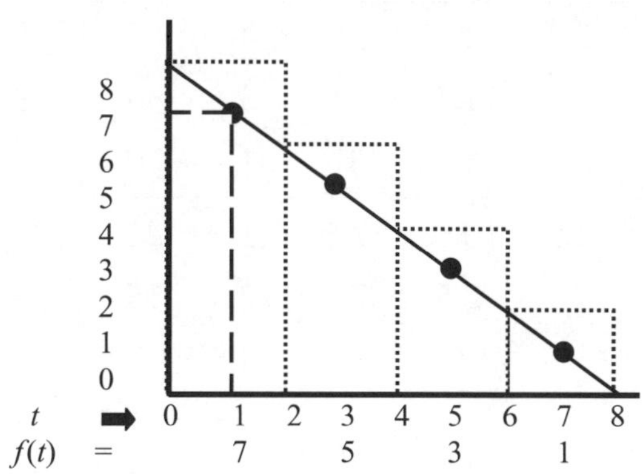

FIGURE 10.26: Graph A of the function, $f(t) = -1\lambda + 8$, with the interval $T = 2$. Note that the values of the sampled function are 7, 5, 3, and 1 at times $t = 1, 3, 5$ and 7, respectively

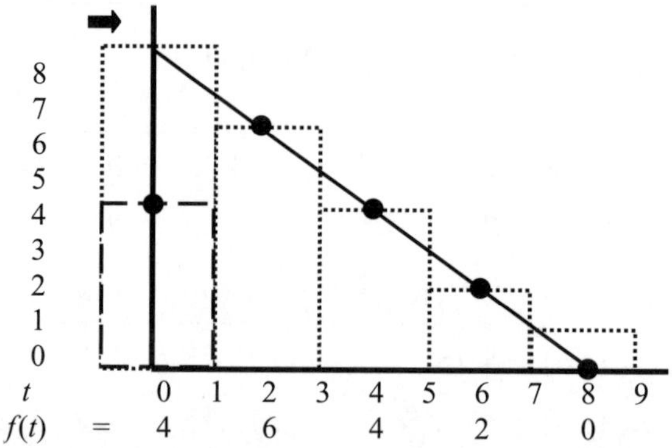

FIGURE 10.27: Graph B of the function, $f(t) = -1\lambda + 8$, with the interval $T = 2$. Note that the values of the sampled function are 4, 6, 4, 2, and 0 at times t = 0, 2, 4, 6, and 8, respectively

However, one may elect to use the interval between $-T/2 < 0 < +T/2$, which then yields slightly different answers as shown by Fig. 10.27. Note that in Fig. 10.27, the midpoints for the interval $T = 2$ are at $t = 1, 3, 5 \ldots$ etc.

Let us convolve the f_1 function with the values of the function $f(t) = -1\lambda + 8$, as shown in Table 10.1 and Fig. (10.26). The calculations and results of the convolution are given in Table 10.2 and are shown in Fig. (10.28). The results are labeled as A, since the second function is composed of the values from Fig. 10.26.

TABLE 10.2: Results A of Convolution with Fig. 10.26

t =	1	2	3	4	5	6				
f_1 =	0.5	1.5	2.5	3	3	3				
f_2 =	7	5	3	1	0					
	3.5	10.5	17.5	21	21	21				
		2.5	7.5	12.5	15	15	15			
			1.5	4.5	7.5	9	9	9		
				0.5	1.5	2.5	3	2	1	
					0	0	0	0	0	0
Col Sum	3.5	13	26.5	38.5	45	47.5	27	11	1	0
$f_1 \times f_2 = (\Sigma)(\Delta T)$	7	26	53	77	90	95	54	22	2	0

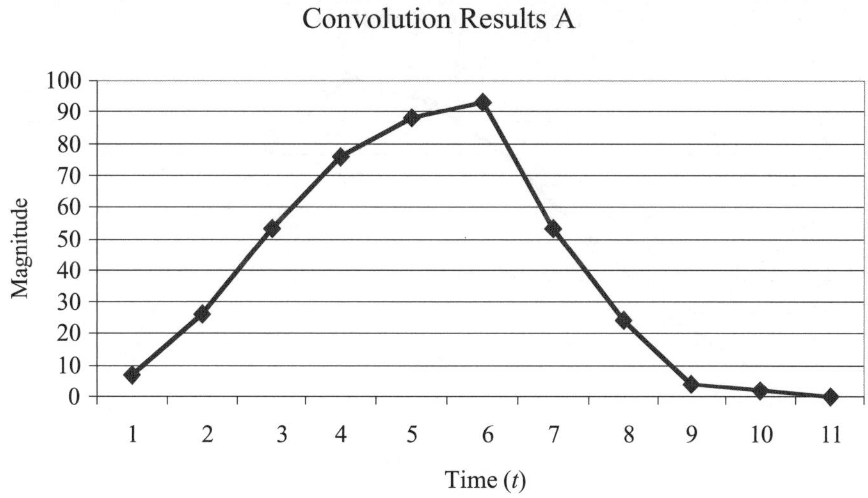

FIGURE 10.28: Graph of Results A

Repeating the convolution of the f_1 function with the values of the function, $f(t) = -1\lambda + 8$, as shown in Table 10.1 and Fig. 10.27. Note the values used in the calculations, which are given in Table 10.3. The results of the convolution are shown in Fig. 10.29. The results are labeled as B, since the second function is composed of the values from Fig. 10.27.

TABLE 10.3: Results B of Convolution with Fig. 10.27

$t =$	1	2	3	4	5	6				
$f_1 =$	0.5	1.5	2.5	3	3	3				
$f_2 =$	4	6	4	2	0					
	2	6	10	12	12	12				
		3	9	15	18	18	18			
			2	6	10	12	12	12		
				1	3	5	6	6		
					0	0	0	0	0	
Col Sum	2	9	21	34	43	47	36	18	0	0
$f_1 \times f_2 = (\Sigma)(\Delta T)$	4	18	42	68	86	94	72	36	0	0

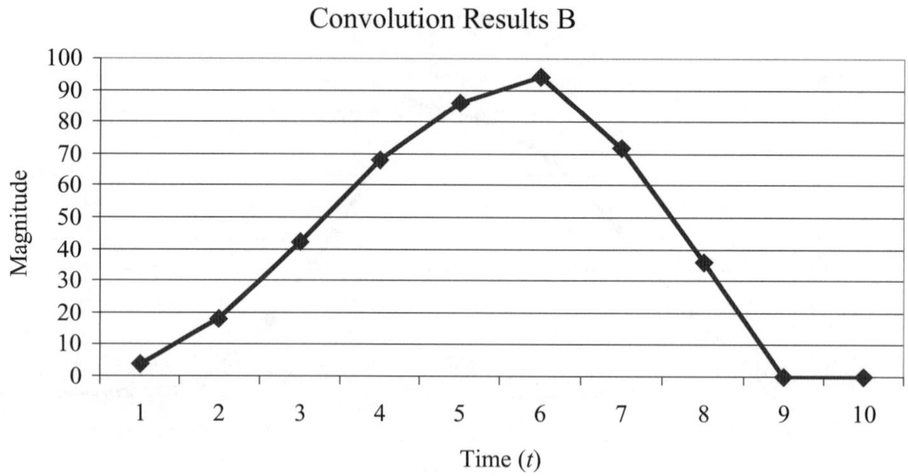

FIGURE 10.29: Graph of Results B

It should be noted that the procedures and results shown in Tables 10.2 and 10.3 and Figs. 10.28 and 10.29 are comparable, but the procedure that locates discontinuities along one of the steps in the approximation gives better results. In the example, the results labeled A (Table 10.2 and Fig. 10.28) are better for the discontinuity at $t = 0$.

10.3 CONVOLUTION ALGEBRA

Convolution is referred to as a "Linear Process" in the arithmetic sense; therefore, let us examine those mathematical properties that make convolution a linear process. Recall the mathematical properties of the "Real Numbering System" whose properties are

1. associative,
2. commutative, and
3. distributive.

Consider the convolution of two functions, one of the functions being the result of a previous convolution process. First, "Convolution is Associative," which means that the order of convolution is not important as indicated by (10.23).

$$f_1(t) \star [f_2(t) \star f_3(t)] = (f_1 \star f_2) \star f_3 = f_1 \star f_2 \star f_3 \tag{10.23}$$

where $f_2 \star f_3 = \int_{-\infty}^{\infty} f_2(\lambda_1) f_3(t - \lambda_1)\, \partial\lambda_1$, and

$$\begin{aligned} f_1 \star [f_2 \star f_3] &= \int_{-\infty}^{\infty} f_2(\tau)\, \partial\tau_2 \left[\int_{-\infty}^{\infty} f_2(\tau_1) f_3(t - \tau_1 - \tau_2) \partial\lambda_1 \right] \\ &= \int \partial\tau_2 \int f_1(\tau_2) f_2(\tau_1) f_3(t - \tau_1 - \tau_2)\, \partial\tau_1 \end{aligned}$$

Second, "Convolution is Commutative," which means that the order of the functions being convolved is immaterial as shown in (10.24). The limits of integration will be different for different orders of the convolved functions.

$$f_1 \star f_2 = f_2 \star f_1 \tag{10.24}$$

The final property is that "Convolution is Distributive," which means that convolution of a function with the sum of two functions can be performed as shown in (10.25).

$$f_1 \star (f_2 + f_3) = f_1 \star f_2 + f_1 \star f_3 \tag{10.25}$$

An interesting property of convolution pertains convolving a function with theoretical delta functions, which results in the original function (see (10.26)). This property is the same as sampling.

$$z(t) \star \delta(t) = \int_{-\infty}^{\infty} z(\lambda)\delta(t - \lambda)\, \partial\lambda = z(t) \tag{10.26}$$

The convolution of a function with a unit step function is given by (10.27).

$$z(t) \times \mu(t) = \int_{-\infty}^{t} z(\lambda)\partial\lambda \tag{10.27}$$

If a function is convolved with the derivative of a delta function, the result is the derivative of the original function as shown by (10.28).

$$z(t) \star \delta'(t) = z'(t) \tag{10.28}$$

where the prime (′) denotes *d/dt*. The first derivative of $\delta(t)$ is called a *doublet.*

The convolution of a step function with the derivative of a delta function will yield the delta function (10.29).

$$\mu(t) \star \delta'(t) = \delta(t) \tag{10.29}$$

It should be noted that the delta function is zero at all other values of time except at the instant of time, t, where the delta function exists. This relation is expressed mathematically as $\delta'(t - t_0) = 0 \text{ for } t \neq t_0$ or as (10.30).

$$\int_{t_1}^{t_2} \delta(t - t_0)\, \partial t = 0 \tag{10.30}$$

For $t_1 < t_0 < t_2$

Other properties of convolution are of interest because of their use in checking validity of a particular computation. For example, in the convolution of two functions, as in (10.31), the area of a convolution product is equal to the product of the areas of the functions. This means that the area under $f_3(t)$ is equal to the area under the function, $f_1(t)$, times the area under the function, $f_2(t)$.

$$f_3(t) = \int_{-\infty}^{\infty} f_1(\tau) f_2(t - \tau)\, d\tau \tag{10.31}$$

Keep in mind that the area is computed by integrating over the interval, $-\infty < t < \infty$. Proof of this property follows in (10.32) through (10.36).

$$\int_{-\infty}^{\infty} f_3(t)\, dt = \int_{-\infty}^{\infty} \left[\int_{-\infty}^{\infty} f_1(\tau) f_2(t - \tau)\, d\tau \right] dt \tag{10.32}$$

$$[\text{Area\_under\_} f_3(t)] = \int_{-\infty}^{\infty} f_1(\tau) \left[\int_{-\infty}^{\infty} f_2(t - \tau)\, dt \right] d\tau \tag{10.33}$$

Rewriting (10.33), as moment generating functions, is shown in (10.34) and (10.35). Equation 10.36 shows that the moment about zero of the convolution function f_3 is equal to the product of the moments about zero of the two functions f_1 and f_2.

$$M_0\left(f_3\right) = \int_{-\infty}^{\infty} f_1\left(\tau\right)\left[\text{area\_under\_}f_2\left(t\right)\right] d\tau \qquad (10.34)$$

$$M_0\left(f_3\right) = A_{f_3(t)} = \left[\text{area\_under\_}f_1\left(t\right)\right] \bullet \left\lfloor A_{f_2(t)}\right\rfloor \qquad (10.35)$$

$$M_0\left(f_3\right) = M_0\left(f_1\right) \bullet M_0\left(f_2\right) \qquad (10.36)$$

Another useful relationship deals with the center of gravity (centroid) of the convolution in terms of the center of gravity of the factors (10.37). If you recall, the center of gravity of a waveform was defined in terms of the energy of the signal as t_0.

$$n = \frac{\int_{-\infty}^{\infty} tf\left(t\right)dt}{\int_{-\infty}^{\infty} f\left(t\right)dt} \qquad (10.37)$$

The nth moment of a function or waveform is defined by the general (10.38), and

$$M_n\left(f\right) = \int_{-\infty}^{\infty} t^n f\left(t\right) dt \qquad (10.38)$$

then the centroid is defined as given by (10.39).

$$n = \frac{M_1\left(f\right)}{M_0\left(f\right)} = \frac{M_1}{\left[\text{area\_under\_the\_waveform}\right]} \qquad (10.39)$$

The first moment of the convolution with infinite limits is given by (10.40).

$$M_1\left(f_3\right) = \int tf_3\left(t\right) dt = \int t\left\lfloor \int f_1\left(\tau\right) f_2\left(t-\tau\right) d\tau \right\rfloor dt \qquad (10.40)$$

The net result in moments is given by expression (10.41):

$$M_1\left(f_3\right) = M_1\left(f_1\right) M_0\left(f_2\right) + M_1\left(f_2\right) M_0\left(f_1\right) \qquad (10.41)$$

And since the moment about zero of the convolution function f_3 is equal to the product of the moments about zero of the two functions f_1 and f_2 , $M_0(f_3) = M_0(f_1) \bullet M_0(f_2)$.

Then the ratio of the first moments to the moment about the origin can be expressed as (10.42) through (10.43).

$$\frac{M_1(f_3)}{M_0(f_3)} = \frac{M_1(f_1)\,M_0(f_2)}{M_0(f_1)\,M_0(f_2)} + \frac{M_1(f_2)\,M_0(f_1)}{M_0(f_1)\,M_0(f_2)} \tag{10.42}$$

After cancellation of common terms, the results are given in 10.43.

$$\frac{M_1(f_3)}{M_0(f_3)} = \frac{M_1(f_1)}{M_0(f_1)} + \frac{M_1(f_2)}{M_0(f_2)} \tag{10.43}$$

Rewriting (10.43) in short notation as $n_3 = n_1 + n_2$.

In the next step, the square of the centroid of the waveform is subtracted from the second moment of the waveform, which results in the square or the radius of gyration that corresponds to the variance in a probability distribution, as described by (10.44):

$$\sigma^2 = \frac{M_2(f)}{M_0(f)} - n^2 \tag{10.44}$$

where $M_2(f)$ is the second moment of a function $f(t)$.

The corresponding expression for the convolution operation is rewritten as (10.45)

$$\sigma_3^2 = \sigma_1^2 + \sigma_2^2 \tag{10.45}$$

The conclusion in probability is that the variance (of the sum of two independent random variables) equals the sum of their variances, since the probability density function of the sum is the convolution of the individual probability density functions.

10.3.1 Deconvolution

If "Convolution" can be thought of as corresponding to polynomial products, then the process of "Deconvolution" should correspond to polynomial division where one factor is known and the second must be recovered. Even though significant progress has been made in numerical methods, the process of deconvolution is very sensitive to errors and is not used in solution of most problems. There is a tendency for the residual (remainder) to become unstable and diverge in magnitude; hence, caution should be taken in using "Time-domain" deconvolution methods.

CHAPTER 11

Digital Filters

This chapter will review the concept of filters before presenting an overview of digital filters. It is not the intent of the chapter to enable the reader to write digital filter programs, but rather to present the engineer with the ability to use application software with filters and to choose the proper filter for specific biomedical applications. Let us start with the basic question, "What are Filters?" By definition, "Filters are circuits that pass the signals of certain frequencies and attenuate those of other frequencies."

11.1 CLASSIFICATION OF FILTERS

The most common classification or types of filters are as follows:

1. *Low-pass*: pass low frequencies; block high frequencies
2. *High-pass*: pass high frequencies; block low frequencies
3. *Band-pass*: pass a band of frequencies
4. *Band-reject*: attenuate a band of frequencies
5. *All-pass filter*: used for phase shifting

In addition filters are also categorized as follows:

1. Butterworth
    a. Low-pass filter
    b. High-pass filter, etc.

2. Chevbychev
    a. Low-pass filter
    b. High-pass filter, etc.

Other classifications that are also used in conjunction with the previous classification are

1. Voltage Controlled Voltage Source (VCVS),
2. Bi Quad,
3. Elliptical, and
4. Digital.

The most common digital filters fall into one of two classes:

1. Finite Impulse Response (FIR) and
2. Infinite Impulse Response (IIR)

Filters are generally described mathematically by their "Transfer Function" in the Time domain as differential equations, or in the Frequency domain as the ratio of two polynomials in "s-domain" or as $j\omega$ in radians. The order of a filter is defined by the degree of the denominator. The higher the order, the closer the filter approaches the ideal filter; however, the tradeoff an engineer must make is that higher order filters require more complex hardware, and consequently, are more expensive (costly).

11.1.1 Low-Pass Butterworth Filter

In many applications, the most common low-pass filter used is referred to as an "All-Pole Filter," which implies that the filter contained "All poles" and "No zeros" in the s-plane. In addition, the all pole, low-pass filter may be called a "Butterworth filter," which means that the filter has an "over damped" response. Figure 11.1 shows the ideal response of a low-pass Butterworth filter with its cutoff frequency at ω_c and the practical response for the same cutoff frequency. It should be noted that the low-pass Butterworth filter

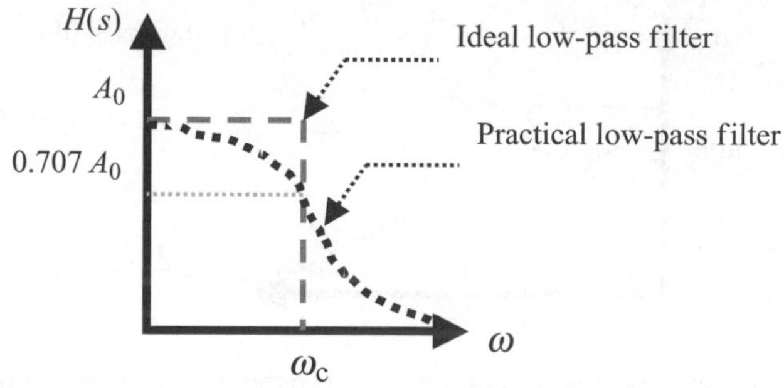

FIGURE 11.1: Butterworth low-pass characteristic transfer function response curves. The ideal low-pass filter is shown as the red trace, and the normal practical response trace is shown in black

magnitude monotonically decreases as frequency increases in both the band-pass and band-stop regions. The flattest magnitude of the filter is in the vicinity of $\omega = 0$, or DC, with its greatest error about the cutoff frequency.

The "Transfer Function" or describing equation for the Butterworth low-pass filter is given by (11.1) for the s-domain and rewritten in (11.2) in terms of frequency.

$$H(s) = \frac{K}{s+1} = \frac{A_{\text{lp}}}{s+j\omega} \tag{11.1}$$

$$H(j\omega) = \frac{A_{\text{lp}}}{\sqrt{1+\left(\frac{\omega}{\omega_c}\right)^2}} \tag{11.2}$$

where A_{lp} = Low-pass gain, and $\frac{\omega}{\omega_c} = \omega_n$

The second-order transfer function for a low-pass Butterworth filter is given by (11.3):

$$H(s) = \frac{A_{\text{lp}}\omega_n^2}{s^2 + 2\zeta\omega_n s + b\omega_n^2}$$

where $$b\omega_c^2 = \frac{1}{R_1 R_2 C_1 C_2} = \frac{1}{T_1 T_2} \tag{11.3}$$

11.1.2 Chebyshev low-pass Filter

The Chebyshev low-pass filter has an underdamped response, meaning that a filter of second order or higher order may oscillate and become unstable if the gain is too high

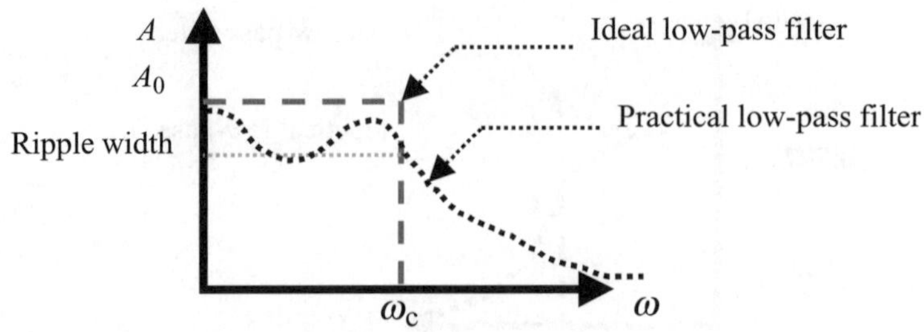

FIGURE 11.2: Chebyshev low-pass characteristic transfer function response curves. The ideal low-pass filter is shown as the dash-line trace, and the normal practical response trace is shown as the dotted-line trace. Note the ripple in the band-pass region

or the input is large. Hence, the term that the "filter rings," meaning it oscillates during it damping phase. The Chebyshev filter response best approaches the "Ideal filter" and is most accurate near the filter cutoff frequency, but at the expense of having ripples in the band-pass region. It should be noted that the filter is monotonic in the stop-band as is the Butterworth low-pass filter. Figure 11.2 shows the ideal response of a Chebyshev low-pass filter with its cutoff frequency at ω_c and the practical response for the same cutoff frequency.

The "Transfer Function" or describing equation for the Butterworth low-pass filter is given by (11.4) in terms of frequency.

$$\left|H(j\omega)\right| = \frac{A_{lp}}{\sqrt{1+\varepsilon^2 C_n^2\left(\frac{\omega}{\omega_c}\right)}} \tag{11.4}$$

where ε is a constant, and Cn is the Chebyshev polynomial, $Cn(x) = \cos[n \cos -1(x)]$

The ripple factor is calculated from (11.5).

$$RW = 1 - \frac{1}{\sqrt{1+\varepsilon^2}} \tag{11.5}$$

Figure 11.3 shows the response of a third-order Chebyshev Voltage Controlled Voltage Source (VCVS) low-pass filter.

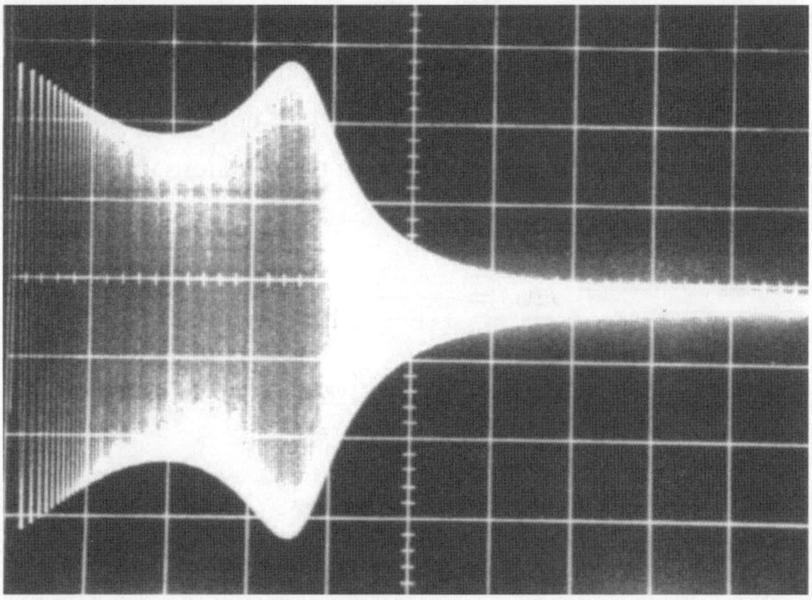

FIGURE 11.3: Chebyshev low-pass filter. Response of a third-order Chebyshev as shown by an oscilloscope

11.1.3 Butterworth High-Pass Filter

High-pass filters are designed to pass high frequencies and block low frequencies. As shown in Fig. 11.4, the Butterworth high-pass Filter magnitude monotonically decreases as frequency decreases, with the flattest magnitude in the high frequencies and the greatest error about the cutoff frequency.

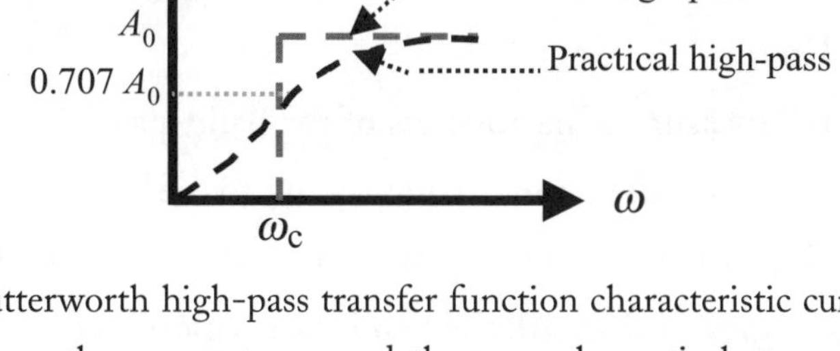

FIGURE 11.4: Butterworth high-pass transfer function characteristic curves. The ideal high-pass filter is shown as the square trace, and the normal practical response trace is shown in dashed-line black

The "Transfer Function" or describing equation for the first-order Butterworth high-pass filter is given by (11.6) for the s-domain and rewritten in (11.7) in terms of frequency. It should be noted that the high-pass filter is not an "All Pole Filter," since the equations show a zero or an s in the numerator of the equations. Hence, in the s-plane a zero exists at the origin, $s = 0$.

$$H(s) = \frac{Ks}{s + \frac{\omega_c}{b}} = \frac{A_{hp}s}{s + \frac{\omega_c}{b}} \tag{11.6}$$

$$H(j\omega) = \frac{A_{hp}s}{s + \frac{1}{R_1 C}} \tag{11.7}$$

where A_{hp} = high-pass gain.

11.1.4 2nd-Order Butterworth High-Pass Filter

The transfer function for a second-order Butterworth high-pass filter is given by (11.8):

$$H(s) = \frac{A_{hp}bs^2}{s^2 + \frac{a}{b}\omega_c s + \frac{\omega_c^2}{b}} \tag{11.8}$$

11.1.5 Band-Pass Filters

Band-pass filters may be of the Butterworth or of the Chebyshev class. Band-pass filters are designed to pass a band of frequencies of bandwidth, B, with the center of the band around a center frequency (ω_0 rad/sec) or $f_0 = \omega_0/2\pi$ Hz. Band-pass filter must designate two cutoff frequencies: a lower cutoff, ωL, and an upper cutoff, ωU.

The frequencies that are passed by the filter are those for which the transfer function, $H(s)$, gain is greater than or equal to $0.707A_0$ as shown in Fig. 11.5.

11.1.5.1 Quality Factor

Quality Factor (Q) is a measure of narrowness of the band-pass filter. By definition, the quality factor, Q, is the ratio of the center frequency, ω_0, to the bandwidth, $B = \omega U - \omega L$ in radians per second; hence, the ratio may be expressed as $Q = \omega_0/B$ or $Q = f_0/B$. If the quality factor, Q, is greater than 10, the band-pass region is considered to be narrow. The gain of the band-pass filter is the magnitude at the center frequency. The transfer function for the band-pass filter is given by (11.9). Band-pass filters must be of second

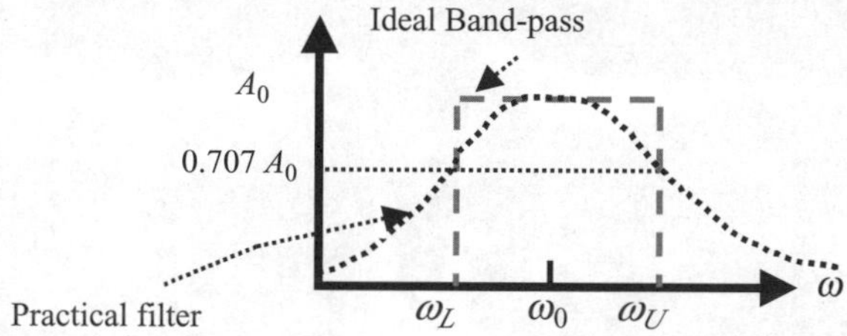

FIGURE 11.5: Band-pass filter transfer function characteristic curves. The ideal band-pass filter is shown as the dashed-line trace, and the normal practical response trace is shown as dotted-line trace

or higher order, since the filters require two slopes (high and low). Note that band-pass filters are not All-Pole filters, since the equation shows a zero in the numerator.

$$H(s) = \frac{A_{\mathrm{BP}}\omega_0 s}{s^2 + \frac{\omega_0}{Q}s + \omega_0^2} \tag{11.9}$$

11.1.6 Band-Reject or "Notch" Filters

Band-reject filters may be of the Butterworth or of the Chebyshev class. Band-reject filters are designed to attenuate or reject a band of frequencies of bandwidth, B, with the center of the band around a center frequency (ω_0 rad/sec) or $f_0 = \omega_0/2\pi$ Hz. As with the band-pass filter, it is necessary to designate the two cutoff frequencies for a band-reject filter; the lower cutoff, ωL, and the upper cutoff, ωU. The frequencies that are passed by the filter are those for which the transfer function, $H(s)$, gain is greater than or equal to $0.707A_0$ as shown in Fig. 11.6.

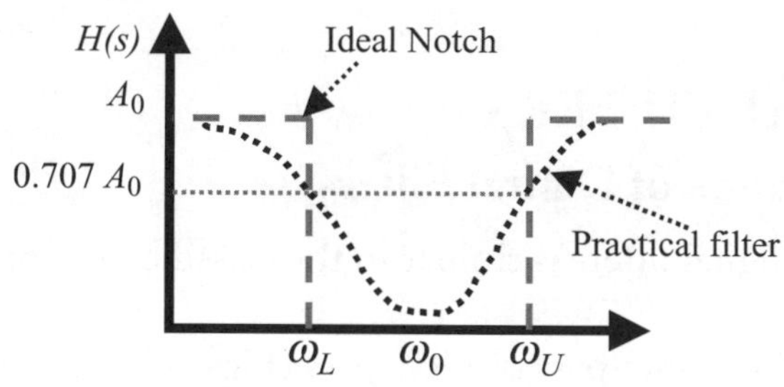

FIGURE 11.6: Band-reject filter transfer function characteristic curves. The ideal band-reject or notch filter is shown as the dashed-line trace, and the normal practical response trace is shown in dotted-line trace

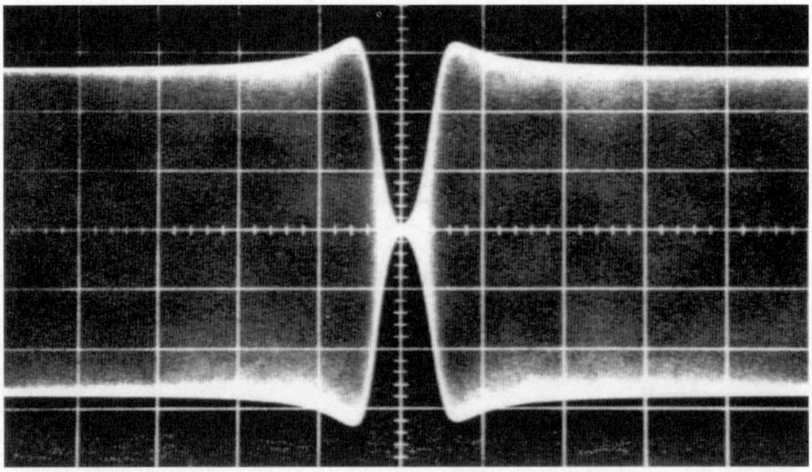

FIGURE 11.7: Chebyshev band-reject filter. Response of a fourth-order Chebyshev notch filter as shown by on an oscilloscope

The transfer function for the band-reject filter is given by (11.10). Band-reject filters must be of second or higher order, since the filters require two slopes (high and low). Note that band-reject filters are not All-Pole filters, since the equation shows a pair of zeros in the numerator (s^2).

$$H(s) = \frac{A_{\mathrm{BR}}(s^2 + \omega_0^2)}{s^2 + \frac{\omega_0}{Q}s + \omega_0^2} \tag{11.10}$$

Figure 11.7 shows the response of a fourth-order Chebyshev band-reject (notch) filter. It should be noted that the maximum attenuation of the band-reject filter is the magnitude at the center frequency, ω_0.

11.2 DIGITAL FILTERS

11.2.1 Classification of Digital Filters

The classification of digital filters is similar to the classification of analog filters

1. Chebyshev filters: low-pass or high-pass (Figs. 11.8 and 11.9)
   a. Underdamped, can oscillate
   b. Ripple factor

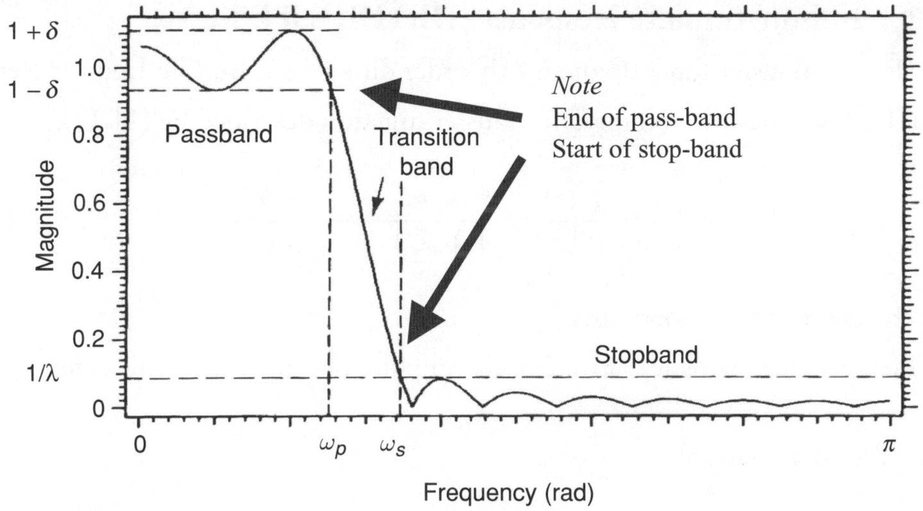

FIGURE 11.8: Low-pass filter. In specifying a digital filter, the end of pass-band and the stop-band parameters must be designated

2. Butterworth filters
   a. Overdamped
   b. Monotonic increasing or decreasing function
3. Elliptical filters

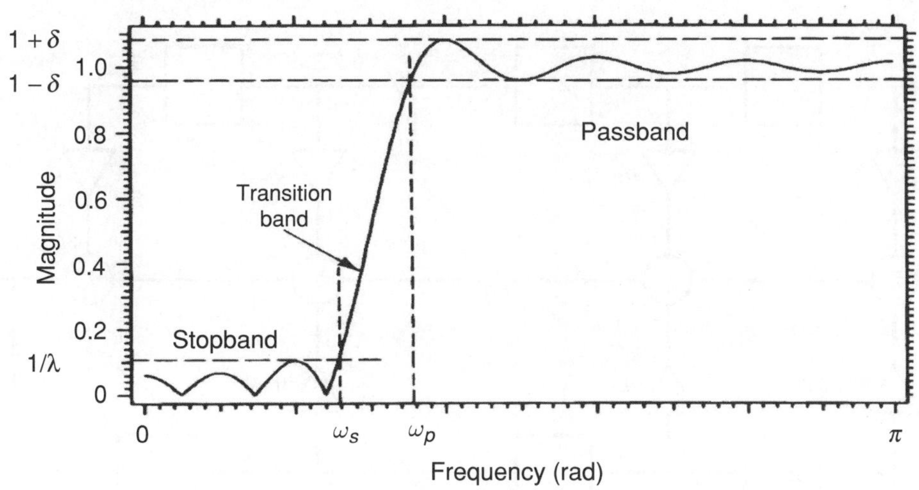

FIGURE 11.9: High-pass filter. In specifying a digital filter, the end of pass-band and the stop-band parameters must be designated. Note the stop-band, the transition band, and the pass-band

11.2.2 Infinite Impulse Response (IIR) Digital Filters

The Pole-zero transfer function of an Lth-order direct form Infinite Impulse Response Filter (IIR-filter) has the z-transform transfer function described by (11.11).

$$H(z) = \frac{B(z)}{A(z)} = \frac{b_o + b_{1Z}^{-1} + \cdots + b_{LZ}^{-L}}{1 + a_{1Z}^{-1} + \cdots + a_{LZ}^{-L}} \tag{11.11}$$

where the zs are the "z-transformation."

General characteristics of the infinite impulse response filters are as follows:

a. Very sharp cutoff

b. Narrow transition band

c. Low-order structure

d. Low-processing time

There are several designs that can be used to implement (11.11), such as the direct, parallel, or cascade design. The "Direct" design is shown in Fig. 11.10, whereas a "Symmetric Lattice" of the IIR design is shown in Fig. 11.11.

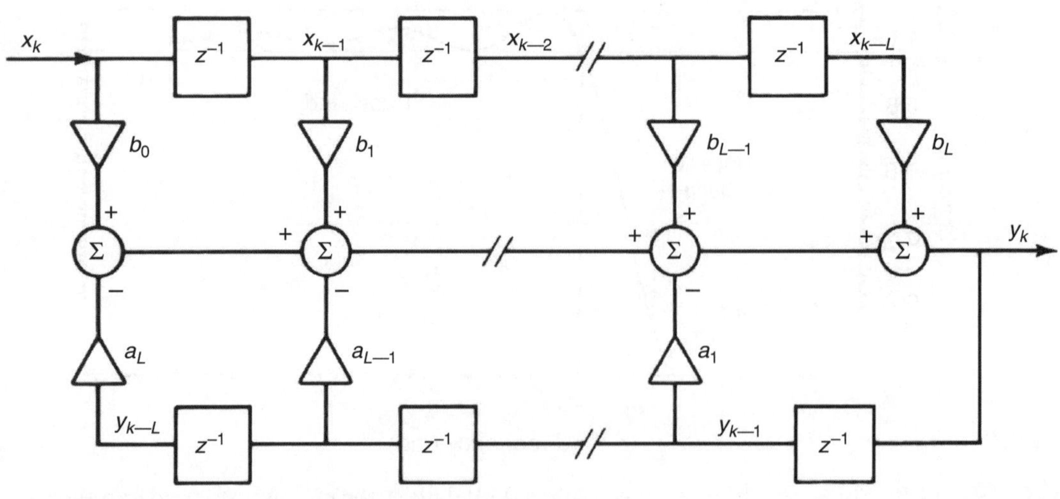

FIGURE 11.10: Direct-form IIR digital filter. Note the numerator coefficients are the b_L terms and the denominator coefficients are the a_L terms

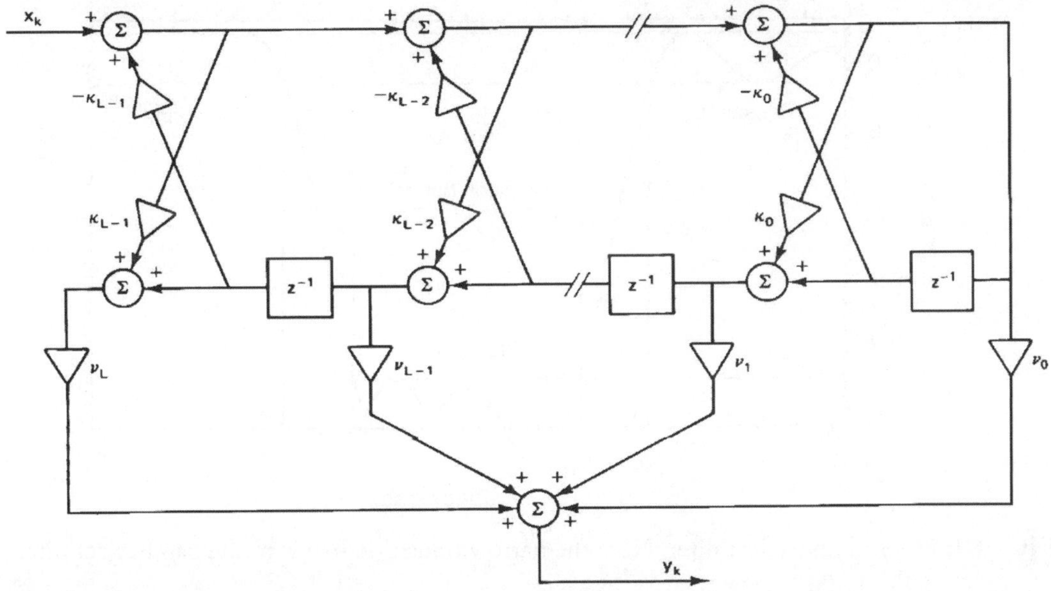

FIGURE 11.11: Symmetric IIR lattice digital filter

IIR filter also have a very narrow transition band, which would be desirable in band-pass and band-reject filters as shown in Figs. 11.12 and 11.13, respectively.

11.2.3 Finite Impulse Response (FIR) Digital Filters

The Finite Impulse Response (FIR) filter maybe derived from the IIR transfer function equation (11.11). If the denominator terms are unity, $A(z) = 1$, the equation for $H(z)$

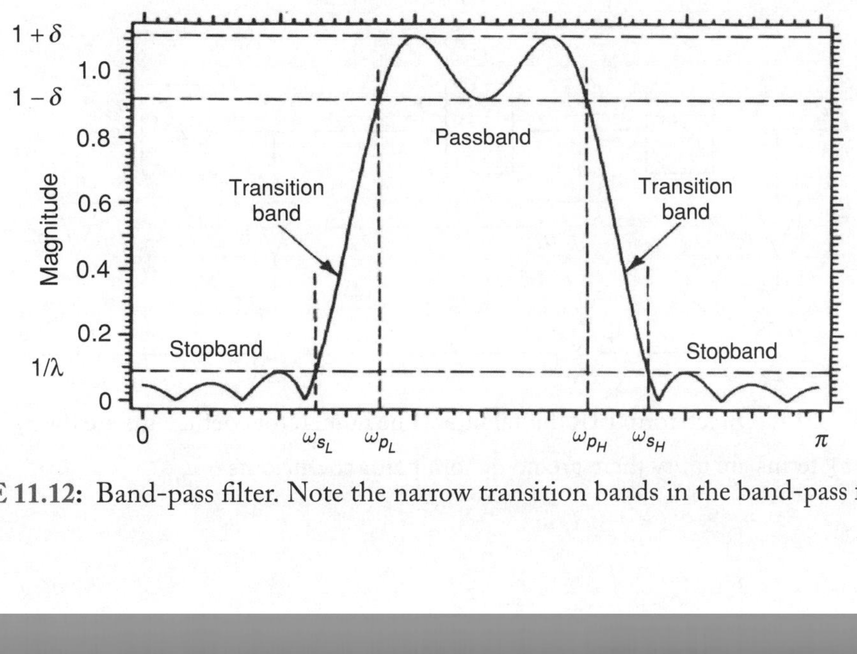

FIGURE 11.12: Band-pass filter. Note the narrow transition bands in the band-pass filter

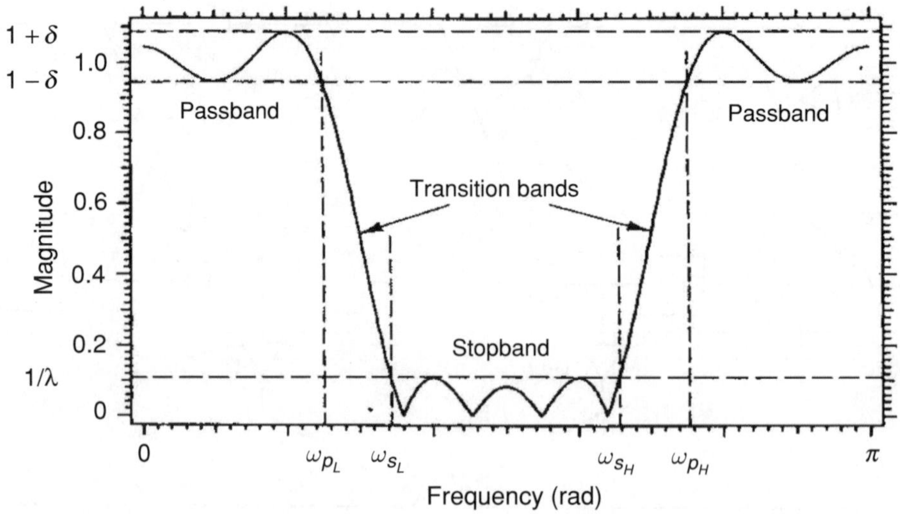

FIGURE 11.13: Band-reject filter. Note the narrow transition bands in the band-reject filter

in the IIR direct version becomes the FIR direct-form implementation equation, as in (11.12).

$$H(z) = B(z) = b_0 + b_{1Z}^{-1} + \cdots + b_{LZ}^{-L} \tag{11.12}$$

The difference equation corresponding to the FIR direct-form implementation equation (11.12) is given by (11.13), and the "Direct" FIR design is shown in Fig. 11.14.

$$y_k = \sum_{n=0}^{L} b_n x_k - n \tag{11.13}$$

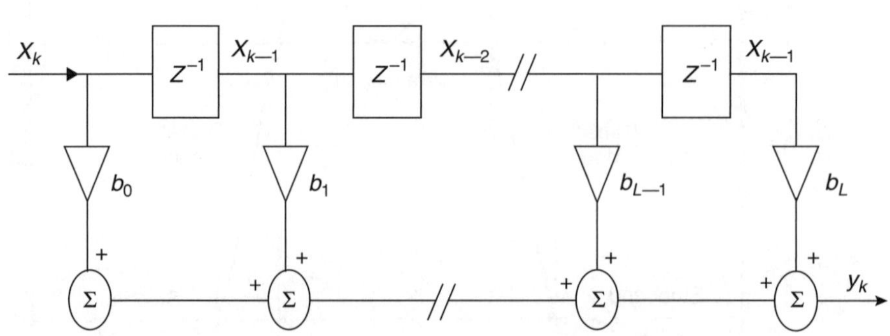

FIGURE 11.14: Direct-form FIR digital filter. The numerator coefficients are the b_L terms, and since the a_L terms are unity there are no denominator coefficients

In summary, FIR filters should be used when phase is important or when flatness without ripples in the pass-band region is important in the analysis of signals. The general characteristics of the FIR filters are listed as follows:

a. many coefficients;

b. designed to have exactly linear phase characteristics; and

c. implementation via fast convolution.

CHAPTER 12

Fourier Series: Trigonometric

The objective of this chapter is for the reader to gain a better understanding of the Basic Fourier Trigonometric Series Transformation and waveform synthesis. If we are interested in studying time-domain responses in networks subjected to periodic inputs in terms of their frequency content, the Fourier Series can be used to represent arbitrary periodic functions as an infinite series of sinusoids of harmonically (not harmoniously) related frequencies as shown in (12.1).

A signal $f(t)$ is periodic with period, T, if $f(t) = f(t+T)$ for all t, then the Fourier Trigonometric Series representation of the periodic function may be expressed as (12.1). It should be noted that both "cosine" and "sine" trigonometric terms are necessary and that an infinite number of terms may be necessary.

$$\begin{aligned} f(t) = a_0 + a_1 \cos\omega_0 t + a_2 \cos 2\omega_0 t + \cdots a_n \cos n\omega_0 t + b_1 \sin\omega_0 t \\ + b_2 \sin 2\omega_0 t + \cdots b_n \sin n\omega_0 t + \cdots n = 1, 2, 3, \ldots \infty \end{aligned} \tag{12.1}$$

where $\omega_0 = \frac{2n\pi}{T}$; $T = \frac{1}{f_0}$; and n is the nth harmonic of the fundamental frequency, ω_0.

12.1 FOURIER ANALYSIS

Recall the chapter on basis functions. In taking the integral transform of a signal, the integral of the product of the signal with any basis function transforms the work from

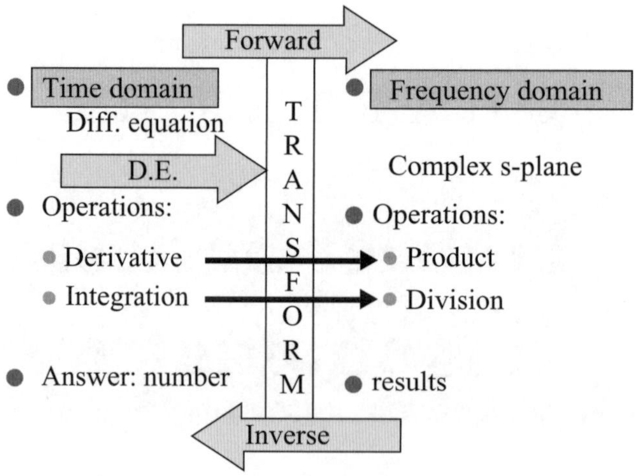

FIGURE 12.1: Transformation from the time domain to the frequency domain

one domain, that is, time domain, into a separate domain, that is, frequency domain with either the La Place or Fourier Basis functions. The reason for working in the transformed domain is to reduce the time-domain operations of integration or differentiation to less complex operations of division or multiplication, respectively.

There are numerous forms of the Fourier Transform. The most used forms include the following:

1. The Fourier Trigonometric Series, which was used prior to the advent of computer to analyze analog data, but had the major drawback of being very time consuming.
2. The Discrete Fourier Transform was used with digitized data, and is considered to be a subset of the LaPlace Transform; however, the Discrete Fourier Transform was also very time consuming (4 minutes to analyze 1024 samples of data).
3. The Fast Fourier Transform became the most popular method used to perform transformation into the frequency domain because it was very efficient and fast.

There are many other Fourier Transforms that will not be studied in this course.

The Fourier analysis consists of two basic operations:

1. The first operation is to determine the values of the coefficients $a_0, a_1, \ldots a_n$, and $b_1, \ldots b_n$.
2. The second operation is to decide how many terms to include in a truncated series such that the partial sum will represent the function within allowable error. It is not possible to calculate an infinite number of coefficients.

12.2 EVALUATION OF THE FOURIER COEFFICIENTS

In evaluating the coefficients, the limits of integration must be set. Since the Fourier Basis functions are orthogonal over the interval t_0 to $(t_0 + T)$ for any t_0, we often use the value $t_0 = 0$ or $t_0 = {}^{-T}/_2$ with the understanding that any period may be used as the period of integration. Thus, we replace the interval of the integral from, t_0 to $(t_0 + T)$, with the limits of integration from, 0 to T.

The next step would be to calculate the a_0 term, which is simply the average value of $f(t)$ over a period; a_0 is often referred to as the DC value of a sinusoid over N complete cycles in the period. Next the complex coefficients, a_n and b_n, are evaluated with the expressions in (12.2), (12.3), and (12.4). To obtain the a_n terms, which constitute the "real-part" of the complex Fourier coefficient, the signal or function is multiplied by the corresponding cosine term, integrated, and then normalized by the fundamental period, T, as shown in (12.3). In a similar manner, to obtain the b_n terms, which constitute the "imaginary-part" of the complex Fourier coefficient, the signal or function is multiplied by the corresponding sine term, integrated, and then normalized by the fundamental period, T, as shown in (12.4).

$$\text{DC Value of } f(t): \quad a_0 = c_0 = \frac{1}{T}\int_{-T/2}^{T/2} f(t)dt = \frac{1}{T}\int_{t_0}^{t_0+T} f(t)dt = \frac{1}{T}\int_0^T f(t)dt \tag{12.2}$$

$$\text{Real Term:} \quad a_n = \frac{2}{T}\int_0^T f(t)\cos n\omega_0 t\, dt \text{ or } \int_{-T/2}^{T/2} \tag{12.3}$$

$$\text{Imaginary term:} \quad b_n = \frac{2}{T}\int_0^T f(t)\sin n\omega_0 t\, dt \tag{12.4}$$

12.3 EQUIVALENT FORM OF THE FOURIER TRIGONOMETRIC SERIES

The shorthand notation for writing the Fourier series is given by (12.5).

$$f(t) = a_0 + a_1 \cos \omega_0 t + a_2 \cos 2\omega_0 t + \cdots a_n \cos n\omega_0 t + \cdots b_1 \sin \omega_0 t$$
$$+ b_2 \sin 2\omega_0 t + \cdots b_n \sin n\omega_0 t + \cdots$$
$$f(t) = a_0 + \sum_{n=1}^{\infty} (a_n \cos n\omega_0 t + b_n \sin n\omega_0 t) \tag{12.5}$$

However, an equivalent expression is to combine the complex coefficients, a_n and b_n, and rewrite the coefficients in polar-coordinate terms of magnitude and phase as in (12.6).

$$c_n \cos(n\omega_0 t + \theta_n) = a_n \cos n\omega_0 t + b_n \sin n\omega_0 t \tag{12.6}$$

where $c_n = \sqrt{a_n^2 + b_n^2}$ and $\theta = \tan^{-1} \frac{b_n}{a_n}$

Then, the Fourier Trigonometric Series representation of the function may be written in the expanded form as in (12.7) or in the compressed form as in (12.8).

$$f(t) = c_0 + c_1 \cos(\omega_0 t + \theta_1) + \cdots c_n \cos(n\omega_0 t + \theta_n) + \cdots \tag{12.7}$$

where c_n is the **magnitude**, and θ_n is the **phase** of the nth harmonic.

$$f(t) = c_0 + \sum_{n=1}^{\infty} c_n \cos(n\omega_0 t + \theta_n) \tag{12.8}$$

12.4 SYMMETRICAL PROPERTIES AS RELATED TO FOURIER COEFFICIENTS

12.4.1 Even Waveform Symmetries

In working with the Fourier Series, it is often advantageous to take advantages of the symmetrical properties of trigonometric functions or other wave forms. If the function $f(t)$ satisfies the condition, $f(t) = f(-t)$, then the function is said to be *even*, as shown in Figs. 12.2 and 12.3. The function will contain only the Fourier coefficients; DC, a_0, and $a_n \cos n\omega_0 t$ terms. Note that the cosine function in Fig. 12.2 is an "even" function as is the square wave function in Fig. 12.3.

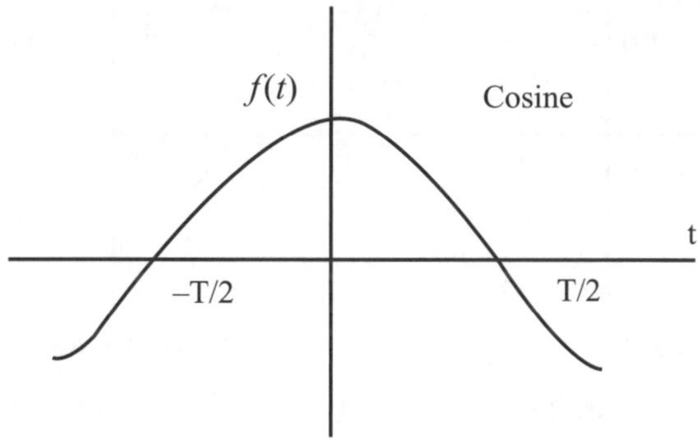

FIGURE 12.2: An even function. The cosine is an even function, since the function is symmetrical about the y-axis, that is, $f(t) = f(-t)$

12.4.2 ODD Waveform Symmetries

Similarly, if $f(t)$ satisfies the condition $f(t) = -f(-t)$, then the function is said to be *odd*, as shown in Fig. 12.4. The function will contain only the Fourier coefficients $b_n \sin n\omega_0 t$ terms. Note that the functions are folded symmetrically about both y- and x-axes. The negative sign before the time, t, folds the function about the y-axis whereas the negative sign before the function, f, inverts the function about the x-axis.

Another useful property to know is the "half-wave" symmetry. If the function $f(t)$ satisfies the condition $f(t) = -f(t \pm \frac{T}{2})$, then the function is said to have *half-wave* symmetry, as shown in Fig. 12.5. The function will contain only the odd complex Fourier coefficients, a_n and b_n, for $n = 1, 3, 5, 7, \ldots$, etc.

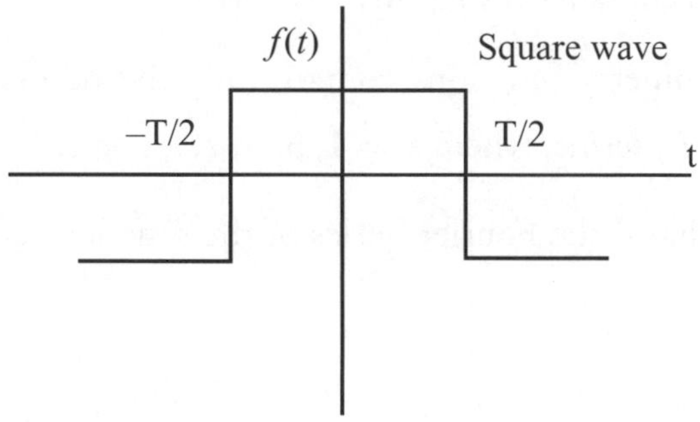

FIGURE 12.3: Another even function. The square wave function is an even function, since the function is symmetrical about the y-axis, i.e., $f(t) = f(-t)$

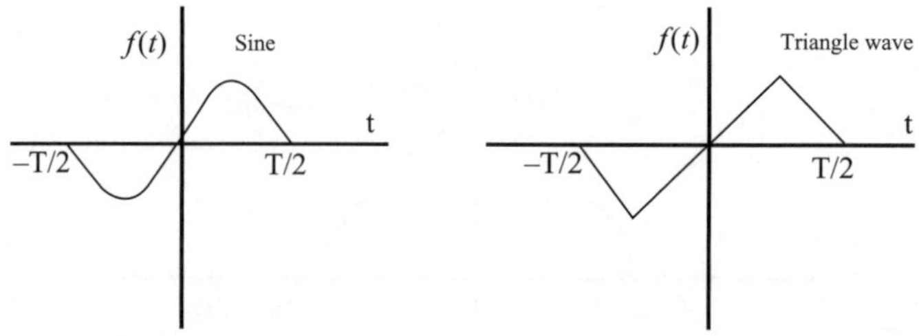

(a) Sine wave (b) Triangular wave

FIGURE 12.4: Odd functions. The sine wave (a) and the triangular wave are considered to be odd functions since they satisfy the condition $f(t) = -f(-t)$

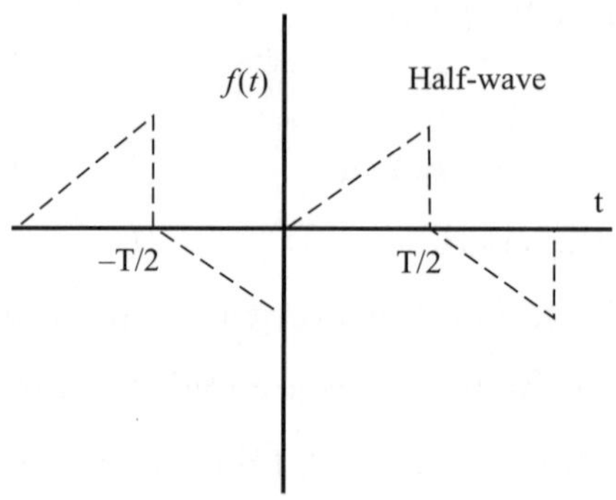

FIGURE 12.5: Half-wave symmetric function. This triangular function is considered to have half-wave symmetry since it satisfies the condition $f(t) = -f(t \pm \frac{T}{2})$

The relationship between the properties of symmetry and the Fourier Series are summarized in Table 12.10 and as follows:

1. Even functions contain only a_0 and $a_n \cos n\omega_0 t$ terms.
2. Odd functions contain only $b_n \sin n\omega_0 t$ terms.
3. Half-wave symmetry functions contain only the odd complex Fourier coefficients, a_n and b_n terms, where $n = 1, 3, 5, 7, \ldots$, etc.

Example Problem: What is the Fourier Series of the function, $v(t)$, in Fig. 12.6?

The function $v(t)$ is given below.

$$v(t) = \begin{cases} V, & 0 < t < {}^{T}\!/_{4} \\ -V, & {}^{T}\!/_{4} < t < {}^{3}\!/_{4}T \\ V, & {}^{3}\!/_{4}T < t < T \end{cases} \tag{12.9}$$

TABLE 12.10 SUMMARY OF SYMMETRY TABLES

TABLE A: Waveform Symmetry

SYMMETRY	CONDITION	ILLUSTRATION	PROPERTY
Even	$f(t) = f(-t)$	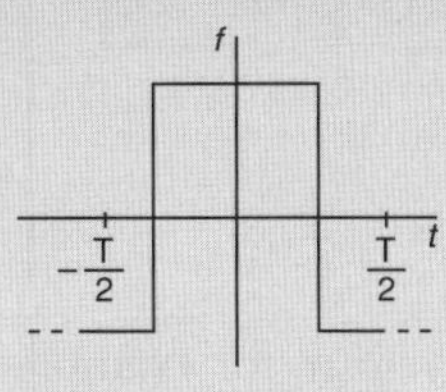	cosine terms only
Odd	$f(t) = -f(-t)$	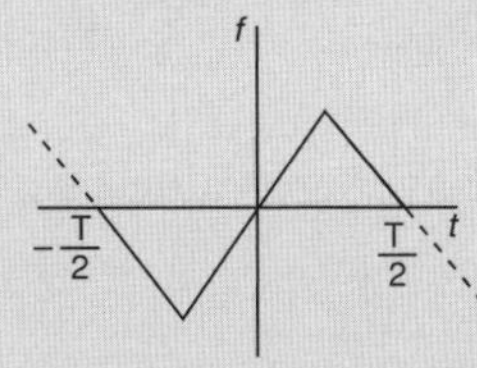	sine terms only
Half-wave	$f(t) = -f(t \pm T/2)$	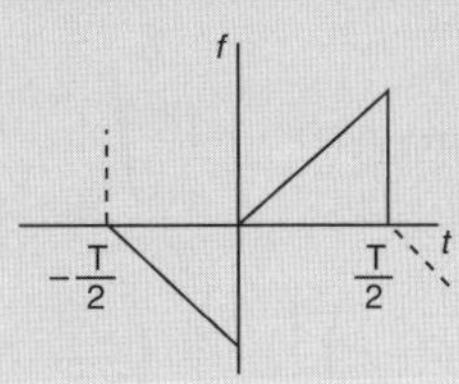	odd n only

TABLE B: Fourier Transform of Symmetrical Waveforms

PROPERTY	a_0	$a_n (n \neq 0)$	b_n
cosine terms only	*	$\frac{4}{T}\int_0^{T/2} f(t) \cos n\omega_0 t \, dt$	0
sine terms only	0	0	$\frac{4}{T}\int_0^{T/2} f(t) \sin n\omega_0 t \, dt$
Odd n only	0	$\frac{4}{T}\int_0^{T/2} f(t) \cos n\omega_0 t \, dt$	$\frac{4}{T}\int_0^{T/2} f(t) \sin n\omega_0 t \, dt$

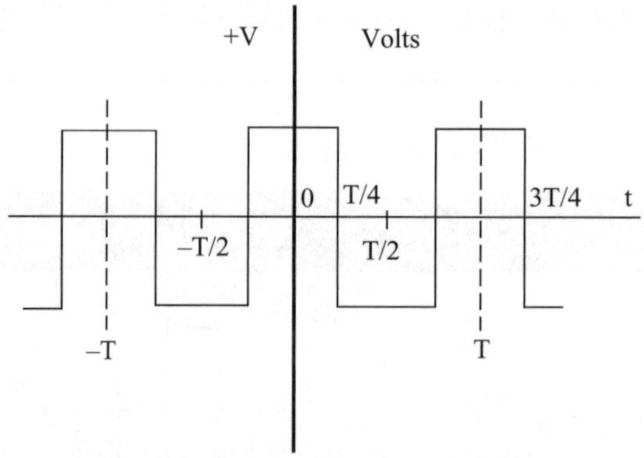

FIGURE 12.6: Square waveform for example problem

By inspection, the average value over one period is zero so that $a_0 = 0$.

Value of a_1 with $n = 1$ is

$$a_n = \frac{2}{T}\left(V\int_0^{T/4}\cos n\omega_0 t\,dt - V\int_{T/4}^{3/4\,T}\cos n\omega_0 t\,dt + V\int_{3/4\,T}^{T}\cos n\omega_0 t\,dt\right)$$

$$a_n = \frac{2V}{T}\left[\left(\frac{1}{n\omega_0}\sin n\omega_0 t\right)\Big|_0^{T/4} - \left(\frac{1}{n\omega_0}\sin n\omega_0 t\right)\Big|_{T/4}^{3/4\,T} + \left(\frac{1}{n\omega_0}\sin n\omega_0 t\right)\Big|_{3/4\,T}^{T}\right]$$

$$a_n = \frac{2V}{\omega_0 T}\left[\left(\sin\frac{\omega_0 T}{4}\right) - \left(\sin\frac{3\omega_0 T}{4} - \sin\frac{\omega_0 T}{4}\right) + \left(\sin\omega_0 T - \sin\frac{3\omega_0 T}{4}\right)\right]$$

Since $\omega_0 T = \frac{2\pi}{T}\cdot T = 2\pi$, then

$$\text{for } n = 1,\ a_1 = \frac{2V}{2\pi}\left[\sin\frac{\pi}{2} - \left(\sin\frac{3\pi}{2} - \sin\frac{\pi}{2}\right) + \left(\sin 2\pi - \sin\frac{3\pi}{2}\right)\right]$$

$$a_1 = \frac{V}{\pi}[1-(-1-1)+(0-(-1))] = \frac{V}{\pi}[1+2+1] = \frac{4V}{\pi}$$

$$a_n = \frac{2}{T}\left(V\int_0^{T/4}\cos n\omega_0 t\,dt - V\int_{T/4}^{3/4\,T}\cos n\omega_0 t\,dt + V\int_{3/4\,T}^{T}\cos n\omega_0 t\,dt\right)$$

$$a_n = \frac{2V}{T}\left[\left(\frac{1}{n\omega_0}\sin n\omega_0 t\right)\Big|_0^{T/4} - \left(\frac{1}{n\omega_0}\sin n\omega_0 t\right)\Big|_{T/4}^{3/4\,T} + \left(\frac{1}{n\omega_0}\sin n\omega_0 t\right)\Big|_{3/4\,T}^{T}\right]$$

Since $2\omega_0 T = 4\pi$, then:

$$\text{For } n = 2,\ a_2 = \frac{2V}{2\omega_0 T}\left[\sin 2\omega_0 \frac{T}{4} - \left(\sin 2\omega_0 \frac{3T}{4} - \sin 2\omega_0 \frac{T}{4}\right)\left(\sin 2\omega_0 T - \sin 2\omega_0 \frac{3T}{4}\right)\right]$$

$$a_2 = \frac{2V}{4\pi}\left[\sin\frac{4\pi}{4} - \left(\sin\frac{4\pi\cdot 3}{4} - \sin\pi\right) + (\sin 4\pi - \sin 3\pi)\right]$$

$$a_2 = \frac{2V}{4\pi}\,[0 - (0-0) + (0-0)] = 0$$

Likewise, $a_n = 0$ for all even values of a_n.

Since $\omega_0 T = 2\pi$, then $n\omega_0 T$ for $n = 3$ is $3\omega_0 T = 6\pi$.

$$\text{For } n = 3,\ a_3 = \frac{2V}{6\pi}\left[\sin\frac{6\pi}{4} - \left(\sin\frac{6\pi\cdot 3}{4} - \sin\frac{6\pi}{4}\right) + \left(\sin 6\pi - \sin\frac{6\pi\cdot 3}{4}\right)\right]$$

$$a_3 = \frac{V}{3\pi}\,[-1 - (+1 - (-1)) + 0 - (+1)]$$

$$a_3 = \frac{V}{3\pi}\,[-1 - 2 - 1] = -\frac{4V}{3\pi}$$

Applying the same procedure for all n, we find that

$$a_n = \begin{cases} +\dfrac{4V}{n\pi}, & n = 1, 5, 9, \ldots \\ -\dfrac{4V}{n\pi}, & n = 3, 7, 11, \ldots \\ 0, & n = \text{even} \end{cases}$$

The imaginary term b_n values are calculated as follows:

$$b_n = \frac{2}{T}\left(V\int_0^{T/4} \sin n\omega_0 t\, dt - V\int_{T/4}^{3/4\,T} \sin n\omega_0 t\, dt + V\int_{3/4\,T}^{T} \sin n\omega_0 t\, dt\right)$$

$$b_n = \frac{2V}{n\omega_0 T}\left[(-\cos n\omega_0 t)\Big|_0^{T/4} - (-\cos n\omega_0 t)\Big|_{T/4}^{3/4\,T} + (-\cos n\omega_0 t)\Big|_{3/4\,T}^{T}\right]$$

$$b_1 = \frac{2V}{n\omega_0 T}\left[-\cos\frac{2\pi}{4} + \cos 0 - \left(-\cos\frac{6\pi}{4} + \cos\frac{2\pi}{4}\right) + \left(-\cos 2\pi + \cos\frac{6\pi}{4}\right)\right]$$

$$b_1 = \frac{2V}{n\omega_0 T}[0 + 1 + 0 - 0 - 1 + 0] = 0$$

$$b_1 = \frac{2V}{2\pi}\,(0) = 0$$

If you continue this process for other integers of n, you will obtain $b_n = 0$ for all values of n.

Before moving into Discrete and Fast Fourier transformations and frequency-domain analysis, let me regress into perhaps the most important function to engineers, physicists, and mathematicians—the **exponential function**—which has the property that the derivative and integral yield a function proportional to itself as in (12.10).

$$\frac{d}{dt}e^{st} = se^{st} \quad \text{and} \quad \int e^{st}\,dt = \frac{1}{s}e^{st} \tag{12.10}$$

It turns out that every function or waveform encountered in practice can always be expressed as a sum of various exponential functions.

12.5 EULER EXPANSION

Euler showed that the expression, $s = -j\omega$, of a complex frequency represents signals oscillating at angular frequency, ω, and that the $\cos \omega t$ could thereby then be represented as (12.11).

$$\cos(\omega t) = \frac{e^{j\omega t} + e^{-j\omega t}}{2} \tag{12.11}$$

In a way, the Discrete Fourier Series is a method of representing periodic signals by exponentials whose frequencies lie along the $j\omega$ axis. Expressing the Fourier Trigonometric Series in an equivalent form in terms of exponentials, or $e^{\pm jn\omega t}$:

$$f(t) = a_0 + \sum_{n=1}^{\infty} (a_n \cos n\omega_0 t + b_n \sin n\omega_0 t)$$

$$\cos n\omega_0 t = \frac{1}{2}\left(e^{jn\omega_0 t} + e^{-jn\omega_0 t}\right)$$

$$\sin n\omega_0 t = \frac{1}{2j}\left(e^{jn\omega_0 t} - e^{-jn\omega_0 t}\right)$$

$$\text{then} f(t) = a_0 + \sum_{n=1}^{\infty}\left(a_n \frac{e^{jn\omega_0 t} + e^{-jn\omega_0 t}}{2} + b_n \frac{e^{jn\omega_0 t} - e^{-jn\omega_0 t}}{2j}\right)$$

Grouping terms, $-j = \frac{1}{j}$:

$$f(t) = a_0 + \sum_{n=1}^{\infty}\left(\left(\frac{a_n - jb_n}{2}\right)e^{jn\omega_0 t} + \left(\frac{a_n + jb_n}{2}\right)e^{-jn\omega_0 t}\right)$$

Redefining coefficients:

$$\tilde{c}_n = \frac{a_n - jb_n}{2}, \quad \tilde{c}_{-n} = \frac{a_n + jb_n}{2}, \quad \text{and} \quad \tilde{c}_0 = a_0$$

Then, the Fourier expression for a function may be written as (12.12):

$$f(t) = \tilde{c}_0 + \sum_{n=1}^{\infty} \left(\tilde{c}_n e^{jn\omega_0 t} + \tilde{c}_{-n} e^{-jn\omega_0 t} \right) \tag{12.12}$$

Letting n range from 1 to ∞ is equivalent to letting n range from $-\infty$ to $+\infty$, including 0.

$$\text{Then } f(t) = \sum_{n=-\infty}^{\infty} \tilde{c}_n e^{jn\omega_0 t}$$

The coefficients $\tilde{c}_n$ can be evaluated in terms of a_n and b_n as shown in (12.13).

$$|\tilde{c}_n| = \frac{1}{2}\sqrt{a_n^2 + b_n^2} = \frac{1}{2}c_n \quad \text{and} \quad \phi_n = \tan^{-1}\frac{b_n}{a_n} \tag{12.13}$$

$$\tilde{c}_n = |\tilde{c}_n| \; e^{j\theta n}$$

$$\tilde{c}_n = \tilde{c}_n^* = |\tilde{c}_n| \; e^{-j\theta n}$$

Note that in (12.13), the coefficients of the exponential Fourier Series are $^1/_2$ the magnitude of the geometric Fourier Series coefficients.

12.6 LIMITATIONS

Overall, the Fourier Series method has several limitations in analyzing linear systems for the following reasons:

1. The Fourier Series can be used only for inputs that are periodic. Most inputs in practice are nonperiodic.
2. The method applies only to systems that are stable (systems whose natural response decays in time).

The first limitation can be overcome if we can represent the nonperiodic input, $f(t)$, in terms of exponential components, which can be done by either the Laplace or

Fourier Transform representation of $f(t)$. Consider the following nonperiodic signal, $f(t)$, which one would like to represent by eternal exponential functions. As a result, we can construct a new periodic function, $f_T(t)$, with period T, where the function $f(t)$ is represented every T seconds. The period, T, must be large enough so there is not any overlap between pulse shapes of $f(t)$. The new function is a periodic function and can be represented with an exponential Fourier Series.

12.7 LIMITING PROCESS

What happens if the interval, T, becomes infinite for a pulse function, $f(t)$? Do the pulses repeat after an infinite interval? Such that $f_T(t)$ and $f(t)$ are identical in the limit, and the Fourier Series representing the periodic function, $f_T(t)$, will also represent $f(t)$?

Discussion on Limiting Process: Let the interval between pulses become infinite, $T = \infty$, in the series, so we can represent the exponential Fourier Series for $f_T(t)$ as in (12.14).

$$\sum_{n=-\infty}^{\infty} F_n \, e^{jn\omega_0 t} \tag{12.14}$$

Where ω_0 is the fundamental frequency and F_n is the term which represents the amplitude of the component of frequency, $n\omega_0$, the coefficient. As T becomes larger, the fundamental frequency, ω_0, becomes smaller, and the amplitude of individual components also become smaller as shown in (12.15). The frequency spectrum becomes denser, but its shape is unaltered.

$$\text{Limit } T \to \infty, \omega_0 \to 0 \tag{12.15}$$

The spectrum exists for every value of ω and is no longer discrete but a continuous function of ω. So, let us denote the fundamental frequency ω_0 as being infinite. The logic is as follows:

Since Limit $T \to \infty$, $\omega_o \to 0$, then $T = \dfrac{2\pi}{\Delta\omega}$,
and TF_n is a function of $jn\Delta\omega$, or $TF = F(jn\Delta\omega)$.

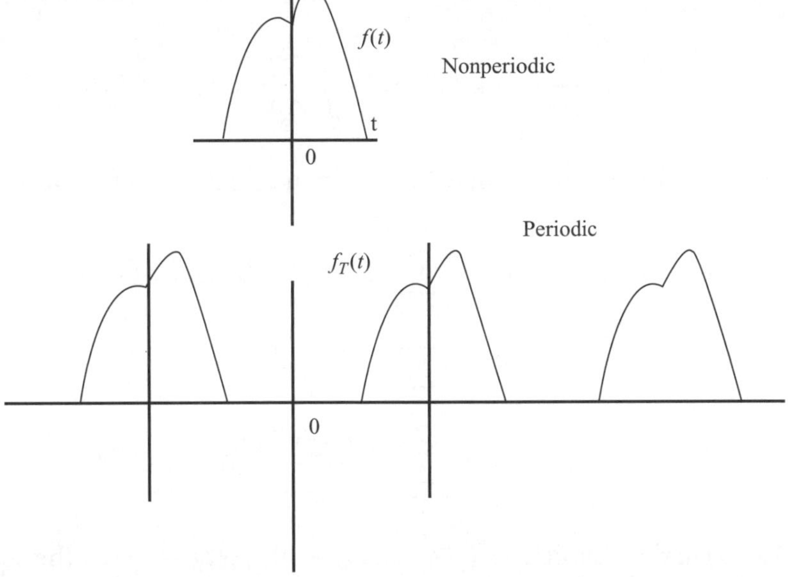

FIGURE 12.7: A transient pulse. Top trace is a transient; bottom trace is periodic

It is an accepted fact that a periodic function, $f_T(t)$, can be expressed as the sum of eternal exponentials of frequencies $0, \pm j\Delta\omega, \pm j2\Delta\omega, \pm 3j\Delta\omega \ldots$ etc. Also, it is known that the Fourier Transform, F_n, amplitude of the component of frequency, $jn\ \Delta\omega$, is expressed as $F\,(jn\Delta\omega)\ {}^{\Delta\omega}\!/_{2\pi}$.

Proof of the Limiting Process: Viewing Fig. 12.7, let us examine the proof of the limiting process.

Let us begin with the general expression (12.16):

$$\text{In the Limit as } T \to \infty : \lim_{T\to\infty} f_T(t) = f(t) \tag{12.16}$$

Then the exponential Fourier Series for $f_T(t)$ is given by (12.17).

$$f_T(t) = \sum_{n=-\infty}^{\infty} F_n\, e^{jn\omega_0 t} \tag{12.17}$$

where $\omega_0 = \frac{2\pi}{T}$ and $F_n = \frac{1}{T}\int_{-T/2}^{T/2} f_T(t)\, e^{-jn\omega_0 t}\, dt$

Since Limit $T \to \infty$, $\omega_0 \to 0$, ω_0 can be denoted by $\Delta\omega$ (see (12.18))

Then $$T = \frac{2\pi}{\omega_0} = \frac{2\pi}{\Delta\omega} \text{ and } TF_n = \frac{1}{T}\int_{-T/2}^{T/2} f_T(t)\, e^{-jn\omega_0 t}\, dt \tag{12.18}$$

Since $TF_n\,(jn\Delta\omega)$, the expression may be rewritten as $TF_n = F\,(jn\Delta\omega)$.

Then (12.17), $f_T(t) = \sum_{n=-\infty}^{\infty} F_n e^{jn\omega_0 t}$, becomes (12.19).

$$f_T(t) = \sum_{n=-\infty}^{\infty} \frac{F\,(jn\Delta\omega)}{T}\, e^{(jn\omega_0 t)t}$$

$$f_T(t) = \sum_{n=-\infty}^{\infty} \left[\frac{F\,(jn\Delta\omega)}{2\pi}\Delta\omega\right] e^{(jn\omega_0 t)t} \tag{12.19}$$

Rewriting (12.19) in the "Limit": as $T \to \infty$, $\omega_0 \to 0$, $f_T(t) \to f(t)$, the equation takes the form of (12.20), thus proving the limit process.

$$f(t) = \operatorname*{Lim}_{T\to\infty} f_T(t) = \left[\operatorname*{Lim}_{\Delta\omega\to 0} \frac{1}{2\pi} \sum F\,(jn\Delta\omega) e^{jn\Delta\omega t}\Delta\omega\right] \tag{12.20}$$

12.8 INVERSE FOURIER TRANSFORM

To return to the "Time Domain" after transforming into the "Frequency Domain," the inverse Fourier transform must be obtained. By definition, the Inverse Fourier Transform is written as (12.21). The process is the same as taking the forward transform, except that in lieu of multiplying the time domain function, $f(t)$, by the Fourier Basis Function, all the Fourier Coefficients form the Frequency domain function, $F_n = F\,(jn\Delta\omega)\,{}^{\Delta\omega}/_{2\pi}$, are multiplied by the inverse Fourier basis function as in (12.19). Particular emphasis is made to point out that the sign of the basis function exponential frequency, $j\omega t$, is positive for the inverse transformation.

$$f(t) = \frac{1}{2\pi}\int_{-\infty}^{\infty} F\,(j\omega)\; e^{j\omega t}\; d\omega \tag{12.21}$$

where $n\Delta\omega$ becomes ω (a continuous variable) and the function $F\,(j\omega) = \operatorname*{Lim}_{\Delta\omega\to 0} F\,(jn\Delta\omega)$.

12.9 SUMMARY

The "Forward Fourier Transform" equations are summarized in (12.22) and (12.23).

$$F_n = \frac{1}{T}\int_{-T/2}^{T/2} f_T(t)\, e^{-jn\omega_0 t}\, dt = \frac{F(jn\Delta\omega)^*}{T} \tag{12.22}$$

then $F(j\omega) = \lim_{T\to\infty} \int_{-T/2}^{T/2} f_T(t)\, e^{-jn\Delta\omega t} dt$ frequency spectrum of $f_T(t)$

$$\text{or for the infinite case: } F(j\omega) = \int_{-\infty}^{\infty} f(t)\, e^{-j\omega t}\, dt \tag{12.23}$$

where in the Limit as $T \to \infty$, $\Delta\omega \to 0$, $f_T(t) \to f(t)$.

The Discrete Inverse Fourier Transform $\lim_{\Delta\omega\to 0} \frac{1}{2\pi} \sum F(j\omega)\; e^{j\Delta\omega t} \Delta\omega$ is, by definition, the integral, which may be expressed as (12.24).

$$f(t) = \frac{1}{2\pi}\int_{-\infty}^{\infty} F(j\omega)\, e^{j\omega t}\, d\omega \tag{12.24}$$

where $n\Delta\omega$ becomes a continuous variable, ω, and the function $F(j\omega)$ is $F(jn\Delta\omega)$.

The result is that $F(j\omega) = \int_{-\infty}^{\infty} f(t)\, e^{-j\omega t}\, dt$ is the presentation of the nonperiodic function, $f(t)$, in terms of exponential functions. The amplitude of the component of any frequency, ω, is proportional to $F(j\omega)$. Therefore, $F(j\omega)$, represents the frequency spectrum of $f(t)$ and is called the *Frequency Spectral of a Function*. Equations (12.23) and (12.24) are referred to as the Fourier Transform pair. Equation (12.23) is the *Direct Fourier Transform* of $f(t)$, and (12.24) is the *Inverse Fourier Transform*.

CHAPTER 13

Fast Fourier Transform

The "Fast Fourier Transform," referred to as the "FFT," is a computational tool that facilitates signal analysis, such as Power Spectral Analysis, by means of digital computers. The FFT is a method for efficiently computing the Discrete Fourier Transform (DFT) of a digitized time series or discrete data samples. The FFT takes advantage of the symmetrical properties of periodic sinusoidal waveforms. Before development of the FFT, transformation of signals into the frequency domain was done by the standard trigonometric Fourier Series computational procedures.

13.1 CONTINUOUS FOURIER TRANSFORM

Before starting on the FFT, let us review or recall from Chapter 12 that the frequency spectrum of an analog signal $x(t)$ may be obtained by taking the Fourier Transform as shown in (13.1):

$$F(\omega) = \int_{-\infty}^{\infty} x(t)e^{-j\omega t}\,dt \tag{13.1}$$

where $x(t)$ and $F(\omega)$ are complex functions.

13.2 DISCRETE FOURIER TRANSFORM

Also recall that the sampled signal may be written either $x(nT)$ or $x(n)$,

$$x(t) \xrightarrow[\text{Interval, } T]{\text{Sampling}} x(nT) \text{ or } x(n)$$

where T-sampling period and n-number of samples, equally spaced with T as the sampling interval.

Then the Discrete Fourier Transform equation is given by (13.2):

$$x(\omega) = \sum_{n=-\infty}^{\infty} x(n)e^{-j\omega n} \tag{13.2}$$

where ω ranges from 0 to 2π and $x(\omega)$ is periodic with 2π

13.3 DEFINITION OF SAMPLING RATE (OR SAMPLING FREQUENCY)

Let us review highlights of Chapter 6 on Sampling Theory before discussing the Fast Fourier Transform (FFT). The basis of the Nyquist sampling theorem is the premise that in order to successfully represent and/or recover an original analog signal, the analog signal must be sampled at least twice the highest frequency in the signal. If the highest frequency in the analog signal is known, then it can be sampled at twice the highest frequency, $f_{n\cdot I}n$ terms of time, the spacing between samples is called time period or sampling interval, T. Thus,

$$T \leq 1(/2f_n) \quad \text{or} \quad f_{\text{sampling}} \geq 2\,f_n = 1/T$$

where T is the sampling period (or sampling interval), and f_{sampling} is the sampling frequency or sampling rate.

It should be noted that the maximum frequency, $f_{n,}$ is also called the *Cutoff Frequency*, the *Nyquist Frequency*, or the *Folding Frequency*.

If the highest frequency in the signal is not known, then the signal should be band-limited with an analog low-pass filter, and then the filtered signal should be sampled at least twice the upper frequency limit of the filter bandwidth. If the low-pass filtering procedure is not followed, then the problem of "Aliasing" arises. The Aliasing problem is usually caused when the spacing between samples is very large, which corresponds to sampling at a rate less than twice the highest frequency in the signal.

Aliasing is often defined as "the contamination of low frequencies by unwanted high frequency signals/noise" (Fig. 13.1), which by itself is an incorrect statement without

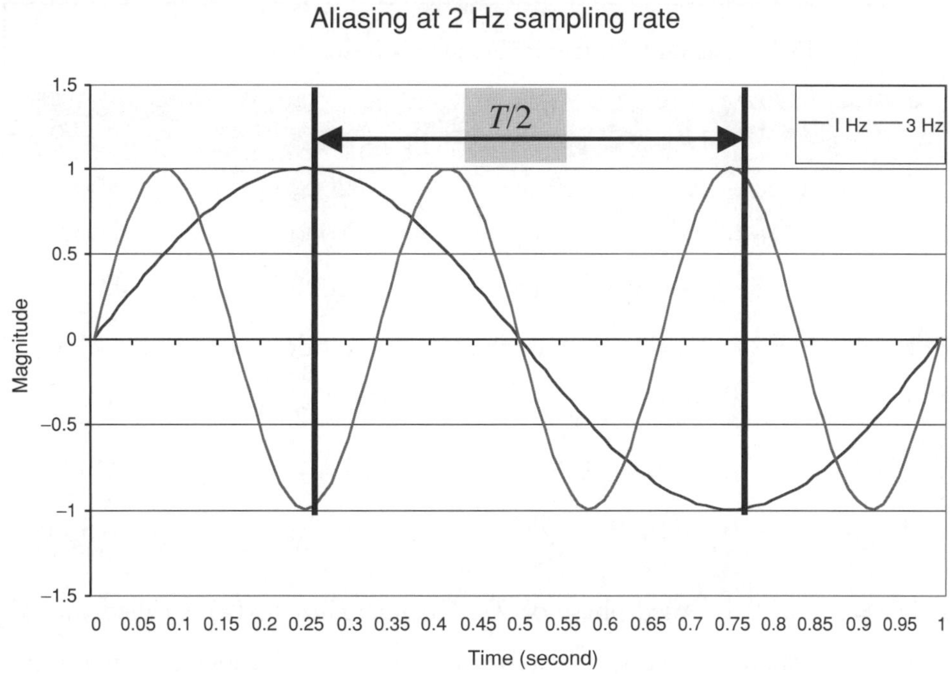

FIGURE 13.1: Aliasing. The example shows that a 3-Hz waveform has the same half period ($T/2$) as a 1-Hz waveform, if the signals are sampled at 2 samples per second

some clarification. Aliasing can be eliminated by proper sampling rate (meaning at least twice higher than the highest frequency in the signal) or by low-pass filtering before sampling the signal at a rate not less than twice the known low-pass filter cutoff frequency if the highest frequency in the signal is not known. What is implied by the statement is that if the period of the sampling window is equal to, or less than, one over twice the highest frequency in the signal, $T \leq 1/(2f_n)$, then in the frequency domain, the spectrum is repeated every $2\pi^* f_n$ Hz. In general, the spectrum repeats over every 2π range from $-\infty$ to $+\infty$. Thus, it would be sufficient to obtain values between 0 and 2π in a DFT.

13.3.1 Short-Hand Notation of the Discrete Fourier Transform (DFT)

The DFT is often written as in (13.3).

$$x(k) = \sum_{n=0}^{N-1} x(n) W_N^{nk} \quad \text{for} \quad k = 0, 1, \ldots, N-1 \tag{13.3}$$

TABLE 13.1: Spectral Window vs. Frequency Resolution

	$f_{\text{sampling rate}}$	N	ΔT
1.	1 sec – 2000 samples/sec	2000	0.0005
2.	1 sec – 1000 samples/sec	1000	0.001
3.	2 sec – 1000 samples/sec	2000	0.05

Since $w_n = 2\pi/N = 2\pi T$. The frequency resolutions are shown below.

1.	0.003141592	$f_{\text{res}} = \dfrac{\omega_0}{2\pi \text{T}}$	=1.0 Hz
2.	0.006283185	$f_{\text{res}} = \dfrac{\omega_n{}^* f_{\text{samp}}}{2\pi}$	=1.0 Hz
3.	0.00314592		=0.5 Hz

where $W_N = e^{-j2\pi/N}$ is called phase or *The Twiddle Factor*, $x(k)$ is called the N-point DFT, and the distance between successive samples in the frequency domain gives the fundamental frequency of $x(t)$.

In terms of normalized frequency units, the fundamental frequency is given by $2\pi/N$. Note in the example given in Table 13.1 that a one-second window results in a 1-Hz frequency resolution regardless of the sampling rate.

By definition, the Frequency Resolution (f_{res}) or the fundamental frequency of the spectral (f_0) is "The minimum frequency that a signal can be resolved in the frequency domain, and is equal to the inverse of the record length or the signal window length being analyzed (in seconds) in time domain" (13.4):

$$f_{\text{res}} = \frac{1}{R.L} = \frac{1}{N\text{T}} \tag{13.4}$$

Record length: The length of the sampled analog signal with a total of N samples is given as $R.L = NT$.

Expanding the DFT equation results in (13.5):

$$\begin{aligned} x(k) = \sum_{n=0N}^{N-1} &\{(\text{Re}[x(n)]\text{Re}[W_N^{nk}] - \text{Im}[x(n)]\text{Im}[W_N^{nk}]) \\ &+ (\text{Re}[x(n)]\text{Im}[W_N^{nk}] + \text{Im}[x(n)]\text{Re}[W_N^{nk}])\} \end{aligned} \tag{13.5}$$

where $k = 0, 1, \ldots, N-1$

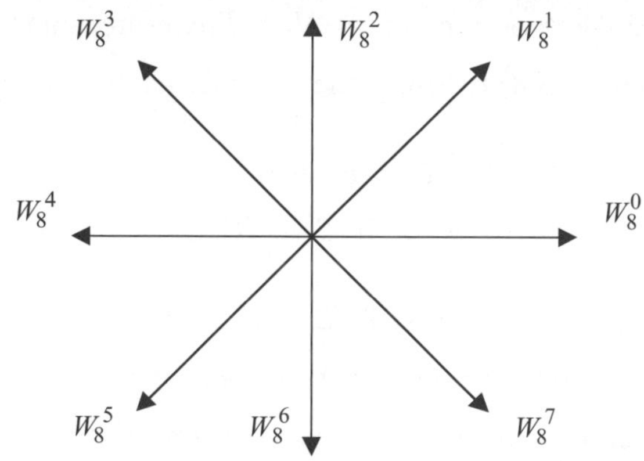

FIGURE 13.2: Symmetry and periodicity of W_N for $N = 8$

For each k, (13.5) needs $4N$ real multiplications and $(4N - 2)$ real additions. The computation of $x(k)$ needs $4N^2$ real multiplications and $N(4N - 2)$ real additions. In terms of complex operations, there are N^2 complex multiplication and $N(N - 1)$ complex additions. As N increases, it can be seen that the number of computations become tremendously large. Hence, there was a need to improve the efficiency of DFT computation and to exploit the symmetry and periodicity properties of the twiddle factor, W_N, as shown in (13.6) and Fig. 13.2.

$$\begin{aligned} \text{Symmetry} : W_N^k &= -W_N^{k+\frac{N}{2}} \\ \text{Periodicity} : W_N^k &= W_N^{N+k} \end{aligned} \tag{13.6}$$

13.4 COOLEY-TUKEY FFT (DECIMIMATION IN TIME)

The Cooley–Tukey FFT was one of the first FFTs developed. Suppose a time series having N samples is divided into two functions, of which each function has only half of the data points ($N/2$). One function consists of the even numbered data points ($x_0, x_2, x_4 \ldots,$ etc,), and the other function consists of the odd numbered data points ($x_1, x_3, x_5, \ldots,$ etc.). Those functions may be written as

$$Y_k = x_{2k}; \quad Z_k = x_{2k+l}, \quad for: \quad k = 0, 1, 2, \ldots, (N/2) - 1$$

Since Y_k and Z_k are sequences of $N/2$ points each, the two sequences have Discrete Fourier transforms, which are written in terms of the odd and even numbered data points. For values of r greater than $N/2$, the discrete Fourier transforms Br and Cr repeat periodically the values taken on when $r < N/2$. Therefore, substituting $r + N/2$ for r in (13.7), one can obtain (13.8) by using the Twiddle Factor, $W = \exp(-2\pi j/N)$.

$$A_r = B_r + W^r C_r \tag{13.7}$$

$$A_{r+\frac{N}{2}} = B_r - W^r C_r \tag{13.8}$$

From (13.7) and (13.8), the first $N/2$ and last $N/2$ points of the discrete Fourier transform of the data x_k can be simply obtained from the DFT of both sequences of $N/2$ samples. Assuming that one has a method that computes DFTs in a time proportional to the square of the number of samples, the algorithm can be used to compute the transforms of the odd and even data sequences using the Ar (13.7) and $Ar + N/2$ (13.8) equations to find Ar. The N operations require a time proportional to $2(N/2)^2$.

Since it has been shown that the computation of the DFT of N samples can be reduced to computing the DFTs of two sequences of $N/2$ samples, the computation of the sequence B_k or C_k can be reduced by decimation in time to the computation of sequences of $N/4$ samples. These reductions can be carried out as long as each function has a number of samples that are divisible by 2. Thus, if $N = 2^n$, n such reductions (often referred to as "Stages") can be made.

For example, the successive decimation (reduction) of a Discrete Fourier Transform will result in a Fast Fourier Transform with fewer computations. In general, $N \log_2 N$ complex additions and at most $(1/2)N \log_2 N$ complex multiplications are required for computation of the Fast Fourier Transform of an N point sequence, where N is a power of 2 ($N = 2^n$).

13.4.1 Derivation of FFT Algorithm

Consider the N-point DFT that can be reduced into two $N/2$ DFTs (even and odd indexed sequences), as $x(n)$ is decomposed into two $N/2$-point sequences (decimation-in-time) and is shown in (13.9).

$$\therefore\ x(k) = \sum_{n=0}^{\frac{N}{2}-1} x(2n) W_N^{2nk} + \sum_{n=0}^{\frac{N}{2}-1} x(2n+1) W_N^{(2n+1)k} \tag{13.9}$$

where: $W_N^2 = \left[e^{-j\frac{2\pi}{N}}\right]^2 = e^{-j\frac{2\pi}{(N/2)}} = W_{\frac{N}{2}}$

Then, (13.9) can then be rewritten as (13.10).

$$\therefore\ x(k) = \sum_{n=0}^{\frac{N}{2}-1} x(2n) W_{\frac{N}{2}}^{nk} + W_N^k \sum_{n=0}^{\frac{N}{2}-1} x(2n+1) W_{\frac{N}{2}}^{nk} \qquad (13.10)$$

where $x_1(k)$ is $= x_1(k) + W_N^k x_2(k)$, $k = 0, 1 \ldots, N$.

The sequences, $x_1(k)$ and $x_2(k)$ can each be further divided into even and odd sequences as shown in (13.11) and (13.12).

$$x_1(k) = \sum_{n=0}^{\frac{N}{2}-1} x(2n)\, W_{N/2}^{nk}, \quad \text{for} \quad k = 0, 1, \ldots, \frac{N}{2} - 1 \qquad (13.11)$$

$$= \sum_{n=0}^{\frac{N}{4}-1} x(4n) W_{N/4}^{nk} + W_{N/2}^{k} \sum_{n=0}^{\frac{N}{4}-1} x(4n+2)\, W_{N/4}^{nk}, \quad \text{for} \quad k = 0, 1, \ldots, \frac{N}{4} - 1 \qquad (13.12)$$

and for the $x_2(k)$ sequence as shown in (13.13) and (13.14):

$$x_2(k) = \sum_{n=0}^{\frac{N}{2}-1} x(2n+1)\, W_{N/2}^{nk}, \quad \text{for} \quad k = 0, 1, \ldots, \frac{N}{2} - 1 \qquad (13.13)$$

$$= \sum_{n=0}^{\frac{N}{4}-1} x(4n+1)\, W_{N/4}^{nk} + W_{N/2}^{k} \sum_{n=0}^{\frac{N}{4}-1} x(4n+3)\, W_{N/4}^{nk}, \quad \text{for} \quad k = 0, 1, \ldots, \frac{N}{4} - 1 \qquad (13.14)$$

Table 13.2 shows the reduction in computations between the DFT and the FFT.

TABLE 13.2: Comparison DFT and FFT Computation with $N = 8$

COMPLEX	DFT	FFT
Multiplications	$N^2 = 64$	$N \log_2 N = 24$
Additions	$N(N-1) = 56$	$N \log_2 1 = 24$
	For $N = 65\,536$	
Complex multiplications	4.3×10^9	1.049×10^6

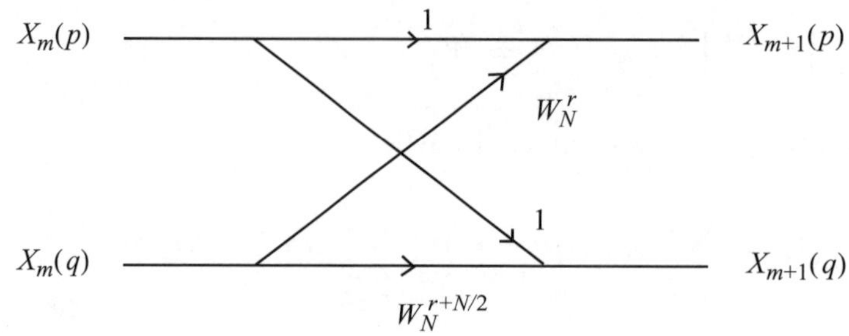

FIGURE 13.3: FFT "Butterfly" signal flow diagram

13.5 THE FFT "BUTTERFLY" SIGNAL FLOW DIAGRAM

To predict the next value in the stage (decimation), the previous stage values are used as inputs, where p may indicate the odd sequence and q the even sequence. The flow graph (Fig. 13.3) is called the "butterfly computation" because of its appearance.

The output of the stage is calculated with (13.15) and (13.16).

$$X_{m+1}(p) = X_m(p) + W_N^r X_m(q) \tag{13.15}$$

$$X_{m+1}(q) = X_m(p) + W_N^{r+\frac{N}{2}} X_m(q) \tag{13.16}$$

However, from symmetry and periodicity properties:

$$W_N^{\frac{N}{2}} = e^{-j\left(\frac{2\pi}{N}\right)\left(\frac{N}{2}\right)} = e^{-j\pi} = -1$$

the reduced Butterfly flow diagram becomes Fig. 13.4 and the equations are rewritten as

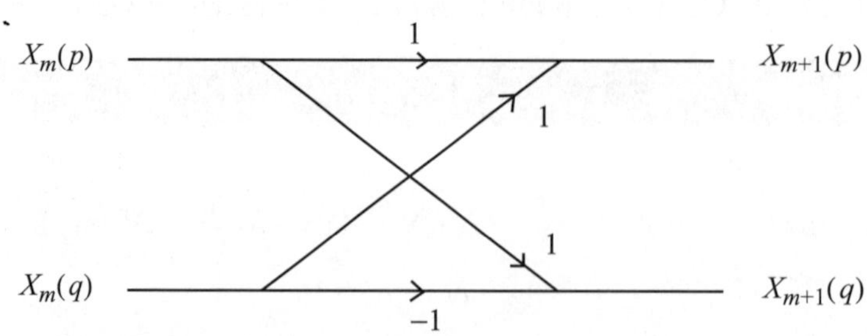

FIGURE 13.4: Reduced Butterfly flow diagram

(13.17) and (13.18).

$$X_{m+1}(p) = X_m(p) + W_N^r X_m(q) \tag{13.17}$$

$$X_{m+1}(q) = X_m(p) - W_N^r X_m(q) \tag{13.18}$$

Bit Reversal: To perform the computation as shown in the Fig. 13.4 flow graph, the input data must be stored in a nonsequential order and is known as the bit-reverse order. For a sequence of $N = 8$, the input data are represented in binary form as

LSB = 0

$X(0) = X(0, 0, 0)$
$X(4) = X(1, 0, 0)$
$X(2) = X(0, 1, 0)$
$X(6) = X(1, 1, 0)$

LSB = 1

$X(1) = X(0, 0, 1)$
$X(5) = X(1, 0, 1)$
$X(3) = X(0, 1, 1)$
$X(7) = X(1, 1, 1)$

Upon observation, one can see that the bits (LSB = 0) of the top section are all even number samples; and LSB = 1 of bottom section are all odd number samples, which are obtained by writing the signal sequences in binary form and bit-reversing each to get the new order. If the input is not bit-reversed, then the output will need to be bit-reversed. The flow graph can be arranged in such a way that both input and output need not be bit-reversed.

13.6 DECIMATION-IN-FREQUENCY

The configuration is obtained by dividing the output sequence into smaller and smaller subsequences. Here the input sequence is divided into first or second halves. According to Fourier integral theorem, a function must satisfy the Dirichlet conditions or every finite interval and the integral from infinity to negative infinity is finite. The conditions are

1) the function should have a finite number of maxima or minima,
2) the function should have a finite number of discontinuities, and
3) the function should have a finite value or $\int_{t}^{t-T} f(t)\,dt$ is less than infinity.

Conditions for the Fourier Series include the following:

1) periodic functions (practical is nonperiodic);
2) can only be applied to stable systems (systems where natural response decays in time, for example, convergent).

Example Problem: Consider the sinusoidal waveform of Fig. 13.5.

According to Nyquist sampling theorem, the waveform in Fig. 13.5 should be sampled at more than two samples per second to successfully reconstruct the signal. For the sake of convenience, consider eight samples per second of the above waveform. Therefore: $N = 8 = 2^n = 2^3$, N is the total number of samples, and $n = 3$ stages; $T = 125$ milliseconds, sampling interval; Record length $= 1$ second (time of recording data $=$ N $\times$ T) Frequency resolution $= 1$ Hz $(1/(N \times T))$

The discrete signal is given by (13.19).

$$X(nT) = \cos\left(2\pi fnT\right) = \cos\left(\frac{2\pi fn}{N}\right) \tag{13.19}$$

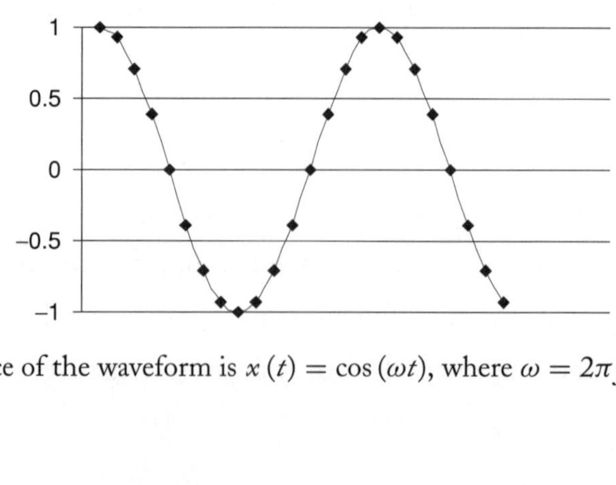

FIGURE 13.5: Trace of the waveform is $x(t) = \cos(\omega t)$, where $\omega = 2\pi f$ and $f = 1$ Hz

Symmetric property of twiddle factor implies:

$$W_N^k = W_N^{k+\frac{N}{2}} = W_8^k = -W_8^{k+4}, \text{ which gives}$$

$$W_8^0 = -W_8^4 \qquad W_8^2 = -W_8^6$$

$$W_8^1 = -W_8^5 \qquad W_8^3 = -W_8^7$$

13.6.1 Butterfly First-Stage Calculations

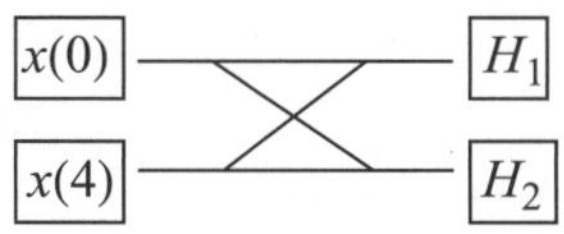

$$H_1 = x(0) + W_8^0 x(4) = (1 + j \times 0) + (1 + j \times 0)(-1 + j \times 0) = 0 + j \times 0$$

$$H_2 = x(0) - W_8^0 x(4) = (1 + j \times 0) - (1 + j \times 0)(-1 + j \times 0) = 2 + j \times 0$$

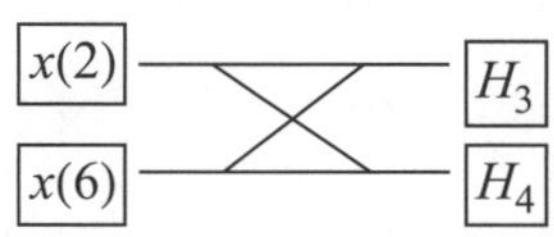

$$H_3 = x(2) + W_8^0 x(6) = (0 + j \times 0) + (1 + j \times 0)(0 + j \times 0) = 0 + j \times 0$$

$$H_4 = x(2) - W_8^0 x(6) = (0 + j \times 0) - (1 + j \times 0)(0 + j \times 0) = 0 + j \times 0$$

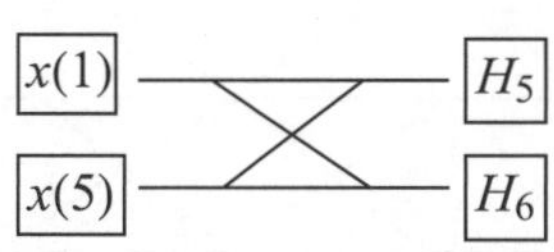

$$H_5 = x(1) + W_8^0 x(5) = (0.707 + j \times 0) + (1 + j \times 0)(-0.707 + j \times 0) = 0 + j \times 0$$

$$H_6 = x(1) - W_8^0 x(5) = (0.707 + j \times 0) - (1 + j \times 0)(-0.707 + j \times 0)$$
$$= 1.414 + j \times 0$$

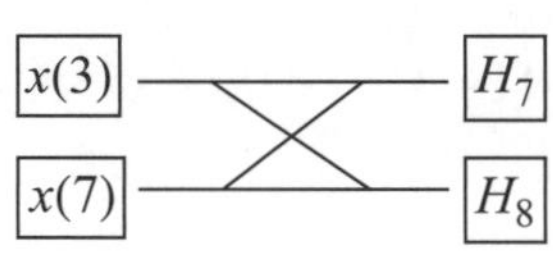

$$H_7 = x(3) + W_8^0 x(7) = (-0.707 + j \times 0) + (1 + j \times 0)(0.707 + j \times 0) = 0 + j \times 0$$

$$H_8 = x(3) - W_8^0 x(7) = (-0.707 + j \times 0) - (1 + j \times 0)(0.707 + j \times 0)$$
$$= -1.414 + j \times 0$$

It should be noted that only $W_8^0 \rightarrow W_8^0$ needs to be calculated for the first stage.

13.6.2 Second-Stage Calculations

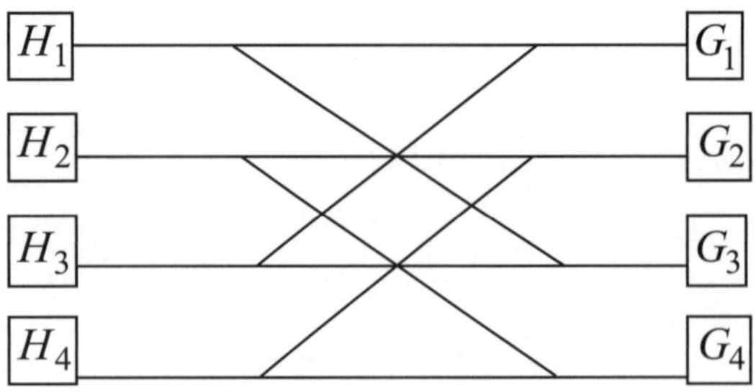

$$G_1 = H_1 + W_8^0 H_3 = (0 + j \times 0) + (1 + j \times 0)(0 + j \times 0) = 0 + j \times 0$$

$$G_2 = H_2 + W_8^0 H_4 = (2 + j \times 0) + (1 + j \times 0)(0 + j \times 0) = 2 + j \times 0$$

$$G_3 = H_1 - W_8^0 H_3 = (0 + j \times 0) - (1 + j \times 0)(0 + j \times 0) = 0 + j \times 0$$

$$G_4 = H_2 - W_8^0 H_4 = (2 + j \times 0) - (1 + j \times 0)(0 + j \times 0) = 2 + j \times 0$$

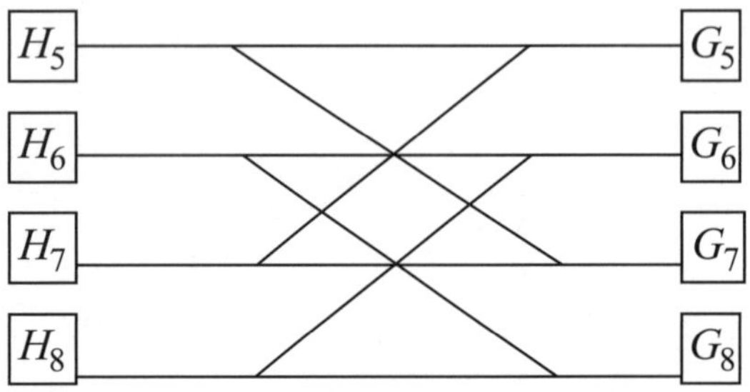

$$G_5 = H_5 + W_8^0 H_7 = (0 + j \times 0) + (1 + j \times 0)(0 + j \times 0) = 0 + j \times 0$$

$$G_6 = H_6 + W_8^0 H_8 = (1.414 + j \times 0) + (1 + j \times 0)(-1.414 + j \times 0)$$
$$= 1.414 - j \times 1.414$$

$$G_7 = H_5 - W_8^0 H_7 = (0 + j \times 0) - (1 + j \times 0)(0 + j \times 0) = 0 + j \times 0$$

$$G_8 = H_6 - W_8^0 H_8 = (1.414 + j \times 0) - (1 + j \times 0)(-1.414 + j \times 0)$$
$$= 1.414 + j \times 1.414$$

13.6.3 Third-Stage Calculations

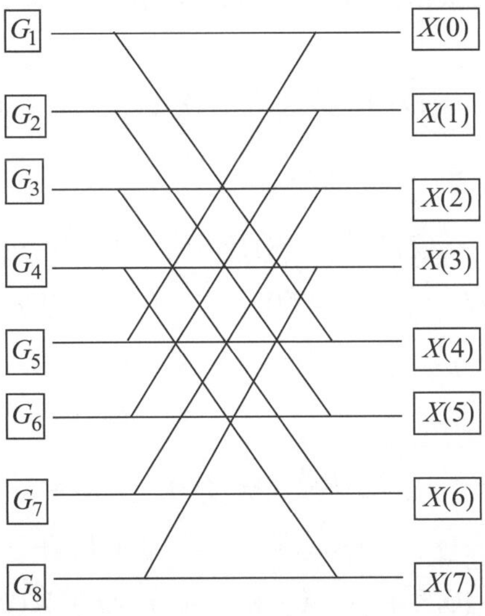

$$X(0) = G_1 + W_8^0 G_5 = (0 + j \times 0) + (1 + j \times 0)(0 + j \times 0) = 0 + j \times 0$$

$$X(1) = G_2 + W_8^1 G_6 = (2 + j \times 0) + (0.707 + j \times 0.707)(1.414 - j \times 1.414)$$
$$= 4 + j \times 0$$

$$X(2) = G_3 + W_8^2 G_7 = (0 + j \times 0) + (0 + j \times 1)(0 + j \times 0) = 0 + j \times 0$$

$$X(3) = G_4 + W_8^3 G_8 = (2 + j \times 0) + (-0.707 + j \times 0.707)(1.414 + j \times 1.414)$$
$$= 0 + j \times 0$$

$$X(4) = G_1 - W_8^0 G_5 = (0 + j \times 0) - (1 + j \times 0)(0 + j \times 0) = 0 + j \times 0$$

$$X(5) = G_2 - W_8^1 G_6 = (2 + j \times 0) - (0.707 + j \times 0.707)(1.414 - j \times 1.414)$$
$$= 0 + j \times 0$$

$$X(6) = G_3 - W_8^2 G_7 = (0 + j \times 0) - (0 + j \times 1)(0 + j \times 0) = 0 + j \times 0$$

$$X(7) = G_4 - W_8^3 G_8 = (2 + j \times 0) - (-0.707 + j \times 0.707)(1.414 + j \times 1.414)$$
$$= 4 + j \times 0$$

Each of the eight samples for $x(nt)$ is given by

$$x(0) = 1 + j \times 0 \qquad x(4) = -1 + j \times 0$$
$$x(1) = 0.707 + j \times 0 \qquad x(5) = -0.707 + j \times 0$$
$$x(2) = 0 + j \times 0 \qquad x(6) = 0 + j \times 0$$
$$x(3) = -0.707 + j \times 0 \qquad x(7) = 0.707 + j \times 0$$

13.6.4 The DFT by Directed Method

In comparison, the direct calculation would be as follows:

$$x(k) = \sum_{n=0}^{N-1} x(n) \times (W_N)^{nk} \qquad k = 0, 1, \ldots N-1 \qquad W_N = e^{-\frac{j2\pi}{N}}$$

$$\begin{aligned} X(0) &= x(0)W^0 + x(1)W^0 + x(2)W^0 + x(3)W^0 + x(4)W^0 + x(5)W^0 + x(6)W^0 + x(7)W^0 \\ &= (1 + j \times 0) + (0.707 + j \times 0) + (0 + j \times 0) + (-0.707 + j \times 0) + (-1 + j \times 0) \\ &\quad + (-0.707 + j \times 0) + (0 + j \times 0) + (0.707 + j \times 0) = 0 + j \times 0 \end{aligned}$$

$$\begin{aligned} X(1) = {} & x(0)W^0 + x(1)W^1 + x(2)W^2 + x(3)W^3 + x(4)W^4 + x(5)W^5 + x(6)W^6 \\ & + x(7)W^7 = (1 + j \times 0) + (.5 + j \times 0) + (0 + j \times 0) + (0.5 + j \times 0) \\ & + (1 + j \times 0) + (0.5 + j \times 0) + (0 + j \times 0) + (0.5 + j \times 0) = 4 + j \times 0 \end{aligned}$$

$$\begin{aligned} X(2) = {} & x(0)W^0 + x(1)W^2 + x(2)W^4 + x(3)W^6 + x(4)W^8 + x(5)W^{10} + x(6)W^{12} \\ & + x(7)W^{14} = (1 + j \times 0) + (0 + j \times 0) + (0 + j \times 0) + (0 + j \times 0) \\ & + (-1 + j \times 0) + (0 + j \times 0) + (0 + j \times 0) + (0 + j \times 0) = 0 + j \times 0 \end{aligned}$$

$$\begin{aligned} X(3) = {} & x(0)W^0 + x(1)W^3 + x(2)W^6 + x(3)W^9 + x(4)W^{12} + x(5)W^{15} + x(6)W^{18} \\ & + x(7)W^{21} = (1 + j \times 0) + (-0.5 + j \times 0) + (0 + j \times 0) + (-0.5 + j \times 0) \\ & + (1 + j \times 0) + (-0.5 + j \times 0) + (0 + j \times 0) + (-0.5 + j \times 0) = 0 + j \times 0 \end{aligned}$$

$$\begin{aligned} X(4) = {} & x(0)W^0 + x(1)W^4 + x(2)W^8 + x(3)W^{12} + x(4)W^{16} + x(5)W^{20} + x(6)W^{24} \\ & + x(7)W^{28} = (1 + j \times 0) + (-0.707 + j \times 0) + (0 + j \times 0) + (0.707 + j \times 0) \\ & + (-1 + j \times 0) + (0.707 + j \times 0) + (0 + j \times 0) + (-0.707 + j \times 0) = 0 \\ & + j \times 0 \end{aligned}$$

$$\begin{aligned} X(5) = {} & x(0)W^0 + x(1)W^5 + x(2)W^{10} + x(3)W^{15} + x(4)W^{20} + x(5)W^{25} + x(6)W^{30} \\ & + x(7)W^{35} = (1 + j \times 0) + (-0.5 + j \times 0) + (0 + j \times 0) + (-0.5 + j \times 0) \\ & + (1 + j \times 0) + (-0.5 + j \times 0) + (0 + j \times 0) + (-0.5 + j \times 0) = 0 + j \times 0 \end{aligned}$$

$$X(6) = x(0)W^0 + x(1)W^6 + x(2)W^{12} + x(3)W^{18} + x(4)W^{24} + x(5)W^{30} + x(6)W^{36}$$
$$+ x(7)W^{42} = (1 + j \times 0) + (0 + j \times 0) + (0 + j \times 0) + (0 + j \times 0)$$
$$+ (-1 + j \times 0) + (0 + j \times 0) + (0 + j \times 0) + (0 + j \times 0) = 0 + j \times 0$$

$$X(7) = x(0)W^0 + x(1)W^7 + x(2)W^{14} + x(3)W^{21} + x(4)W^{28} + x(5)W^{35} + x(6)W^{42}$$
$$+ x(7)W^{49} = (1 + j \times 0) + (0.5 + j \times 0) + (0 + j \times 0) + (0.5 + j \times 0)$$
$$+ (1 + j \times 0) + (0.5 + j \times 0) + (0 + j \times 0) + (0.5 + j \times 0) = 0 + j \times 0$$

13.6.5 The DFT by FFT Method

$$X(k) = \sum_{n=0}^{N-1} x(n)\, W_N^{nk} = \sum_{n=0}^{7} x(n)\, W_8^{nk}; \qquad W_8 = e^{\frac{-j2\pi}{8}}; \quad k = 0, 1, \ldots 7$$
$$= \sum_{n=0}^{3} x(2n)\, W_4^{nk} + W_8^k \sum_{n=0}^{3} x(2n+1)\, W_4^{nk}; \qquad W_4 = e^{\frac{-j2\pi}{4}}; \quad k = 0, 1, \ldots 3$$
$$= \sum_{n=0}^{1} x(4n)\, W_2^{nk} + W_4^k \sum_{n=0}^{1} x(4n+2)\, W_2^{nk}$$
$$+ W_8^k \left[\sum_{n=0}^{1} x(4n+1)\, W_2^{nk} + W_4^k \sum_{n=0}^{1} x(4n+3)\, W_2^{nk} \right]; W_2 = e^{\frac{-j2\pi}{2}}; \quad k = 0, 1$$

Periodicity property of Twiddle factor used in the FFT implies:

$$W_N^K = W_N^{N+K} \quad \text{or} \quad W_8^K = W_N^{8+K}$$

Therefore, one needs to calculate the twiddle factor for $W_8{}^0 \rightarrow W_8^7$.

13.7 SUMMARY

In summary, the usefulness of the FFT is the same as the DFT in Power Spectral Analysis or Filter Simulation on digital computers; however, the FFT is fast and requires fewer computations than the DFT. Decimation-in-time is the operation of separating the input data series, $x(n)$ into two $N/2$ length sequences of even-numbered points and of odd-numbered points, which can be done as long as the length is an even number, i.e., 2 to any power. Results of the decimation in the FFT are better shown with the "Butterfly"

Index	Real (a_n)	Imaginary(jb_n)	
0	2	0	DC offset
1	0.5	0.95	Fundamental (f_0)
2	3	2.1	2nd Harmonic
3	1.1	−6.2	$3 \times f_0$
4	0.25	−6.7	Center fold
5	1.1	6.2	Conjugate $3 \times f_0$
6	3	−2.1	Conjugate $2 \times f_0$
7	0.5	−0.95	Conjugate f_0

FIGURE 13.6: Output values for an eight-sample FFT

signal flow diagram with the number of complex computation Stages as, $V = \text{Log}2N$, since $N = 2V$. In general, computation of a subsequent stage from the previous butterfly is given by the basic butterfly equations, which reduce the number of complex operations from N^2 for the DFT to $N\text{Log}2N$ for the FFT; where $N = 2^m$, and m is a positive integer. The output of the FFT is shown in Fig. 13.6. Note that the first index output is the DC value of the signal. Note that the spectrum folds about the center (index 4) and the output values for indices, 5 through 7, are the conjugate of the output values for indices, 1 through 3.

CHAPTER 14

Truncation of the Infinite Fourier Transform

Problems in the accuracy of the Fourier Transform occur when adherence to the "Dirichlet Conditions" have not been met. Recall the Dirichlet condition requirements on the signal to be transformed as

1. a finite number of discontinuities,
2. a finite number of maxima and minima, and
3. that the signal be absolutely convergent (14.1)

$$\int_{o}^{T} |f(t)|\, dt < \infty \tag{14.1}$$

It is impractical to solve an infinite number of transform coefficients; therefore, the engineer must decide how many Fourier coefficients are needed to represent the waveform accurately. How to make a decision on where to truncate the series is a very important part of signal processing.

Truncation means that all terms after the nth term are dropped, resulting in n finite number of terms to represent the signal. However, not including all the coefficients results in an error referred to as the "Truncation error." The truncation error, ε_n, is defined as the difference between the original function $f(t)$ and the partial sum, $s_n(t)$ of the inverse transformed truncated Fourier terms.

$$\varepsilon_n = f(t) - s_n(t)$$

where ε_n is the truncation error. The partial sum is often expressed as the "mean-squared error" as in (14.2).

$$E_n = \frac{1}{T}\int_0^T [\varepsilon_n(t)]^2 dt \tag{14.2}$$

To understand truncation of a series, it is important to know what is desirable in truncation. Since accuracy is one of the most desirable specifications in any application, it would be desirable to attain acceptable accuracy with the least number of terms. Without proof in limits and convergence, let us define convergence as the rate at which the truncated series approaches the value of the original function, $f(t)$. In general, the more rapid the convergence, the fewer terms are required to obtain desired accuracy.

The convergence rate of the Fourier Transform is directly related to the rate of the decrease in the magnitude of the Fourier Coefficients. Recall the Fourier transform coefficients of a square wave as having the following magnitudes:

$$|a_3| = \frac{1}{3};\quad |a_5| = \frac{1}{5};\quad |a_7| = \frac{1}{7};\quad |a_9| = \frac{1}{9};\ \cdots;\quad |a_n| = \frac{1}{n}$$

Note that the sequence is the reciprocal of the nth term to the 1st power.

Now let us examine the Fourier transform of the triangular waveform shown in Fig. 14.1.

The Fourier trigonometric series representation is given by (14.3).

$$v(t) = \frac{8V}{\pi^2}\left(\sin\omega_0 t - \frac{1}{3^2}\sin 3\omega_0 t + \frac{1}{5^2}\sin 5\omega_0 t + \cdots + \frac{1}{n^2}\sin n\omega_0 t\right) \tag{14.3}$$

Note that the magnitude of the coefficients are the square of the nth term.

$$|a_3| = \frac{1}{3^2};\quad |a_5| = \frac{1}{5^2};\quad |a_7| = \frac{1}{7^2};\quad |a_9| = \frac{1}{9^2};\cdots;\quad |a_n| = \frac{1}{n^2}$$

It is noted that the Fourier coefficients for a triangular waveform diminish faster (rate of convergence) than the Fourier coefficients for a rectangular waveform. To help in

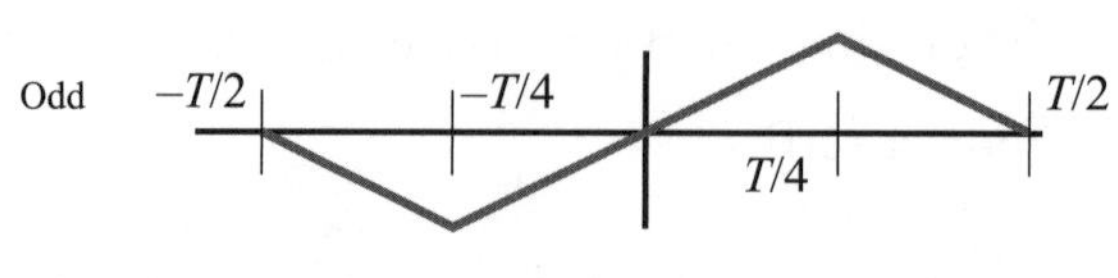

FIGURE 14.1: One cycle of a triangular waveform with odd symmetry

TABLE 14.1: Convergence Table

$f(t)$	JUMP-IN	IMPULSE-IN	DECREASING a_n	WAVEFORM
Square wave	$f(t)$	$f'(t)$	$1/n$	Discontinuities
Triangle wave	$f'(t)$	$f''(t)$	$1/n^2$	
Parabolic/ sinusoid	$f''(t)$	$f'''(t)$	$1/n^3$	
...				
– – –	$f^{k-1}(t)$	$f^k(t)$	$1/n^k$	Smoother

determining the rate of convergence, a basic expression was formulated called the Law of Convergence.

The law covering the manner in which the Fourier coefficients diminish with increasing "n" is expressed in the number of times a function must be differentiated to produce a jump discontinuity. For the kth derivative, the convergence of the coefficients will be on the order of $1/n^{k+1}$. For example, if that derivative is the kth, then the coefficients will converge at the rate shown in (13.4).

$$|a_n| \leq \frac{M}{n^{k+1}} \quad \text{and} \quad |b_n| \leq \frac{M}{n^{k+1}} \tag{13.4}$$

where M is a constant, which is dependent on $f(t)$.

There are two ways to show convergence. Let us look at Table 14.1, the Convergence Table. The left most column of the table denotes the type of waveform with the right most column describing the waveform as going from with discontinuities to smoother. The second and third columns labeled "Jump-in" and "Impulse-in," respectively. *Jump-in* means that discontinuity occurs in the function after successive differentiation, where the discontinuities occur in the *square-like waveform*. *Impulse-in* means what derivative (successive differentiation) of the function will result in an *impulse train*. The fourth column gives the general expression for convergence of the coefficients. From the table one may conclude that the smoother the function, the faster the convergence.

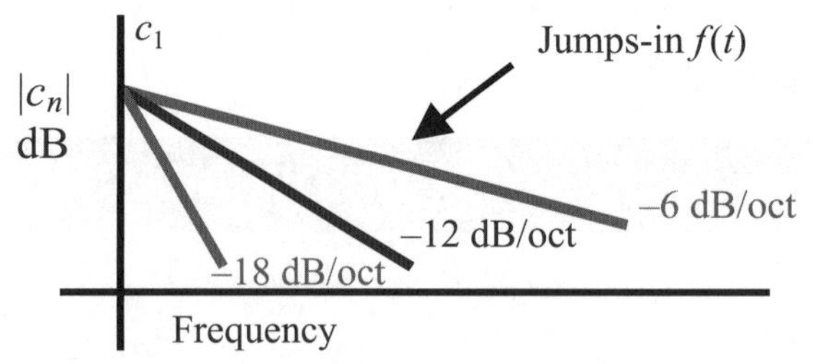

FIGURE 14.2: Graphs of convergence rates

Often, the rates of decrease in the Fourier coefficients with the number of terms, n, are expressed in the logarithmic terms of decibels per octave by expressing the absolute magnitude of the Fourier coefficients, $|c_n|$ in decibels and plotting the coefficients against frequency as shown in Fig. 14.2.

14.1 PRACTICAL APPLICATIONS

So how realistic are discontinuities in the real world, specifically physiological signals in clinical and biomedical engineering? Unfortunately, discontinuities in medicine are real. Discontinuities in the human and animal physiological waveforms may be found in the following signals:

1. Electroencephalograms (EEG); Seizure waveforms
2. Spike waveforms in EEG and Electromyogram (EMG)
3. Saccadic movements in the Electro-occulograms (EOG)

It is important to understand what problems discontinuities present in signal processing. It is well known that as more coefficients are added to the waveform approximation, the accuracy improves everywhere except in the immediate vicinity of a discontinuity. Even in the "Limit" as $n \to \infty$, the discrepancy (error) at the point of discontinuity becomes approximately 9%. This error is known as the *Gibbs phenomenon.*

One may think of the Gibbs phenomenon as the response to a square pulse, which is the characteristic overshoot and damped oscillatory decay in nonlinear systems (filters). What the discontinuities and the Gibbs phenomenon indicate is that one can expect about a 9% error at each region of discontinuity in a signal. It should be noted that discontinuities may occur at the beginning and end points of the analysis window.

Can one correct for the discontinuities and the Gibbs phenomenon? There are a few approaches that may be used. For example, one could use the low-pass filter the signal to remove the high-frequency components, thus smoothing the signal, for example, rounding off the QRS in the electrocardiogram (ECG) signal. Or, one could avoid any analysis in the region of discontinuity by "Piece-meal" or segment analysis.

14.2 SUMMARY

Keep in mind that all Fourier analysis are approximations due to truncation in the number of Fourier coefficients; therefore, from "lessons learned," one should do the best approximation possible, knowing that 100% accuracy is not attainable. The smoother the signal, the faster it will converge, thus requiring fewer Fourier coefficient terms. Discontinuities either at the edges or in the middle of the signal will result in greater discrepancy (error). Remember that the Gibbs phenomenon indicates that approximately a 9% error will occur in the region of any discontinuity.

CHAPTER 15

Spectral Analysis

15.1 INTRODUCTION

Spectral analysis is the process by which we estimate the energy content of a time-varying function (or signal) as a function of frequency. For signals with very little frequency content, the spectral density function can be represented by a "line spectrum." The variations in the frequency resolution of spectra are due to different window lengths. As the signal becomes more complex (in terms of frequency content), the power spectrum resembles a more continuous function, hence, spectral density functions. The sharp peaks that correspond to the raw estimates of the energy for the main frequency components will be present; however, a certain amount of signal conditioning (i.e., smoothing) is necessary prior to this final estimation of the "Smoothed Power Spectral Estimate." Notice that at times, the text will refer to the Spectral Density Function as the "Power Spectrum" (singular) or "Power Spectra" (plural). Units for the power spectra are in power per frequency band (spectral resolution, $\Delta f = 1/T$).

The harmonic analysis of a random process will yield some frequency information, whereas the power spectrum is a useful tool in analysis of a random process. So let us examine what information can be obtained from the power spectrum. The power spectrum has four major uses. The first is to get an overview of the function's frequency distribution. Second is for comparison of distribution via statistical testing or by discriminate analysis. Third, the power spectrum can be used to obtain estimates the parameters of interest. And last, but probably the most important application is that power spectra can be used to show hidden periodicities. In summary, spectral estimates are used primarily to

1. show hidden *Periodicities* in frequency domain;
2. obtain descriptive statistics, for example, distribution, mean, mode, bandwidth of the spectral from a random signal;
3. get an overview of the frequencies in a function, $F(\omega)$;
4. in obtaining parameter estimation and/or feature extraction, that is, clinical EEG bands; and
5. in classification testing and discrimination analysis.

There are several methods for calculating the power spectra. The most commonly used spectral estimation methods are calculated via

1. the autocorrelation function,
2. the direct or fast fourier transform,
3. the autoregression, which is briefly included in this chapter, but this method of spectral estimation will not be covered in the course, since an autoregression course is offered by the statistics department.

It should be noted, however, that there are other methods. The power spectrum (periodogram), most commonly used by engineers, is an estimator of the "raw power spectrum" and must undergo smoothing to clearly reveal informative features (preferred by statisticians). Autoregressive spectral estimation is a powerful tool that estimates the power spectrum without employing any smoothing techniques.

Before commencing with the spectral analysis of a signal, the signal has to be qualified as deterministic (periodic) or random. The random data must be tested for normality of distribution, independence, and stationarity. If any of these statistical properties are not met, the data is a nonstationary process. If the data is nonstationary, appropriate steps should be followed to make the data independent and stationary before proceeding with spectral analysis; however, nonstationary processes are beyond the scope of this book.

15.2 SPECTRAL DENSITY ESTIMATION

A basic question one might ask is whether the Fourier Series or transform is applied to a voltage signal, how did the unit become power? To answer this question, let us consider the function $v(t)$ defined over the interval $(-\infty, +\infty)$. If $v(t)$ is periodic or almost periodic, it can be represented by the Fourier series as in (15.1).

$$v(t) = a_0 + \sum_{1}^{\infty} (a_n \cos nwt + b_n \sin nwt) \tag{15.1}$$

where $\omega = 2\pi f$, f represents the frequency, and the period is $T = 1/f$. The complex Fourier coefficients a and b are then represented by equations in (15.2).

$$\begin{aligned} a_0 &= \frac{1}{2T}\int_0^T v(t)\,dt \\ a_n &= \frac{1}{2T}\int_0^T v(t)\,\cos(nw_0 t)\,dt \\ b_n &= \frac{1}{2T}\int_0^T v(t)\,\sin(nw_0 t)\,dt \end{aligned} \tag{15.2}$$

The magnitude of the complex coefficients are obtained from (15.3), which has units of "Voltage/Hertz."

$$c_n = \sqrt{(a_n + b_n)(a_n - b_n)} \tag{15.3}$$

The Fourier coefficients are not in units of energy or power. By using Parseval's Theorem, the power spectrum is obtained from (15.4).

$$c_n^2 = a_n^2 + b_n^2 \tag{15.4}$$

The squared magnitude of the complex coefficients from (15.4) has units of "Voltage Squared/Hertz," "Average energy or power per hertz."

Parseval's Theorem may be stated as the following equation form (15.5).

$$\int_{-\infty}^{+\infty} |v(t)|^2\,dt = \frac{1}{2\pi}\int_{-\infty}^{+\infty} |X(\omega)|^2\,d\omega \tag{15.5}$$

where $v(t)$ and $X(\omega)$ are the Fourier Transform pair.

The left-hand side of (15.5), is defined as the total energy in the signal $v(t)$.

A periodic signal contains real and imaginary components of power. However, determination of even symmetry (real power only) and odd symmetry (imaginary power only) simplifies calculations because b_n and a_n cancel for each case, respectively. Fourier Series Analysis is a very powerful tool, but it is not applicable to random signals.

15.2.1 Power Spectra Estimation Via the Autocorrelation Function

The classic method used to efficiently obtain the estimated Power Spectra from random processes before the advent of the Fast Fourier Transform was by taking the finite Fourier transform of the correlation function. The process required one to calculate the autocorrelation function (or cross-correlation) first, and then to take the Fourier Transform of the resulting correlation function.

The autocorrelation function (R_{xx}) and cross-correlation function (R_{xy}) will not be discussed in this section since the material was presented in Chapter 9. Hence, let us assume that the correlation function exists and that its absolute value is finite. The Fourier Transform (FT) is defined as shown in (15.6) for the autocorrelation and (15.7) for the cross-correlation:

$$S_{xx}(\omega) = \int_{-\infty}^{\infty} R_{xx}(\tau)e^{-j\omega\tau}\,d\tau \tag{15.6}$$

$$S_{xy}(\omega) = \int_{-\infty}^{\infty} R_{xy}(\tau)e^{-j\omega\tau}\,d\tau \tag{15.7}$$

where $S_{xx}(\omega)$ is the autospectral density function for an input function $x(t)$, and $S_{xy}(\omega)$ is the cross-spectral density function for the input $x(t)$ and output $y(t)$.

The cross-spectral density function is estimated by taking the cross-correlation of two different functions in the time domain, that is, the input signal to a system and the output signal from the same system, and then taking the Fourier transform of the resulting cross-correlation function, S_{xy}. The Inverse Fourier Transforms are then

$$R_{xx}(\tau) = \frac{1}{2\pi}\int_{-\infty}^{\infty} S_{xx}(\omega)e^{j\omega t}\,d\omega \tag{15.8}$$

$$R_{xy}(\tau) = \frac{1}{2\pi} \int_{-\infty}^{\infty} S_{xy}(\omega) e^{j\omega t} d\omega \tag{15.9}$$

Equations (15.6) through (15.9) are called the "Wiener–Kinchine Relation."

The Fourier transform of the correlation functions contain real and imaginary components of energy, meaning that the spectrum is a two-sided spectrum with half of the energy containing negative frequencies, that is, $-\omega$; hence, the following equivalent equations are defined in (15.10) and (15.11).

$$S_{xx}(-\omega) = \overline{S}_{xx}(\omega) = S_{xx}(\omega) \tag{15.10}$$

$$S_{xy}(-\omega) = \overline{S}_{xy}(\omega) = S_{xy}(\omega) \tag{15.11}$$

where the bar over the letter S denotes the complex conjugate.

The $S(-\omega)$ and $S(\omega)$ are the estimate of the "two-sided spectral density function," in which the magnitude of the FT for a frequency ω in the real plane is equal to the magnitude of the FT of ω in the imaginary plane. In the case when the autocorrelation function is evaluated over the interval from 0 to ∞, the autospectra may be obtained by using the symmetry properties of the function. An alternate equation is given in (15.12), with its inverse transform as (15.13).

$$S_{xx}(\omega) = 2 \int_{0}^{\infty} R_{xx}(\tau) \cos(\omega\tau) d\tau \tag{15.12}$$

$$R_{xx}(\tau) = \frac{1}{\pi} \int_{0}^{\infty} S_{xx}(\omega) \cos(\omega\tau) d\omega \tag{15.13}$$

For the case of the "one-sided spectral density function" (Fig. 15.1), in which only the real frequency components of the power spectrum are represented, (15.14) and (15.15) are the same as doubling the energy from one side of the two-sided spectra.

$$G_{xx}(\omega) = 2S_{xx}(\omega) \tag{15.14}$$

$$G_{xy}(\omega) = 2S_{xy}(\omega) \tag{15.15}$$

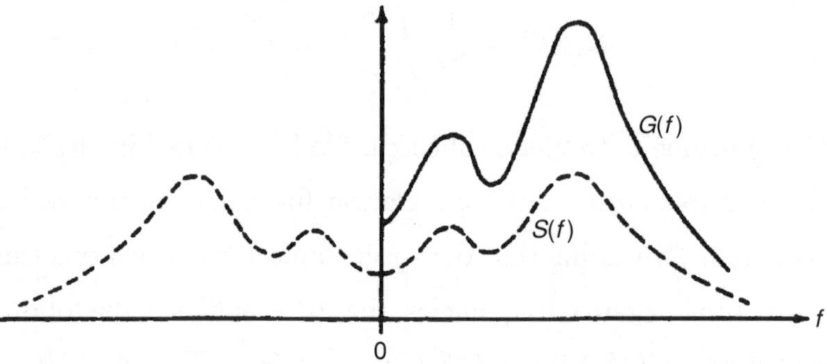

FIGURE 15.1: The two-sided and one-sided spectral density functions

15.2.2 Power Spectra Estimation Directly from the Fourier Transform

With computers and the Fourier transform, it is easier to obtain the spectral estimate directly from the either discrete Fourier transform (DFT) or the fast Fourier transform (FFT). The FFT has become the most popular and widely used method for spectral estimation. The first step in the process requires one to obtain the Fourier coefficients, and then to multiply each coefficient of the autospectrum by its conjugate to get spectral power. The cross-spectral density function is obtained by taking the Fourier transform of the two functions, $x(t)$ and $y(t)$, then the coefficients of one function, $X(\omega)$ are multiplied by the conjugate of the coefficients of the second function, $Y(\omega)$. The equations for the autospectum ($\hat{G}_{xx}$) and the cross-spectrum ($\hat{G}_{xy}$) are given by (15.16) and (15.17), respectively. The "^" symbol is used to denote a "raw" spectrum.

$$\hat{G}_{xx} = F_1(\omega) \times \overline{F_1}(\omega) \tag{15.16}$$

where the "bar" denotes the conjugate of the function.

$$\hat{G}_{xy} = F_1(\omega) \times \overline{F_2}(\omega) \tag{15.17}$$

An alternate method that works well, but only with the autospectrum, is to obtain the Fourier coefficients and calculate the sum of the real term squared, a_n^2, and the imaginary term squared, b_n^2, as in (15.18). Equation (15.18) should not be used to calculate the cross-spectrum.

$$\hat{G}_{xx} = a_n^2 + b_n^2 \tag{15.18}$$

15.3 CROSS-SPECTRAL ANALYSIS

Cross-spectral analysis is the process for estimating the "*Energy*" content of "Two" time-varying functions (signals) as a function of frequency. The cross-spectrum has many practical uses in engineering among which the major applications are in

1. determining output/input relationship,
2. obtaining the transfer function of a system from its input and output spectra,
3. determining bode plots from the spectra
   a. Magnitude, $H(\omega)$
   b. Phase angle, $\theta(\omega)$,
4. estimating transfer function parameter, and
5. determining "Coherence" or *Goodness Fit* of the Transfer Function (Model).

The equation for the "one-sided cross-spectral function" from the cross-correlation method is given by (15.19).

$$G_{XY}(f) = 2S_{XY}(f) \quad 0 \leq f < \infty \tag{15.19}$$

The expanded form of (15.19) may be expressed as (15.20).

$$G_{XY}(f) = \int_{-\infty}^{\infty} R_{XY}(\tau)e^{-j2\pi f\tau}\,d\tau = C_{XY}(f) + jQ_{XY}(f) \tag{15.20}$$

Complex terms of the cross-spectrum are

1. $C_{xy}(f)$ is called the *coincident spectral density function* (*Co-spectrum*), and
2. $Q_{xy}(f)$ is called the *quadrature spectral density function* (*Quad-spectrum*).

Since the cross-spectrum has two terms, a real and an imaginary term, the cross-spectrum can be represented in complex polar notation as in (15.21).

$$G_{XY}(f) = \left|G_{XY}(f)\right| e^{-j\theta_{XY}(f)} \quad \text{for} \quad 0 \leq f < \infty \tag{15.21}$$

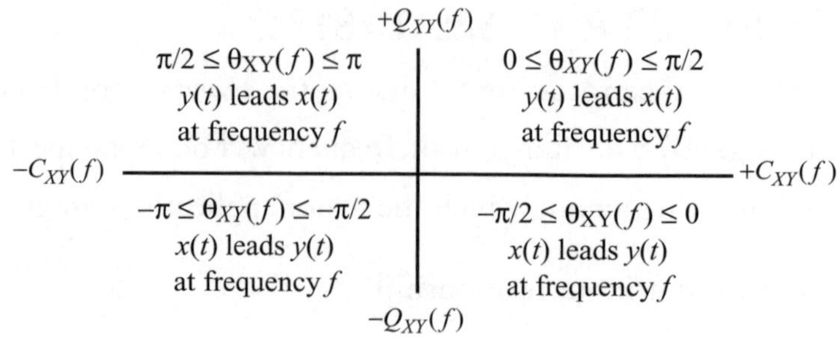

FIGURE 15.2: The relationship between the phase angle to cross-spectral terms

The magnitude of the cross-spectrum is calculated from (15.22), and the phase is calculated from (15.23).

$$|G_{XY}(f)| = \sqrt{C_{XY}^2(f) + Q_{XY}^2} \tag{15.22}$$

$$\theta_{XY}(f) = \tan^{-1}\frac{Q_{XY}(f)}{C_{XY}(f)} \tag{15.23}$$

Special note should be made of the fact that the autospectrum does not retain any phase information whereas the cross-spectrum contains phase information. Keep in mind that engineers look for relationships in both magnitudes or in phase between an input stimulus and the output response of a system. Phase relations are often expressed as leading or lagging a reference signal. The relationship between the phase angle to cross-spectral terms is shown in Fig. 15.2. Since phase is calculated with a tangential function, the relationship is expressed in quadrants.

15.4 PROPERTIES OF SPECTRAL DENSITY FUNCTIONS

For both autospectra and cross-spectra, the Power Spectral Density functions are "*Even*" Functions and the Power Spectra are always "Positive" valued functions, $G_{xx} > 0$.

15.5 FACTORS AFFECTING THE SPECTRAL DENSITY FUNCTION ESTIMATION

There are several factors that adversely affect the spectral density function estimation, regardless of the method of estimation that is employed, if these factors are not taken into account.

1. *Filtering*: Analog filtering is the first major consideration for the acquisition and processing of data. The engineer must determine the appropriate cutoff frequencies within the signal. Both high- and low-frequency cutoff should be determined and appropriate analog filters with sufficient attenuation in the bandstop regions should be employed. The high-pass filter should be used to remove the DC component of the signal and to remove trends. An antialias filter (low-pass filter) should be employed to eliminate "Aliasing" of the data. It is important to use an analog antialias filter prior to sampling of the data; any subsequent digital filtering should be done after sampling.

2. *Frequency resolution*: Frequency resolution is defined as the smallest unit of frequency in the spectrum. The "Fundamental Frequency" or resolution of the spectrum ($\Delta f_0 = 1/T$) is equal to the inverse of time the signal segment (length of time in seconds: T) that is being analyzed by Spectral Analysis. The segment time is also referred to as the window of observation or analysis. The energy in the spectrum of the signal is identified as harmonics of the fundamental frequency, that is, multiples of the fundamental frequency.

3. *Digitizing and the Nyquist criterion*: Probably the most important rule to follow in spectral analysis is the Nyquist Criteria in digital sampling of the signal. The sampling rate must be twice the highest frequency of the signal bandwidth (W). Sampling at a frequency lower than the Nyquist frequency causes aliasing of the spectrum, which will then yield an erroneous spectral estimator.

4. *Windowing*: A "Window" function should be applied prior to obtaining the estimation of the "Spectral Density Function." Windowing is a process by which the sampled data points are "weighted" in the time domain. The effect is smoothing of the signal at the beginning and end of the record to prevent the occurrence of discontinuities and Gibb's Phenomenon in the data. Improper windowing produces "Leakage" in the spectrum, which adversely affects estimation of the Spectral Density Function. There are numerous window functions that can be applied. The problem lies in deciding which window function is the best to apply on the signal being studied.

5. *Smoothing*: To obtain a fairly good estimator, statisticians would insist on some "Smoothing" function, since the Spectral Density Function or Periodogram is a raw power spectrum estimate that contains high-frequency components. Smoothing is a technique equivalent to low-pass filtering to eliminate the high-frequency components from the spectrum for the final smoothed spectral density estimate.

15.6 ADVANCED TOPIC

15.6.1 Brief Description of Autoregressive Spectral Estimation

The concept of smoothing serves as a transition to the Autoregressive (AR) Spectral Estimation Method. The AR Spectral Estimator, also known as the Maximum Entropy Spectral Estimator, is becoming more popular in the engineering community as an accurate estimator of the spectral density function.

To perform the AR Spectral Estimation, it is necessary to do the following procedural steps in order.

1. Fit AR model to data
2. Calculate AR coefficients
3. Enter the AR coefficients into formula for spectrum of AR process.
4. Estimate spectrum of AR process

The AR model of order p is represented by (15.24):

$$x_t = \alpha_0 + \alpha_1 x_{t-1} + \alpha_2 x_{t-2} + \cdots + \alpha_p x_{t-p} + z_t \tag{15.24}$$

The order of the model is chosen using Akaike's Information Criterion (AIC). The coefficients are then calculated from (15.25).

$$\hat{\alpha}_k = \frac{\sum_{t=1}^{N-1} (x_t - \bar{x})(x_{t-1} - \bar{x})}{\sum_{t=1}^{N-1} (x_t - \bar{x})^2} \tag{15.25}$$

where $\bar{x}$ is the mean of the sample record, N is the total number of data points, $x(t)$ is the point at time t, and $\hat{\alpha}_k$ is the approximate estimator of α.

Another approximation that has often been used is given by (15.26).

$$\sum_{t=1}^{N-1}(x_t - \bar{x})^2 = \sum_{t=1}^{N}(x_t - \bar{x})^2 \tag{15.26}$$

If (15.26) is used then $\alpha_k \cong r_k$, where r_k is the kth correlation coefficient. Then the sample correlation coefficients are substituted into the Yule–Walker equations (15.27), and the matrix equations are solved for α, the AR coefficients.

$$\begin{bmatrix} 1 & r_1 & & & r_{p-1} \\ r_2 & 1 & r_1 & & r_{p-2} \\ - & - & - & - & - \\ - & - & - & - & - \\ - & - & - & - & r_1 \\ r_{p-1} & - & - & - & 1 \end{bmatrix} x \begin{bmatrix} \alpha_1 \\ \alpha_2 \\ - \\ - \\ - \\ \alpha_p \end{bmatrix} = \begin{bmatrix} r_1 \\ r_2 \\ - \\ - \\ - \\ r_p \end{bmatrix} \tag{15.27}$$

Since the Fourier transform is given by (15.28), the autospectral density function is estimated by using an AR model of order p as given in (15.29).

$$F(\omega) = \sum_{k=-\infty}^{+\infty} \gamma_k e^{-j\omega k} \tag{15.28}$$

$$X(\omega) = \frac{\sigma^2}{\pi}\left(1 + \sum_{k=1}^{p} \alpha_k e^{-j\omega k} + \sum_{k=1}^{p} \alpha_k e^{j\omega k}\right) \tag{15.29}$$

where the autocovariance function is $\gamma(k) = \sigma^2 \alpha_k$.

It should be noted that the frequency ω varies within the interval $(0, \pi)$, where π represents the Nyquist Frequency.

15.6.2 Summary

The cross-spectrum shows the relationship between two waveforms (input and output) in the "Frequency Domain." The cross-spectrum indicates what frequencies in the output are related to frequencies in the input. The cross-spectrum consists of complex terms, which are termed the *coincident spectral density function* [*Co-spectrum, Cxy(f)*] and the *quadrature spectral density function [Quad-spectrum Qxy(f)]*. The cross-spectrum is calculated in the same manner as the autospectrum, via direct Fourier Transform of the two signals or via the cross-correlation function. As with the autospectrum, the "Raw" cross-spectrum must be *smoothed*, and then *normalized* to get a good estimation of the cross-spectrum.

15.7 SUGGESTED READING

1. Bendat, J., and Piersol, A. G. *Random Data Analysis and Measurement Procedures.* New York: Wiley & Sons, Inc., 1986.
2. Chatfield, C. *The Analysis of Time Series: An Introduction.* London: Chapman & Hall, 1980.
3. Lynn, P. A. *An Introduction to the Analysis and Processing of Signals.* New York: Wiley & Sons, Inc., 1973.
4. McIntire, D. A. "A Comparative Case Study of Several Spectral Estimators," In *Applied Time Series Analysis*, D. F. Findley, ed., Academic Press, 1978.

CHAPTER 16

Window Functions and Spectral Leakage

The purpose of window functions is to minimize a phenomenon called "Spectral Leakage," which is the result of selecting a finite time interval and a nonorthogonal trigonometric basis function over the interval of analysis. Only frequencies that coincide with the basis function will project on a single basis vector, such as shown in Fig. 16.1. Signals, which include frequencies not on the basis set, will not be periodic in the observing interval (often referred to as the "Observation or Observing Window"). The periodic extension of a signal with a period that is not equal to the fundamental period of the window (as in Fig. 16.2) will produce discontinuities at the boundaries of the observing interval. It is these discontinuities that cause the frequency leakage of the signal across the spectrum.

The sine function in Fig. 16.2 does not fit perfectly in the window, since its value at the beginning is not the same as the value at the end of the observation interval [4]. Since the window is supposed to contain a periodic function, one should note that in the next cycle, the value must go from 1 to 0 at the same instant of time. Hence, the discontinuity at the edge of the window will result in the spectral power of a specific frequency leaking out of the main lobe or specific spectral bandwidth to adjacent spectral side lobes. Leakage into the side lobes may occur in either side (direction) of the main lobe. The decrease of power from the main lobe adds to the power in the sideband harmonics resulting in an erroneous estimate of power in the sideband harmonics.

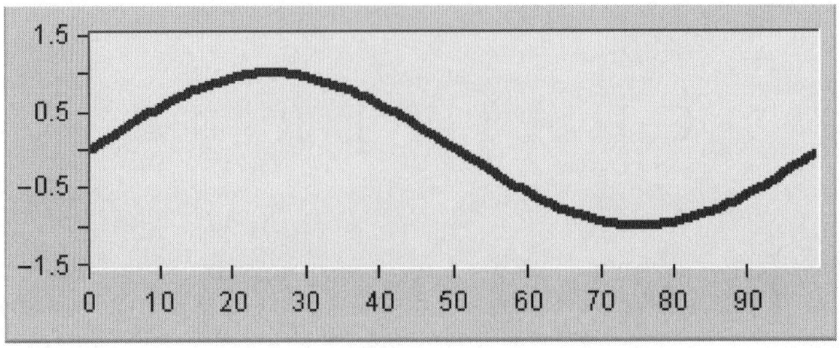

FIGURE 16.1: Sinusoidal waveform. The sine function fits perfectly in the window, since its value is zero (0) at the beginning and end of the observation interval

Outward leakage can be easily demonstrated by some simple examples. Figure 16.3 shows the Power Spectral Density (PSD) of a sinusoidal with a period that is equal to the observation window length of 1 second. Note that there is only one lobe at 1 Hz.

Figure 16.4 shows the Power Spectral Density (PSD) of a sinusoidal with a period that is not equal to the observation window length of 1 second. Note the main-lobe spectral energy at 0.59 Hz with side lobes at 0.41, 0.50, 0.68, and 0.77 Hz.

In dealing with random physiological data, the possibility of an integral fundamental frequency repetition is very limited; hence, regardless of what sampling window interval is used, leakage will be present in the spectra.

16.1 GENERALITIES ABOUT WINDOWS

The window functions (often called restoring functions) are basically weighting functions applied to the time domain data. Weighting function selection can be made early in the design process because the choices of FFT algorithm and weighting function are independent of each other. Choice of a weighting function to provide the specified

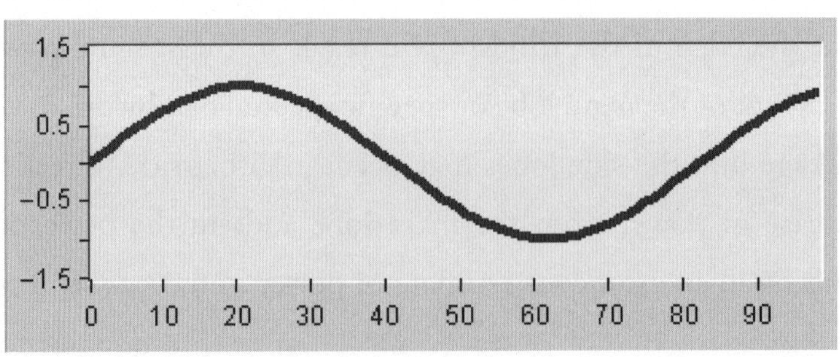

FIGURE 16.2: Discontinuity in sinusoidal waveform

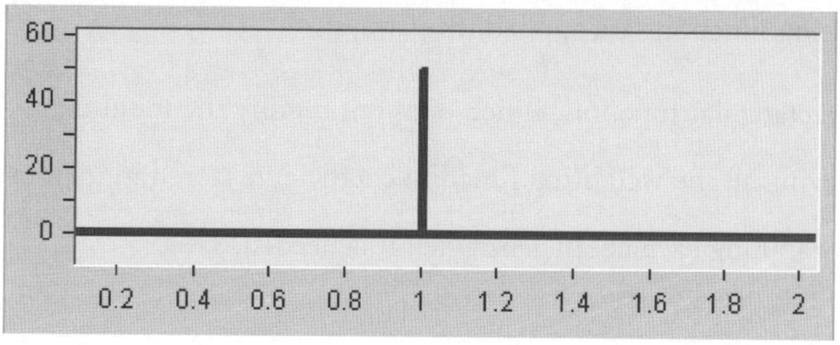

FIGURE 16.3: Power spectral density of the trace in Fig. 16.1. Note the main lobe at 1 Hz without any side lobes

side-lobe level is done without concern for the FFT algorithm that will be used because

1. they work for any length FFT,
2. they work the same for any FFT algorithm, and
3. they do not alter the FFT ability to distinguish two frequencies (resolution).

Window functions, $w(t)$, have the following properties:

1. The $w(t)$ function has to be real, even and nonnegative.
2. The transformed function has to have a main lobe at the origin of the fundamental frequency, and side lobes at both sides to ensure symmetry.
3. If the nth derivative has an "impulse" character, then the peak of the side lobes of the transformed function will decay asymptotically as 6^*n dB/octave (least estimate).

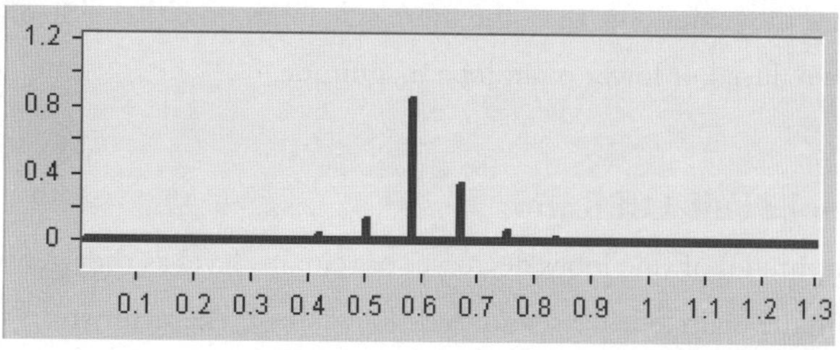

FIGURE 16.4: Power spectral density of the trace with discontinuity shown in Fig. 16.2. Note the main lobe of the spectra at 0.59 Hz with side lobes at 0.41, 0.50, 0.68, and 0.77 Hz

Weighting functions are applied three ways:

1. as a rectangular function, which does not modify the input data,
2. by having all the weighting function coefficients stored in memory, and
3. by computing each coefficient when it is needed.

16.2 PERFORMANCE MEASURES

The choice of a weighting function depends on which of the features of the narrowband DFT filters (*resolution*) are most important to the application. Those features are performance measures of the narrowband filters to analytically compare weighting functions. All these measures, except frequency straddle loss, refer to individual filters. Frequency straddle loss is associated with how filters work together [3].

16.2.1 Highest Side-Lobe Level

Highest side-lobe level (in decibels) is an indication on how large the effect of the side lobes is. Side lobes are a way of describing how a filter responds to signals at frequencies that are not in its main lobe, commonly called its passband. Each FFT filter has several side lobes. With rare exception, the highest side lobe is closest in frequency to the main lobe and is the one that is most likely to cause the passband filter to respond when it should not. For a signal with the maximum point of the main lobe at 0 dB, rectangular window the first side lobe is about -13 dB, relative amplitude. A good leakage suppressor should have its first side lobe much lower than -13 dB. Since side lobes are the result of leakage from the main lobe, then the more side lobes on either side of the main lobe will result in a reduced or lower main lobe magnitude.

16.2.2 Sidelobe Fall-Off Ratio

The peak (amplitude) of side lobes decreases or remains level as they get further away in frequency from the main lobe passband. Side lobe fall-off characteristic is very important since it indicates the rate at which the side lobes are decreasing in magnitude. The rectangular window has a decreasing rate of −6 dB/octave. This means that the next side lobe will be encountered at $-13 - 6 = -19$ dB. If the side lobes reduce rapidly, then the main lobe of the FFT bands will lose less energy into the sidelobes and the spectra will be less erroneous. Side lobe fall-off performance measure is important for applications with multiple signals that are close in frequency.

16.2.3 Frequency Straddle Loss

Frequency straddle loss is the reduced output of a DFT filter caused by the input signal not being at the filter's center frequency. Frequencies seldom fall at the center of any filter's passband. When a frequency is halfway between two filters, the response of the FFT has its lowest amplitude. For a rectangular weighting function, the frequency response halfway between two filters is 4 dB lower than if the frequency were in the center of a filter. Each of the other weighting functions in this chapter has less frequency straddle loss than the rectangular one. This performance measure is important in applications where maximum filter response is needed to detect the frequencies of interest.

16.2.4 Coherent Integration Gain

Coherent integration gain (often referred to as *Coherent Gain* or *Processing Loss*) is the ratio of amplitude of the DFT filter output to the amplitude of the input frequency. N-point FFTs have a coherent gain of N for frequencies at the centers of the filter passbands. Since most weighting function coefficients are less than 1, the coherent gain of a weighted FFT is less than N. While weighting functions reduce the coherent integration gain, the combination of this reduction and the improved straddle loss results in an overall signal response improvement halfway between two filters. Like frequency straddle loss, this performance measure is important in applications where maximum filter response is needed to detect the frequencies of interest. To restore the main lobes to their original magnitudes, the FFT coefficients may be multiplied by the inverse of the coherent gain. Window coherent gains are compared as shown in Table 16.1 to the rectangular window gain, which is 1 or 0 dB.

TABLE 16.1: Windows and Figures of Merit

WINDOW	HIGHEST SIDE-LOBE LEVEL (DB)	SIDE-LOBE FALL-OFF (DB/OCT)	COHERENT GAIN	EQUIV. NOISE BW (BINS)	3.0-DB BW (BINS)	SCALLOP LOSS (DB)	WORST CASE PROCESS LOSS (DB)	6.0-DB BW (BINS)	OVERLAP CORRELATION (PCNT) 75% OL	OVERLAP CORRELATION (PCNT) 50% OL
Rectangle	−13	−6	1.00	1.00	0.89	3.92	3.92	1.21	76.0	60.0
Triangle	−27	−12	0.50	1.33	1.28	1.82	3.07	1.78	71.9	25.0
Cos*(X) Hanning										
$\alpha = 1.0$	−23	−12	0.64	1.23	1.20	2.10	3.01	1.65	75.5	31.8
$\alpha = 2.0$	−32	−16	0.50	1.50	1.44	1.42	3.16	2.00	65.9	16.7
$\alpha = 3.0$	−39	−24	0.42	1.73	1.66	1.08	3.47	2.32	56.7	8.5
$\alpha = 4.0$	−47	−30	0.38	1.94	1.86	0.86	3.75	2.59	48.6	4.3
Hamming	−43	−6	0.64	1.36	1.30	1.78	3.10	1.81	70.7	23.5
Riesz	−21	−12	0.67	1.20	1.16	2.22	3.01	1.59	76.5	34.4
Riemann	−26	−12	0.59	1.30	1.26	1.89	3.03	1.74	73.4	27.4
De La Valle Poussin	−53	−24	0.38	1.92	1.82	0.90	3.72	2.55	48.3	5.0
Tukey										
$\alpha = 0.25$	−14	−18	0.88	1.10	1.01	2.96	3.39	1.36	74.1	44.4
$\alpha = 0.50$	−15	−18	0.75	1.22	1.15	2.24	3.11	1.57	72.7	36.4
$\alpha = 0.75$	−19	−18	0.63	1.36	1.31	1.73	3.07	1.80	70.4	25.1
Bohman	−46	−24	0.41	1.79	1.71	1.02	3.54	2.38	54.5	7.4
Poisson										
$\alpha = 2.0$	−19	−6	0.44	1.30	1.21	2.09	3.23	1.09	69.9	27.8
$\alpha = 3.0$	−24	−6	0.32	1.05	1.45	1.46	3.64	2.08	64.8	15.1
$\alpha = 4.0$	−31	−6	0.25	2.08	1.75	1.03	4.21	2.58	40.4	7.4

Hanning Poisson										
$\alpha = 0.5$	−35	−18	0.43	1.61	1.54	1.26	3.33	2.14	61.3	12.6
$\alpha = 1.0$	−39	−18	0.38	1.73	1.64	1.11	3.50	2.30	56.0	9.2
$\alpha = 2.0$	NONE	−18	0.29	2.02	1.87	0.87	3.94	2.65	44.6	4.7
Cauchy										
$\alpha = 3.0$	−31	−6	0.42	1.48	1.34	1.71	3.40	1.90	61.6	20.6
$\alpha = 4.0$	−35	−6	0.33	1.76	1.50	1.36	3.83	2.20	48.8	13.2
$\alpha = 5.0$	−30	−6	0.28	2.06	1.68	1.13	4.28	2.53	38.3	9.0
Gaussian										
$\alpha = 2.5$	−42	−6	0.51	1.39	1.33	1.69	3.14	1.86	67.7	20.0
$\alpha = 3.0$	−55	−6	0.43	1.64	1.55	1.25	3.40	2.18	67.5	10.6
$\alpha = 3.5$	−69	−6	0.37	1.90	1.79	0.94	3.73	2.52	47.2	4.9
Dolph Cheryshev										
$\alpha = 2.5$	−50	0	0.53	1.39	1.33	1.70	3.12	1.85	69.6	22.3
$\alpha = 3.0$	−60	0	0.48	1.51	1.44	1.44	3.23	2.01	64.7	16.3
$\alpha = 3.5$	−70	0	0.45	1.62	1.55	1.25	3.35	2.17	60.2	11.9
$\alpha = 4.0$	−80	0	0.42	1.73	1.65	1.10	3.48	2.31	55.9	8.7
Kaiser Bssel										
$\alpha = 2.0$	−46	−6	0.49	1.50	1.43	1.46	3.20	1.99	65.7	16.9
$\alpha = 2.5$	−57	−6	0.44	1.65	1.57	1.20	3.38	2.20	59.5	11.2
$\alpha = 3.0$	−69	−6	0.40	1.80	1.71	1.02	3.56	2.39	53.9	7.4
$\alpha = 3.5$	−82	−6	0.37	1.93	1.83	0.89	3.74	2.57	48.8	4.8

(*cont.*)

TABLE 16.1: (*Continued*)

WINDOW	HIGHEST SIDE-LOBE LEVEL (DB)	SIDE-LOBE FALL-OFF (DB/OCT)	COHERENT GAIN	EQUIV. NOISE BW (BINS)	3.0-DB BW (BINS)	SCALLOP LOSS (DB)	WORST CASE PROCESS LOSS (DB)	6.0-DB BW (BINS)	OVERLAP CORRELATION (PCNT) 75% OL	OVERLAP CORRELATION (PCNT) 50% OL
Barcilon Temes										
$\alpha = 3.0$	−53	−6	0.47	1.56	1.49	1.34	3.27	2.07	63.0	14.2
$\alpha = 3.5$	−58	−6	0.43	1.67	1.59	1.18	3.40	2.23	58.6	10.4
$\alpha = 4.0$	−68	−6	0.41	1.77	1.68	1.05	3.52	2.36	54.4	7.6
Exact Blackman	−51	−6	0.46	1.57	1.52	1.33	3.29	2.13	62.7	14.0
Blackman	−58	−18	0.42	1.73	1.68	1.10	3.47	2.35	56.7	9.0
Minimum 3 Sample Blackman-Harris	−67	−6	0.42	1.71	1.66	1.13	3.46	1.81	57.2	9.6
Minimum 4 Sample Blackman-Harris	−92	−6	0.36	2.0	1.90	0.83	3.85	2.72	46.0	1.8
61 dB 3 Sample Blackman-Harris	−61	−6	0.45	1.61	1.56	1.27	3.34	2.19	61.0	12.6
74 dB 4 Sample Blackman-Harris	−74	−6	0.40	1.79	1.74	1.03	3.56	2.44	53.9	7.4
4 Sample Kaiser Bessel $\alpha = 3.0$	−69	−6	0.40	1.80	1.74	1.02	3.56	2.44	53.9	7.4

16.2.5 Equivalent Noise Bandwidth

Equivalent noise bandwidth is the ratio of the input noise power to the noise power in the output of an FFT filter times the input data sampling rate. Every signal contains some noise. That noise is generally spread over the frequency spectrum of interest, and each narrowband filter passes a certain amount of that noise through its main lobe and sidelobes. White noise is used as the input signal and the noise power out of each filter is compared to the noise power into the filter to determine the equivalent noise bandwidth of each passband filter. In other words, equivalent noise bandwidth represents how much noise would come through the filter if it had an absolutely flat passband gain and no sidelobes. It should be noted that leakage to the side lobes not only decreases the magnitude of the main lobe but also increases the frequency resolution (spreading) of the main lobe. The width of the spread is compared to the rectangular window, which has a main lobe width of 1. Table 16.1 shows the equivalent bandwidth spread to be greater than 1 for all window functions.

16.2.6 Three-Decibel Main-Lobe Bandwidth

The standard definition of a filter's bandwidth is the frequency range over which sine waves can pass through the filter without being attenuated more than a factor of 2 (3 dB) relative to the gain of the filter at its center frequency. The narrower the main lobe, the smaller the range of frequencies that can contribute to the output of any FFT filter. This means that the accuracy of the FFT filter, in defining the frequencies in a waveform, is improved by having a narrower main lobe.

16.3 WINDOW FUNCTIONS AND WEIGHTING EQUATIONS

16.3.1 The Rectangle Window

In this section, some of the most used window models to reduce spectral leakage are presented. Some terms will be used to relate the effects of the particular window function to the rectangular window. Recall that the reduction of the side-lobe leakage introduces leakage from the expansion of the main lobe. Also some gain is lost because of the main

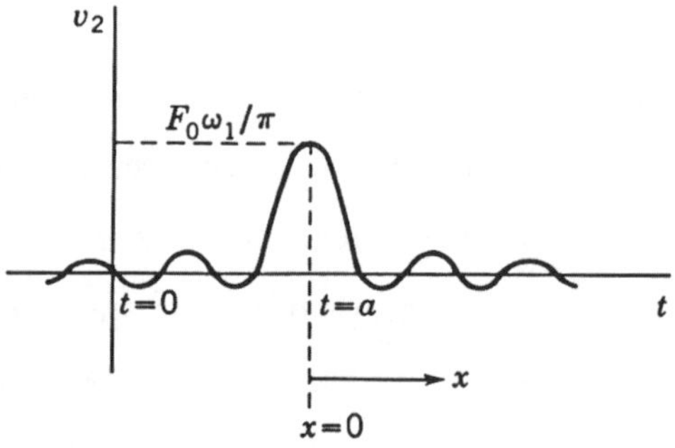

FIGURE 16.5: The Sinc function. Note the ripples

lobe spreading. More noise is introduced in the spectrum, which is a characteristic that is most often forgotten. So the question an engineer must consider is "What is the best solution or what is the best window?" According to the characteristics of the data, the analyst must select a function that produces the least error and complications.

Let us begin with the most common window, which is the "Rectangular window." The simplest way to model a finite record length is through the usage of the rectangular function (often called boxcar). The function equals one in the sampling interval and it is zero everywhere else. The Window Weighting function is given by (16.1).

$$w(n) = 1; \quad \text{For} \quad n = 0 \; to \; N - 1 \tag{16.1}$$

Consider a signal, $x(t)$, with a Fourier Transform, $X(f)$, and that the signal is defined in some finite period of time zero (0) to $+T$. The observed signal is multiplied by the window Weighting Function $w(t)$ in the time domain. The resulting function is the product of $w(t)$ and $x(t)$. The major limitation of the new function is that it is zero outside the time interval (observation window).

The Fourier transform of the window function $w(t)$ is the well-known *SINC* function. The *SINC*(x) is defined as $\sin(x)/x$, which is given in Fig. 16.5.

16.3.2 The Rectangular Window and the Gibbs Phenomenon

The rectangular function truncates a signal in a discontinuous manner at its edges. The Fourier transform approximates the function with a linear combination of sines and cosines. The Gibbs phenomenon expresses the fact that the *SINC* function oscillation is always present when discontinuities exist. If the discontinuity is smoothed with a window function, the fit will suppress the leakage, which will result in a smaller error. As the number of linear combinations in the transformation approaches infinity, the MSE in the region of the discontinuity approaches a steady error value of 0.09 (9%).

When the Fourier coefficients are squared, the negative parts of the waveform become positive. In the spectrum, the squared ripples of the Sinc function are the sidelobes. One should note that in Fig. 16.5 the following are true:

1. The SINC function is centered at the main lobe frequency "*MAXIMA*," which is an indication that the majority of the information remains around the fundamental frequency (f_0 Hz) or its multiples (harmonics).
2. The transformed function $X(f)$ is zero at the crossover points, which introduces "SMEARING." Instead of having information centered on 0, the information is centered in a RANGE around f_0.

The spectrum of the Rectangular Window function is shown in Fig. 16.6.

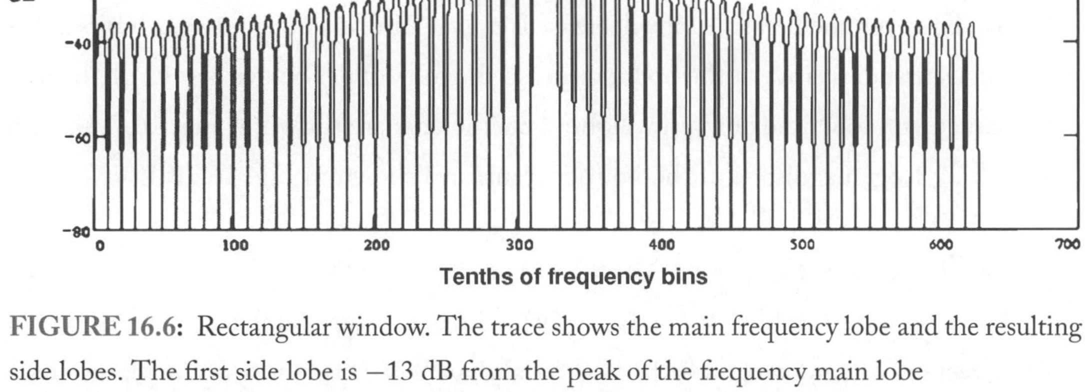

FIGURE 16.6: Rectangular window. The trace shows the main frequency lobe and the resulting side lobes. The first side lobe is −13 dB from the peak of the frequency main lobe

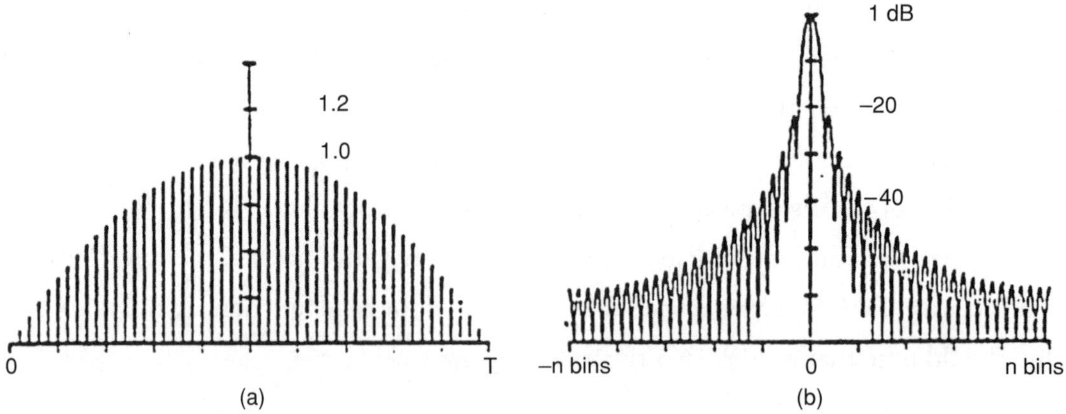

FIGURE 16.7: Parzen/Rietz window. Trace (a) is the time domain weighted coefficients. Trace (b) shows the main frequency lobe and the resulting side lobes. The first sidelobe is −21 dBs from the peak of the frequency main lobe

Characteristics of the rectangular window performance are −13 dB of the first side lobe from the height of the main lobe with a −6 dB/octave fall-off rate, 1.00 coherence gain, and 3.92-dB scallop loss (3.92-dB worst case process loss).

16.3.3 Parzen/Rietz Window

The Parzen/Rietz window is the simplest continuous polynomial window. The mathematical model is shown in (16.2).

$$x(i) = 1 - \left| \frac{(i - 0.5(N-1))}{0.5(N-1)} \right| \tag{16.2}$$

The Parzen/Rietz window gives a smoother representation than the Hamming window; however, the first side lobe decrease is only −21 dB. The window exhibits a discontinuous first derivative at the boundaries, so the roll-off per octave is −12 dB/octave. The Parzen window time domain weighted coefficients are shown in Fig. 16.7(a), with its resulting Log-magnitude of the Fourier transform shown in Fig. 16.7(b).

16.3.4 The Tukey Family of Windows

The Tukey window can be described as a cosine lobe of width, (alpha/2)(N), convolved with a rectangle window of width, (1.0 to alpha/2)/N. The term *alpha* represents a parameter, and according to this value we have a family of windows. The usual value of alpha is 0.10 or "10% Tukey Window" (often called a ten percent window) [1]. Another name for the family of these windows is cosine-tapered windows. The main idea behind any window is to smoothly set the data to zero at the boundaries, and yet minimize the reduction of the process gain of the windowed transform. There are several Tukey windows for different values of alpha ranging from 0.10 to 0.75.

The mathematical model for the ten percent Tukey window is given by equations in (16.3). Since the formula is taken from our signal processing program, the variable "alpha" has a value of 0.10.

$$\begin{aligned} w(i) &= 0.5\left(1-\cos\left(\frac{10\pi i}{10}\right)\right) \quad \text{for } i = 00 \text{ to } 0.1N \\ w(i) &= 1 \quad \text{for } i \text{ between } i = 0.1N \text{ to } 0.9N \\ w(i) &= 0.5\left(1-\cos\left(\frac{10\pi i}{10}\right)\right) \quad \text{for } i = 0.9N \text{ to } N \end{aligned} \tag{16.3}$$

Notice that for the last and first ten percent of data the window function is the same. Higher alpha values tend to increase the main lobe spread with less leakage results. For an alpha of 0.10, the Tukey window has the first side-lobe level at −14 dB with a roll-off rate of −18 dB. It is interesting that the first side-lobe level and the roll-off rate characteristics remain the same for alpha values up to 0.25. The Tukey window time domain weighted coefficients for alpha values of 0.25, 0.5, and 0.75 are shown in Figs. 16.8(a), 16.9(a), and 16.10(a), respectively. The resulting Log-magnitude of the Fourier transform for alpha values of 0.25, 0.5, and 0.75 are shown in Figs. 16.8(b), 16.9(b), and 16.10(b), respectively [1] and [2].

16.3.5 Hanning Windows

The Hanning family of windows is also known as the "Cosine Windows," since the functions used are true cosines. For the readers, historical information, the name "Hanning" does not exist, since the founder of these functions was an Austrian meteorologist named "Hann." The *ing* ending was added because of the popularity of the cosine window models. The mathematical model for the Hanning window is given by equations in (16.4).

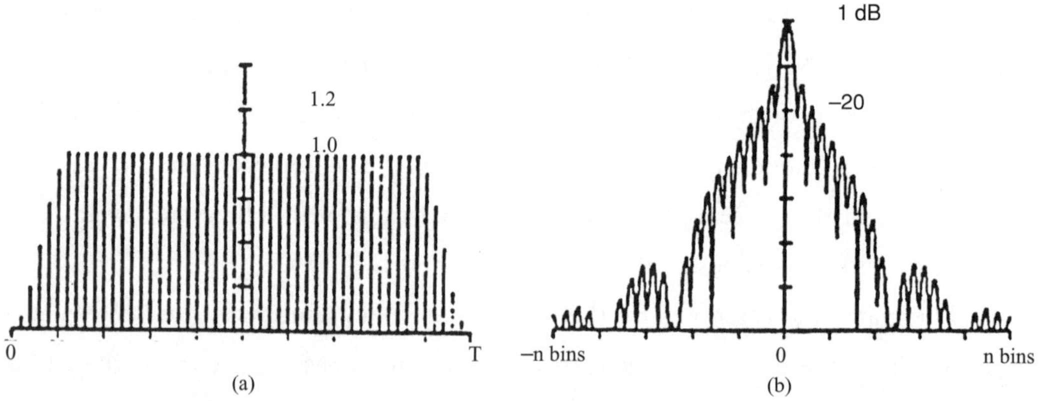

FIGURE 16.8: Twenty-five percent (25%) Tukey window. For alpha of 0.25, the trace (a) is the time domain weighted coefficients, whereas the trace (b) shows the main frequency lobe and the resulting side lobes. The first side lobe is −14 dBs from the peak of the frequency mainlobe with a −18-dB per octave side lobe fall-off rate

The variable "alpha" is an exponent of the cosine with a range from 1 to 4.

$$w(i) = 1 - \cos^{\text{alpha}}(x) \tag{16.4}$$

As alpha takes larger values, the windows are becoming smoother and smoother, and the first side lobe is smaller. The family of Hanning window time domain weighted

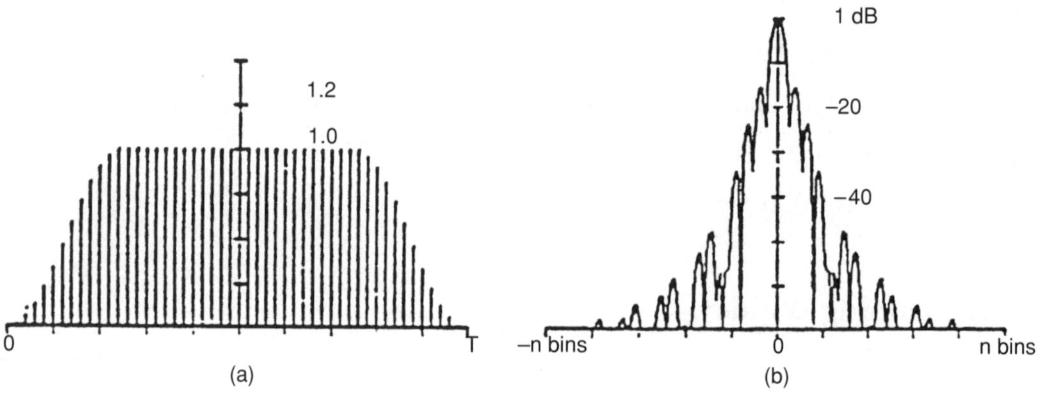

FIGURE 16.9: Fifty percent (50%) Tukey window. For alpha of 0.50, the trace (a) is the time domain weighted coefficients, whereas the trace (b) shows the main frequency lobe and the resulting side lobes. The first side lobe is −15 dBs from the peak of the frequency main lobe with a −18 dB per octave sidelobe fall-off rate

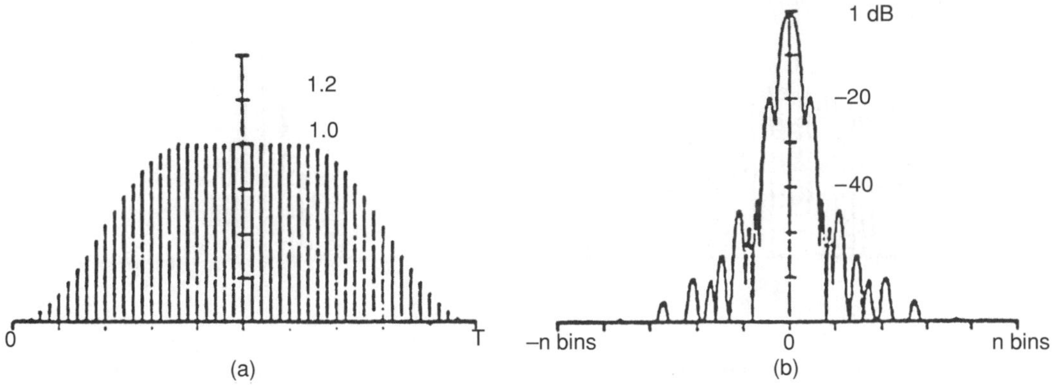

FIGURE 16.10: Seventy-five percent Tukey window. For alpha of 0.75, the trace (a) is the time domain weighted coefficients, whereas the trace (b) shows the main frequency lobe with the first side lobe at −19 dB and a −18 dB per octave side lobe fall-off rate

coefficients for alpha values of 1, 2, 3, and 4 are shown in Figs. 16.11(a), 16.12(a), 16.13(a), and 16.14(a), respectively. The resulting Log-magnitude of the Fourier transform for alpha values of 1, 2, 3, and 4 are shown in Figs. 16.11(b), 16.12(b), 16.13(b), and 16.14(b), respectively. Notice in the figures what effect increasing the value of alpha from 1 to 4 has on the first side-lobe value and the side-lobe roll-off rate. Note that for each increase of alpha by 1 the absolute magnitude of the side-lobe roll-off rate increases by 6 dB.

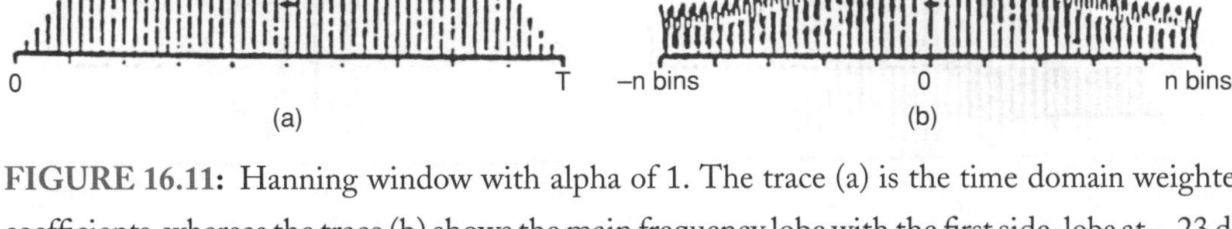

FIGURE 16.11: Hanning window with alpha of 1. The trace (a) is the time domain weighted coefficients, whereas the trace (b) shows the main frequency lobe with the first side-lobe at −23 dB and a −12 dB per octave side-lobe fall-off rate

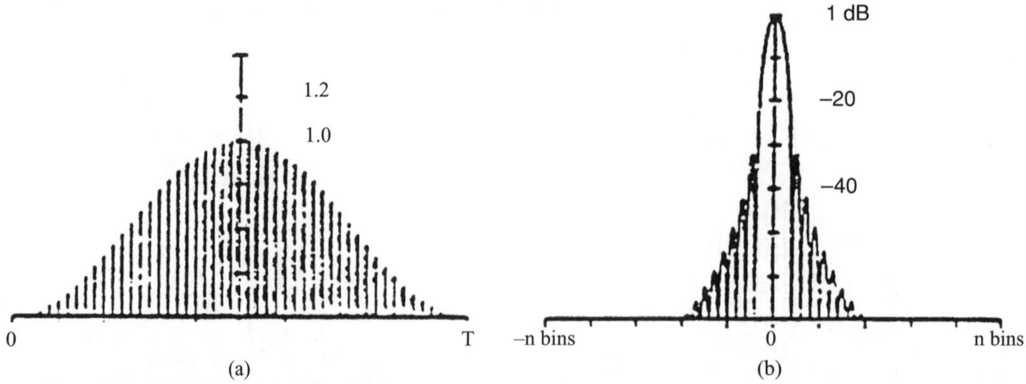

FIGURE 16.12: Hanning window with Alpha of 2. The trace (a) is the time domain weighted coefficients, whereas the trace (b) shows the main frequency lobe with the first side lobe at −32 dB and a −18 dB per octave side-lobe fall-off rate

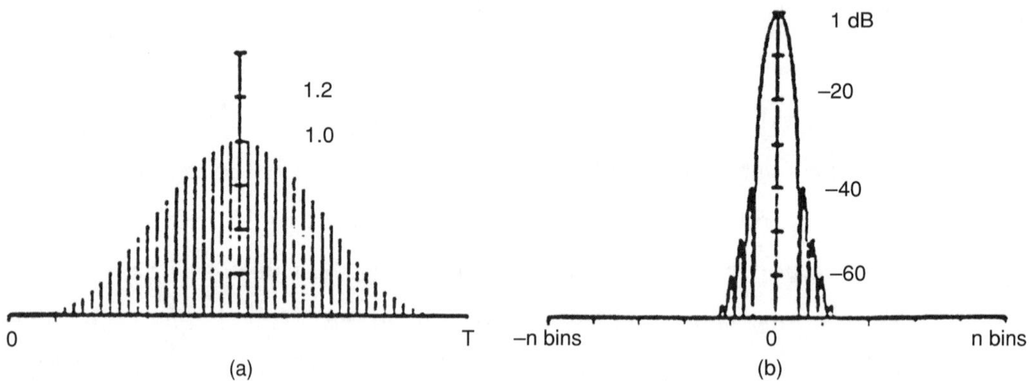

FIGURE 16.13: Hanning window with alpha of 3. The trace (a) is the time domain weighted coefficients, whereas the trace (b) shows the main frequency lobe with the first side lobe at −39 dB and a −24 dB per octave side-lobe fall-off rate

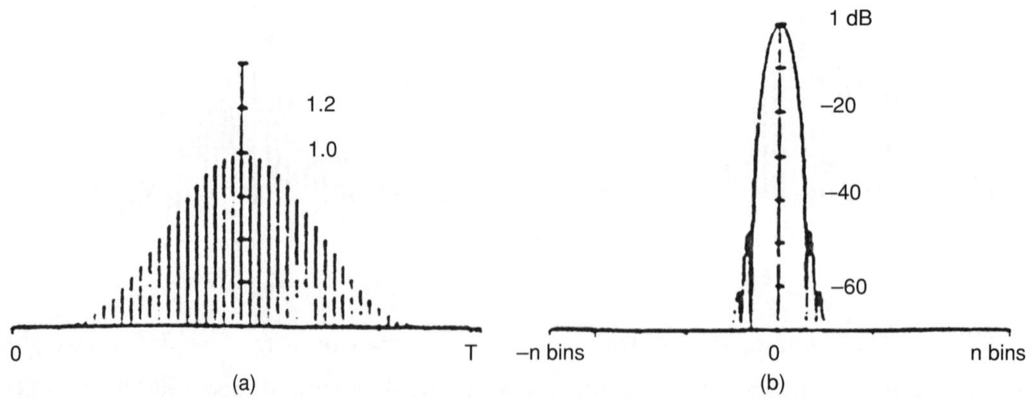

FIGURE 16.14: Hanning window with alpha of 4. The trace (a) is the time domain weighted coefficients, whereas the trace (b) shows the main frequency lobe with the first side lobe at −47 dB and a −30 dB per octave side-lobe fall-off rate

From trigonometric identities, an alternate method for expressing the cosine window is to write the expression in terms of the "Sine" function, when the data are not symmetrical about the origin. For example if $\alpha = 2$ the cosine model becomes as shown in by (16.5).

$$\sin^2(x) = 1 - \cos^2(x) \tag{16.5}$$

Hence, the basic cosine model could be expressed as (16.6).

$$w(n) = \sin^{\text{alpha}}\left(\frac{n\pi}{N}\right) \text{ for alpha values of 2 to 4} \tag{16.6}$$

16.3.6 Welch Window

This window is a modification of the Parzen window with some better characteristics especially in the coherent gain; however, the Welch window produces greater main lobe leakage due to a wider spreading. The mathematical model for the Welch window is given by (16.7). Notice the polynomial similarity between Parzen and Welch expressions.

$$w(i) = 1 - \left(\frac{i - 0.5(N-1)}{0.5(N-1)}\right)^2 \tag{16.7}$$

This type of window is not good for signals with closely related spectral components. The main side-lobe effect can be worse than the leakage from the rectangular window.

16.3.7 Blackman Window

The name "Blackman Window" was given to honor Dr. Blackman, a pioneer in the harmonic signal detection. There are several windows under his name, differing in the amount of variables used. For simplicity only the "Exact Blackman Window" will be described, since this window has superb tone discrimination characteristics (used in acoustics).

The expression for the Exact Blackman Window is given by (16.8).

$$w(n) = 0.42 + 0.50\cos\left[\frac{2\pi}{N}n\right] + 0.08\left[\frac{2\pi}{N}2n\right] \tag{16.8}$$

where $n = -N/2, \ldots., -1, 0, +1, \ldots., +N/2$

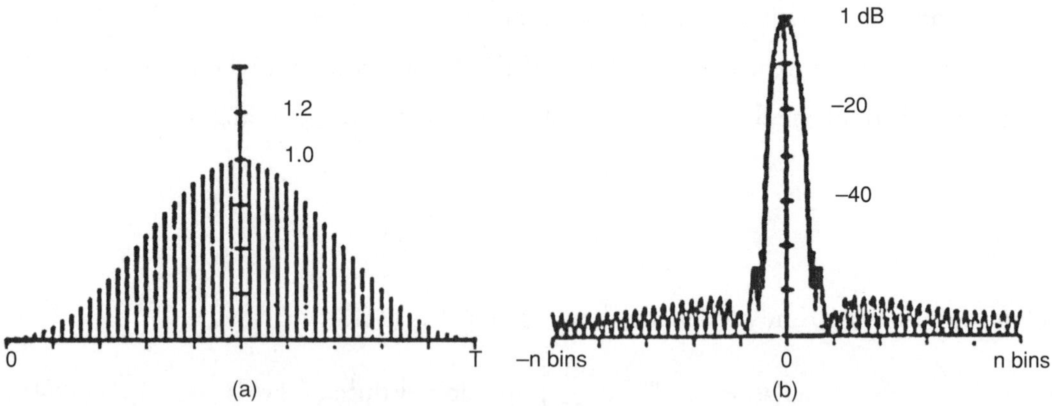

FIGURE 16.15: The Exact Blackman Window. Trace (a) is the time domain weighted coefficients. Trace (b) shows the main frequency lobe and the resulting side lobes. The first sidelobe is −51 dB from the peak of the frequency main lobe

Notice that the third term deals with the second harmonic. If we add the terms at $n = 0$ then $w(n) = 0$, something that we want to see. The side-lobe level of the Exact Blackman Window is −51 dB almost 3.5 times lower than the normal rectangular window. The roll-off for the Exact Blackman Window is −18 dB/octave. These characteristics are impressive, since the second side-lobe is at a level of −69 dB. What these numbers indicate is that the second harmonic leakage is about 1200 times less than the maximum value at the center frequency f_0. For a normal power spectrum, most analysts are not interested below −60 dB; hence, a value of −69 dB is exceptionally good (the decibel values are referred to normalized spectra). The disadvantage of the Exact Blackman Window is the coherent gain, which unfortunately is less than 0.5, closer to 0.47. However, this error can be compensated in the frequency domain by multiplying with the inverse of the coherent gain.

The Exact Blackman Window time domain weighted coefficients are shown in Fig. 16.15(a) with its resulting Log-magnitude of the Fourier transform shown in Fig. 16.15(b).

16.4 SUMMARY

Leakage is the result of making an infinitely long function finite in length. A smearing effect is the result of convolving the infinite period of the signal, with the SINC function.

The effects of leakage can be suppressed or reduced with the application of special functions called "Windows." As many as 58 window functions have been developed with various optimizations in mind. In practice, most engineers will apply a Hanning or Hamming window; however, it is not recommended that only one window be used for all applications, but rather that several windows should be tried for the best results in the particular situation.

16.5 REFERENCES

1. Bendat, J., and Piersol, A. *Randon Data Analysis and Measurement Procedures*, 2nd Edition, Wiley, pp. 393–400.
2. Gabel, R., and Roberts, R. *Signals and Linear Systems*, pp. 253–370.
3. Flannery, B., Tenkolsky, S., and Veterling, W. *Numerical Recipes: The Art of Scientific Computing*. Wesley Press, pp. 381–429.
4. Van Valkenburg, M. E. *Network Analysis*, pp. 452–495.

16.6 SUGGESTED READING

1. Bergland, G. D. "A Guided Tour of the Fast Fourier Transform," IEEE Spectrum, pp. 41–52, July 1969.
2. Harris, F. "The use of Windows for Harmonic Analysis with DFT," Proceedings of IEEE, Vol. 6, No. 1, January 1978.
3. Geckinki, N., and Yanuz, D. *IEEE Transactions on Acoustics Speech and Signal Processing*. Vol. 26, No. 6, December 1978.
4. Nudall, A. "Some Windows With Very Good Sidelobe Behavior," IEEE Transactions on Acoustics Speech and Signal Processing, Vol. 29, No. 1, February 1981.

CHAPTER 17

Transfer Function Via Spectral Analysis

17.1 INTRODUCTION

This chapter describes the method for estimation of the frequency response of a linear system, using spectral density functions. The autocorrelation and cross-correlation functions are discussed to provide background for the development of their frequency domain counterparts, the autospectral and cross-spectral density functions. The coherence function is also discussed as a means of calculating the error in the transfer function estimation. Applications will be restricted to single input/single output models, where it is assumed that systems are constant-parameter, linear types.

To meet control system specifications, a designer needs to know the time response of the controlled variable in the system. Deriving and solving the differential equations of the system can obtain an accurate description of the system's response characteristics; however, this method is not particularly practical for anything but very simple systems. With this solution, it is very difficult to go back and decide what parameters need to be adjusted if the performance specifications are not met. For this reason, the engineer would like to be able to predict the performance of the system without solving its associated differential equations and he/she would like for his/her analysis to show which parameters need to be adjusted or altered in order to meet the desired performance characteristics.

The transfer function defines the operation of a linear system and can be expressed as the ratio of the output variable to the input variable as shown in Fig. 17.1.

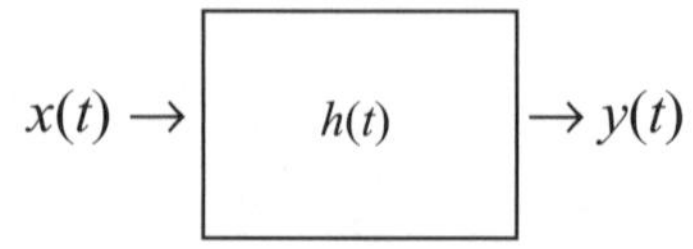

FIGURE 17.1: Transfer function from input–output

The common approach to system analysis is to determine the system's transfer function through observation of its response to a predetermined excitation signal. Evaluation of the system characteristics can be performed in the time or frequency domains.

17.2 METHODS

The engineer has several methods available to him/her for evaluating the system's response. The Root Locus Method is a graphical approach in the time domain, which gives an indication of the effect of adjustment based on the relation between the poles of the transient response function, and the poles and zeros of the open loop transfer function. The roots of the characteristic equation are obtained directly to give a complete solution for the time response of the system.

In many cases it is advantageous to obtain system performance characteristics in terms of response at particular frequencies. Two graphical representations for transfer function derivations are the Bode and Nyquist Methods. The Bode Plot Method yields a plot of the magnitude of the output/input ratio versus frequency in rectangular or logarithmic coordinates. A plot of the corresponding phase angle versus frequency is also produced. The Nyquist Plot Method is very similar: it plots the output/input ratio against frequency in polar coordinates. The amplitude and phase data produced with these methods are very useful for obtaining an estimate of the system's transfer function. They can be determined experimentally for a steady-state sinusoidal input at various frequencies, from which the magnitude and phase angle diagrams are then derived directly, leading to the synthesis of the transfer function.

Analytical techniques that fit rational polynomial transfer functions using digital computers are very popular as they speed the evaluation process considerably. In the next section, two functions, which provide analytical methods for measuring the time-domain properties of signal waveforms, will be developed.

17.3 AUTOCORRELATION

The autocorrelation function gives an average measure of the time-domain properties of a signal waveform. It is defined as (17.1):

$$R_{xx}(\tau) = \lim_{T_0 \to \infty} \frac{1}{T_0} \int_{-T_0/2}^{T_0/2} f(t) f(t+\tau)\, dt \tag{17.1}$$

This is the average product of the signal, $f(t)$, and a time-shifted version of itself, $f(t+\tau)$. The expression above applies to the case of a continuous signal of infinite duration. In practice, the intervals must be finite and it is necessary to use a modified version as given by (17.2).

$$R_{xx}(\tau) = \int_{-\infty}^{\infty} f_1(t)\ f_1(t+\tau)\, dt \tag{17.2}$$

The autocorrelation function may be applied to deterministic as well as random signals. Each of the frequency components in the signal $f(t)$ produces a corresponding term in the autocorrelation function having the same period in the time-shifted variable, τ, as the original component has in the time variable, t. The amplitude is equal to half of the squared value of the original. The phase shifts of each signal produce a single cosine term in the autocorrelation as shown in Fig. 17.2.

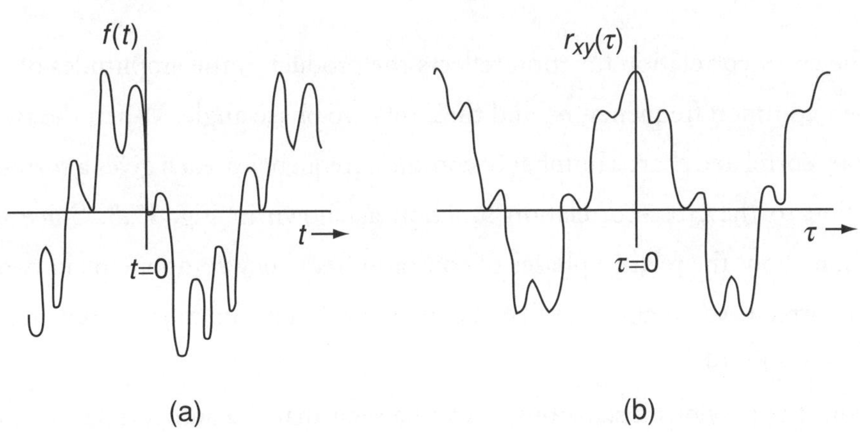

FIGURE 17.2: Autocorrelation. Trace (b) is the autocorrelation of the signal shown in (a)

The fact that the autocorrelation is composed of cosines implies that it must be an even function of τ. When $\tau = 0$, all of the cosine functions reinforce each other to give a peak positive value for the autocorrelation. Whether this peak value is ever attained for other shifts of τ depends on whether the components of the signal are harmonically related or not. The peak value of the autocorrelation function at $\tau = 0$ is simply the mean square value, or average power of the signal, and is given by (17.3) [1].

$$R_{xx}(\tau)\Big|_{\tau=0} = R_{xx}(0) = \lim_{T_0 \to \infty} \int_{-T_0/2}^{T_0/2} [f(t)]^2\, dt \tag{17.3}$$

17.4 THE CROSS-CORRELATION FUNCTION

The cross-correlation function is essentially a time-averaged measure of shared signal properties and is defined as (17.4).

$$R_{xy}(\tau) = \lim_{T_0 \to \infty} \frac{1}{T_0} \int_{-T_0/2}^{T_0/2} f_1(t) f_2(t+\tau)\, dt \tag{17.4}$$

where τ is a time shift on one of the signals. Since signals to be compared must be of finite duration, the function is modified as shown in (17.5).

$$R_{xy}(\tau) = \int_{-\infty}^{\infty} f_1(t)\; f_2(t+\tau)\, dt \tag{17.5}$$

The cross-correlation function reflects the product of the amplitudes of $f_1(t)$ and $f_2(t)$, their common frequency, ω, and their relative phase angle. When the two signals being cross-correlated share a number of common frequencies, each gives a corresponding contribution to the cross-correlation function as shown in Fig. 17.3. Since it retains information about the relative phases of common frequency components in two signals, the cross-correlation function, unlike the autocorrelation function, is not normally an even function of τ (4).

From the previous discussions, it can be seen that the autocorrelation and cross-correlation functions provide all analytical means, which may evaluate the magnitude

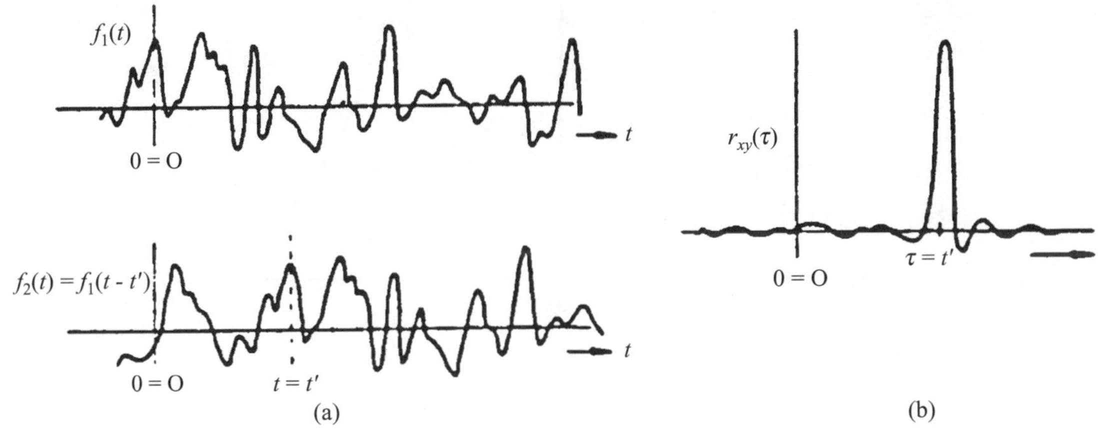

FIGURE 17.3: Cross-correlation function. Trace (b) is the cross-spectral estimate of the two traces in (a). The bottom trace in (a) is the time shifted upper trace

and phase of a system's time response. The next step is to look at the equivalents of these functions in the frequency domain where computation will be simplified.

17.5 SPECTRAL DENSITY FUNCTIONS

Spectral density functions can be derived in several ways. One method takes the direct Fourier transform of previously calculated autocorrelation and cross-correlation functions to yield the two-sided spectral density functions given in (17.6).

$$S_{xx}(f) = \int_{-\infty}^{\infty} R_{xx}(\tau)\, e^{-j2\pi f\tau}\, d\tau$$
$$S_{yy}(f) = \int_{-\infty}^{\infty} R_{yy}(\tau)\, e^{-j2\pi f\tau}\, d\tau$$
$$S_{xy}(f) = \int_{-\infty}^{\infty} R_{xy}(\tau)\, e^{-j2\pi f\tau}\, d\tau \tag{17.6}$$

These integrals always exist for finite intervals. The quantities $S_{xx}(f)$ and $S_{yy}(f)$ are the autospectral density functions of signals $x(t)$ and $y(t)$ respectively, and $S_{xy}(f)$ is the cross-spectral density function between $x(y)$ and $y(t)$. These quantities are related

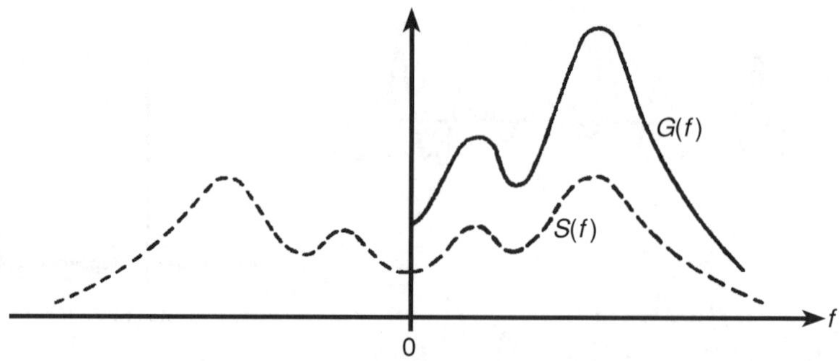

FIGURE 17.4: One- and two-sided autospectrum

as in (17.7).

$$S_{yy}(f) = |H(f)|^2 \, S_{xx}(f)$$
$$S_{xy}(f) = H(f) \, S_{xx}(f) \tag{17.7}$$

In terms of physically measurable one-sided spectral density functions where f varies over $(0, \infty)$, the results are given in (17.8):

$$G_{xx}(f) = 2S_{xx}(f)$$
$$G_{yy}(f) = 2S_{yy}(f)$$
$$G_{xy}(f) = 2S_{xy}(f) \tag{17.8}$$

Figure 17.4 shows the relation between the two-sided autospectral density function and the one-sided representation.

The autospectral and cross-spectral density functions are also shown in (17.9).

$$G_{yy}(f) = |H(f)|^2 \, G_{xx}(f)$$
$$G_{xy}(f) = H(f) \, G_{xx}(f) \tag{17.9}$$

Note that the top equation of (17.9) is a real-valued relation which contains only the transfer function gain factor of the system, $|H(f)|$. The lower equation is a complex-valued relation which can be broken down into a pair of equations to give both the gain factor, $|H(f)|$, and the transfer function phase factor, □ (f) of the system. Recall that the FFT of the cross-correlation yields the complex cross-spectra as given by the following equation. Note that the cross-spectrum is a complex function with real and imaginary functions, where $C_{xy}(f)$ is the coincident spectral density function (cospectrum) and $Q_{xy}(f)$ is the quadrature spectral density function (quad-spectrum) as shown in (17.10).

$$G_{xy}(f) = 2\int R_{xy}(\tau)\, e^{-j2\pi f\tau}\, d\tau = C_{xy}(f) - j\, Q_{xy}(f) \tag{17.10}$$

In complex polar notation, the cross-spectral density becomes 17.11.

$$\begin{aligned} G_{xy}(f) &= \left|G_{xy}(f)\right|\, e^{-j\theta_{xy}(f)} \\ H(f) &= \left|H(f)\right|\, e^{-j\phi(f)} \end{aligned} \tag{17.11}$$

where

$$\begin{aligned} \left|G_{xy}(f)\right| &= \left(C_{xy}^2(f) + Q_{xy}^2(f)\right)^{1/2} \\ \theta_{xy}(f) &= \tan^{-1}\left(\frac{Q_{xy}(f)}{C_{xy}(f)}\right) \end{aligned}$$

Thus, by knowing the autospectra of the input $G_{xx}(f)$ and the cross-spectra $G_{xy}(f)$ one can determine the transfer function, $|H(f)|$. However, the magnitude of the transfer function does not provide any phase information. Nevertheless, the complete frequency response function with gain and phase can be obtained when both $G_{xx}(f)$ and $G_{xy}(f)$ are known [2]. Figure 17.5 illustrates these relationships.

An alternative direct transform method is widely used and does not require computation of the autocorrelation and cross-correlation functions beforehand. It is based on finite Fourier transforms of the original data records. For any long, but finite records of length, T, the frequency domain equation is written as (17.12).

$$Y(f) = H(f)X(f) \tag{17.12}$$

where $X(f)$ and $Y(f)$ are finite Fourier transforms of $x(t)$ and $y(t)$ respectively. It follows that (17.12) can be rewritten as (17.13).

$$\begin{aligned} Y^*(f) &= H^*(f)X^*(f) \\ \left|Y(f)\right|^2 &= \left|H(f)\right|^2 \left|X(f)\right|^2 \\ X^*(f)Y(f) &= H(f)\left|X(f)\right|^2 \end{aligned} \tag{17.13}$$

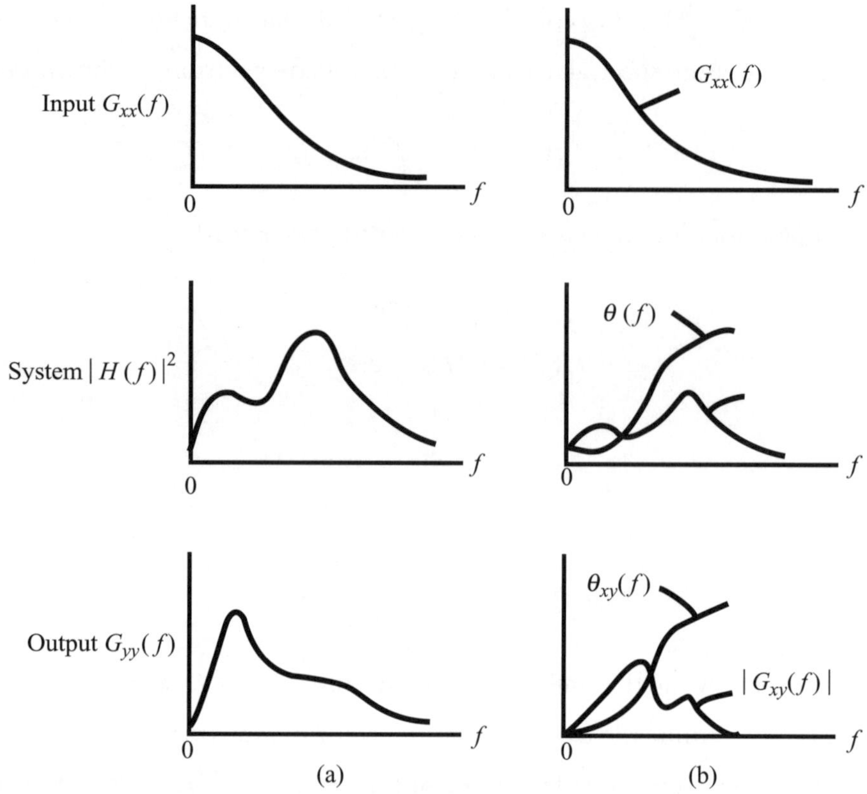

FIGURE 17.5: Input/output spectral relationships for linear systems. Traces in column (a) are autospectra while traces in column (b) are cross-spectra

With some algebraic manipulation, (17.14) can be obtained.

$$\begin{aligned} G_{yy}(f) &= \left|H(f)\right|^2 G_{xx}(f) \\ G_{xy}(f) &= H(f)\, G_{xx}(f) \end{aligned} \qquad (17.14)$$

The complex conjugation of the lower equation yields (17.15).

$$G^*_{xy}(f) = G_{yx}(f) = H^*(f)\, G_{xx}(f) \qquad (17.15)$$

where

$$\begin{aligned} G_{xy}(f) &= \left|G_{xy}(f)\right| e^{j\theta_{xy}(f)} \\ H^*(f) &= \left|H(f)\right| e^{j\phi(f)} \end{aligned}$$

Thus, the phase factor can be determined from (17.16).

$$\frac{G_{xy}(f)}{G_{yx}(f)} = \frac{H(f)}{H^*(f)} = e^{-j2\phi(f)} \tag{17.16}$$

and the complete response function becomes (17.17).

$$G_{yy}(f) = H(f)[H^*(f)G_{xx}(f)] = H(f)G_{yx}(f) \tag{17.17}$$

For the ideal single input/single output model, the frequency response is given by (17.18).

$$H(f) = \frac{G_{xy}(f)}{G_{xx}(f)} \tag{17.18}$$

The magnitude of the transfer function is commonly expressed in logarithmic form, as decibels. The logarithmic of the magnitude of the transfer function $G(f)$ expressed in decibels as (17.19).

$$20 \log \left| H(f) \right| \quad \text{in} \quad \text{dB} \tag{17.19}$$

Since the transfer function is a function of frequency, the log magnitude is also a function of frequency [1 and 3].

17.6 THE COHERENCE FUNCTION

System error is defined as the ideal or desired system output value minus the actual system output. The ideal value establishes the desired performance of the system. Coherence provides a measure of the linear correlation between two components of a process at a particular frequency. The coherence function between the input $x(t)$ and the output $y(t)$ is a real-valued quantity defined by (17.20).

$$\gamma_{xy}^2(f) = \frac{\left|G_{xy}(f)\right|^2}{G_{xx}(f)} G_{yy}(f) \tag{17.20}$$

where

$$0 \le \gamma_{xy}^2 \le 1 \quad \text{for all} \quad f$$

For a constant parameter linear system with a single, clearly defined input and output, the coherence function will be unity. If $x(t)$ and $y(t)$ are completely unrelated, the coherence function will be zero. Coherence function values greater than 0 but less than 1 may be due to the presence of noise in the measurements, or there may be nonlinearities in the system relating $x(t)$ and $y(t)$, or it may be that the output $y(t)$ may be a function of or related to other endogenous inputs in addition to the input signal $x(t)$. For linear systems, the coherence function can be interpreted as the fractional portion of the mean square value at the output $y(t)$ that is contributed by $x(t)$ at frequency f. Conversely, the quantity $[1 - \gamma_{xy}^2(f)]$ is a measure of the mean square value of $y(t)$ not accounted for by $x(t)$ at frequency f.

17.7 SUMMARY

The ability to derive the transfer function for a system's response is an integral part of the design process. Of the graphical and analytical techniques available to the engineer, the spectral density functions prove to be powerful computational tools for this purpose. In addition, through the coherence function, the spectral density functions provide an effective means for evaluating error in the transfer function model.

17.8 REFERENCES

1. Bendat, J. S., and Piersol, A. G., *Random Data: Analysis and Measurement Procedures*, 2nd Edition, John Wiley and Sones, 1986.
2. D'Azzo, J. J., and Houpis, C., *Linear Control System Analysis and Design: Conventional and Modern*. McGraw-Hill, 1981.
3. Lessard, C. S. "Analysis of Nystagmus Response to Pseudorandom Velocity $\tau = 0$ Input," In *Computer Methods and Programs in Biomedicine*, Vol. 23, 1986, pp. 11–18.
4. Lynn, P. A. *An Introduction to the Analysis and Processing of Signals*. Halsted Press, 1973.

CHAPTER 18

Coherence Function from Spectral Analysis

18.1 INTRODUCTION

The coherence function is a mathematical tool used to determine the cause-and-effect relationship between the input and the output of a system. The coherence function is used in modeling and transfer function estimation to determine the "goodness fit" of the model or transfer function to the data. In addition, the coherence function is used to determine the "random error" of the estimates of functions which use spectral analysis. Several examples given and a description of the coherence function in the context of a correlation coefficient is included for clarity. Like all scientific tools, the coherence function has its limitations and biases. These are explained in general, and specific references are cited. Two experimental examples are given showing that the coherence function can be used in primary analysis of an input–output system or at the end of a series of steps used in analyzing a more sophisticated type input–output system.

Signal-processing problems involve determining how various signals are modified in amplitude, phase, or frequency as the signals pass through some type of system. The system may be an airplane, various electromechanical devices, machine, or the human body. The block diagram model in Fig. 18.1 shows how a signal may be altered by a system.

In Fig. 18.1, the system may modify the input signal so that the output signal $Y(f)$ has different characteristics than the input signal $X(f)$. Subsystems may be connected

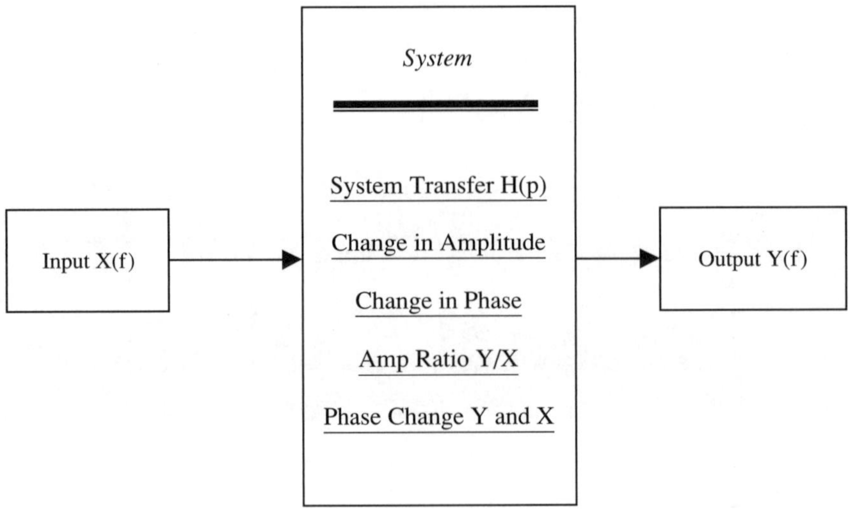

FIGURE 18.1: System modification of input signal

to other subsystems in such a way that each subsystem will modify its respective input such that the total output is changed. In addition, the system itself may generate some type of noise or interference that is only an artifact, which has nothing to do with the actual function of the system. Noise may cause alteration in the output which may make the output appear as having been modified by the system. Another type of system may have multiple inputs as illustrated in Fig. 18.2.

In systems with multiple inputs, one would be evaluating causability. The signal-processing objective is to determine how much each source of the input signal though the system transfer, $H(f)$, is responsible for the output. Other types of systems involve

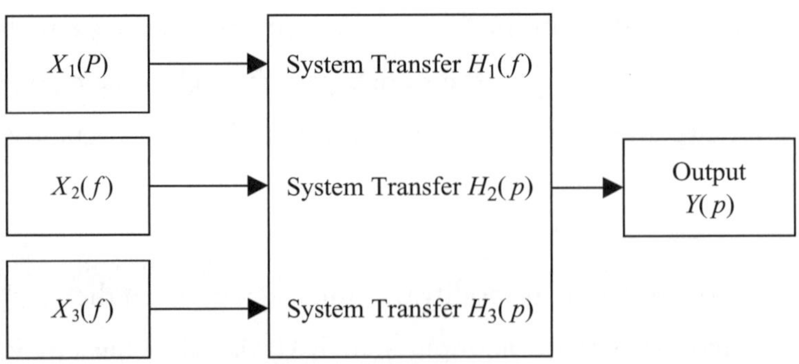

FIGURE 18.2: Multiple input, single output system

multiple inputs with multiple output systems or single input with multiple outputs systems [1]. The relationships between the input transfer function of the system and the output are analyzed by several techniques [2].

This chapter will deal primarily with coherence function of single input, single output systems only. The coherence function is a function of frequency. Its maximum values occur at the frequencies where the greatest transfer of energy within a system takes place.

18.2 DESCRIPTION OF THE COHERENCE FUNCTION

The simplest explanation of the coherence function would be that it is similar to the correlation coefficient, r[3]. The correlation coefficient, r, is a measure of the linear relationship between two variables, with one being the independent variable and the other being the dependent variable. Mathematically, the correlation coefficient, r, is shown in terms of variance as given by (18.1) through (18.5), where (18.1) through (18.3) are the mathematical definitions of terms.

$$\sigma_X^2 = \frac{1}{n}\sum_{i=1}^{n} XiXi = \text{Variance of } X \tag{18.1}$$

$$\sigma_y^2 = \frac{1}{n}\sum YiYi = \text{Variance of } Y \tag{18.2}$$

$$yx = \frac{1}{n}\sum YiXi = \text{Covariance of } X \text{ and } Y \tag{18.3}$$

$$r_{xy} = \hat{\rho}_{xy} = \frac{s_{xy}}{s_x s_y} = \frac{\sum_{t-1}^{N}(x_t-\overline{x})(y_t-\overline{y})}{\left[\sum_{t-1}^{N}(x_t-\overline{x})^2\sum_{t-1}^{N}(y_t-\overline{y})^2\right]^{1/2}} \tag{18.4}$$

or

$$= \frac{\sum_{t-1}^{N} x_t y_t - N\overline{x}\,\overline{y}}{\left[\left(\sum_{t-1}^{N} x_t^2 - N\overline{x}^2\right)\left(\sum_{t-1}^{N} y_t^2 - N\overline{y}^2\right)\right]^{1/2}} \tag{18.5}$$

The correlation coefficient, r, value has a range from -1 to $+1$. The correlation coefficient squared term r^2 is the coefficient of determination, which is often referred to as the "Goodness Fit." The correlation coefficient squared shows the percentage of variation in the output or dependent variable that is accounted for from the input or independent variable. One word of caution when using this comparison is the correlation coefficient squared r^2 between one dependent and multiple independent variables is that the r^2 does not represent the linear relationship. There may be linear, quadratic, or cubic relationship contained within the correlation coefficient squared r^2 term.

The coherence function uses almost identical calculations when comparing two spectra, where

G_{xx} (of autospectrum 1) is analogous to (18.1),
G_{yy} (of autospectrum 2) is analogous to (18.2), and
G_{xy} (of both cross-spectrums) is analgous to (18.3).

The coherence function , γ^2, is equivalent to the correlation coefficient squared, r^2. It should be noted that the correlation coefficient, r^2 is a single value measure of "Goodness," the coherence is a function with values related to frequency analysis of continuous and/or sampled random data. Therefore, the coherence function is evaluated at specific analysis frequencies. From (18.6), it can be noted that it is necessary to calculate the autospectra of both the input and output signals, as well as, the cross-spectra of the input–output signals.

$$\gamma_{xy}^2(f) = \frac{\left|\hat{G}_{xy}(f)\right|^2}{\hat{G}_{xx}(f)\hat{G}_{yy}(f)} \tag{18.6}$$

In Fig. 18.1, a single input–output system is shown with transfer function $H(f)$. The coherence will show causality (unlike the correlation coefficient) between the input and the output. Another advantage over the correlation or cross correlation, which is used in signal processing, is that the coherence function is a function of frequency. The maximum value of one (1) for the coherence function occurs when all of the output (response) is directly attributed to the input (stimulus), whereas at lowest value, which is equal to 0, of the coherence function occurs at frequencies where little or no transfer of energy from input to output is taking place. If an output signal is interfered with by noise, the method of suppressing these depends on the frequencies at which they occur; therefore, the coherence function not only shows the causality between the input and the output but also may be of use in helping solve the interference or noise problems within a system.

In summary, the coherence is a frequency domain function with observed values ranging from 0 to 1. At each frequency where the coherence function is performed, it represents the fraction of the power output related to input. If the coherence function is less than 1, then there are three possible explanations:

1. there is noise in the system or
2. the system has some nonlinearity generating energy at other frequencies or
3. there are other inputs into the system that have not be accounted for.

The coherence function can be calculated using the Fast Fourier Transform [1]. If only one transform is performed on a power spectrum, the coherence function will always be unity. Therefore, it is necessary to average a number of transforms to get a good estimate of the function.

18.3 MISAPPLICATION OF THE COHERENCE FUNCTION

Let us look at an example of when a single coherence function by itself may not give desirable results to a noise problem. Assume that there are multiple inputs to a system as shown in Fig. 18.3.

The coherence function will be 1 if each suspect input is evaluated separately, while the other inputs are turned off. If two or more sources are related to or caused by the primary source, the coherence may not reveal the secondary cause, if both suspect inputs are wired to the same power source. This problem is similar to the "spurious correlation" and "multicolinearity" problems in statistics [4,7]. This problem may be solved by "multiple coherences" [6].

Bendat and Piersol [1] note there may be bias in the coherence function analysis of a frequency response, if there are other inputs correlated with the "input of interest," if

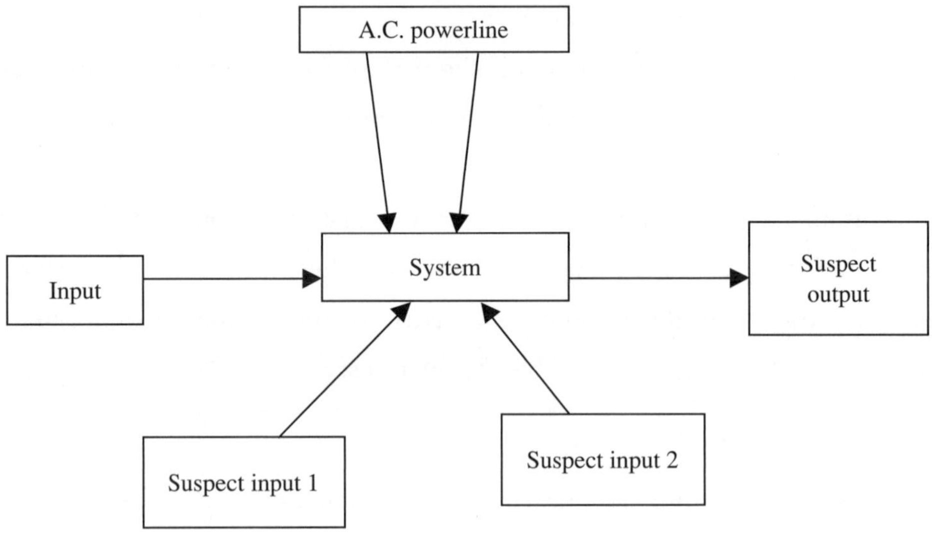

FIGURE 18.3: Spurious correlation in the analysis of a system

there is bias in the autospectral and cross-spectral density measurements, and if there are nonlinear and/or time-varying parameters within the system. Mathematical explanations of these biases are shown by (9.62) through (9.92) in Bendat and Piersol [1].

A confidence interval may be used with the coherence function. By definition, a confidence interval of 95%, for example, is "in repeated sampling 95% of all intervals will include the true fixed value of the parameter of interest" [4], then it is a true random interval.

18.4 EXAMPLES OF THE USE OF COHERENCE FUNCTIONS

The first example of how a coherence function might be used is demonstrated [5 and 6] to measure the vibration spectrum of a small printed circuit board shown in Fig. 18.4.

In this example, the ratio of acceleration/force was measured by connecting an accelerometer to the output and a pseudorandom signal to the input. In this case, the modal vibration resonances were measured and converted to the power spectrum. At 100 Hz, the coherence decreased from 1 to 0.83, indicating that the signal-to-noise ratio

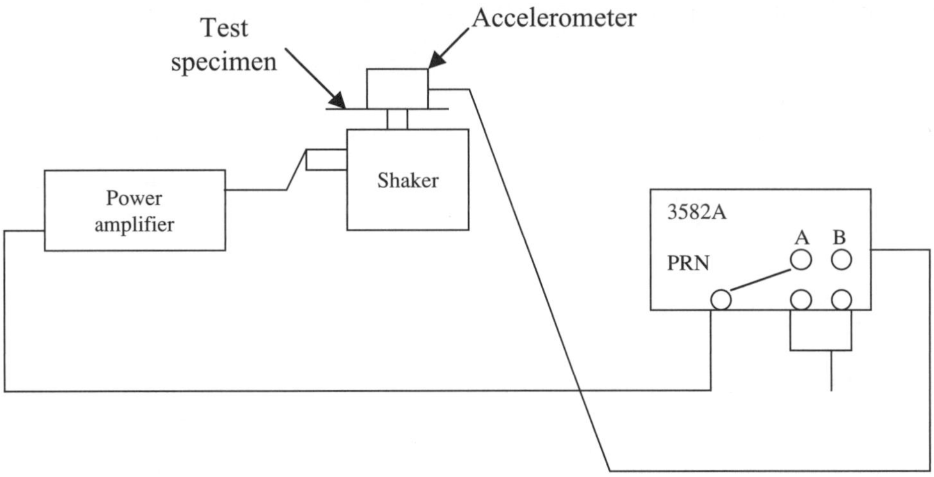

FIGURE 18.4: Test setup. The possible results in the evaluation would be (a) normal transfer function response or (b) transfer function is contaminated by vibration from loose hardware during testing of specimen

was 0.83/(1 − 0.83), so the coherence function is an alternative method of measuring the signal-to-noise ratio or an input–output system.

Lessard [3] used the coherence function to determine the cause-and-effect relationship between the output of the vestibular system and pseudorandom angular velocity input. The subject was subjected to pseudorandom angular velocity the saccade of the horizontal eye movement was converted to a signal. A full analysis of the output frequencies was performed. The coherence function analysis was performed as the last step in the analysis. This project was done with the objective of determining the nature and causes of space sickness.

18.5 PROBLEMS THAT LIMIT THE APPLICATION OF COHERENCE

There are numerous problems that limit the application. The major problems will be listed without discussion. Among these problems are

1. inaccurate source measurements,
2. source measurement interference,

3. correlation among sources,
4. input–output time-delay errors,
5. input measurement noise,
6. input–output nonlinearities,
7. input–output feedback,
8. reverberation effects, and
9. common source periodicities.

18.6 CONCLUSION

The coherence function, like many of the other mathematical operations used in signal processing, is susceptible to abuses by researchers who intentionally misapply its use or is not well informed with regard to its limitations. The coherence function should be used to establish a relationship between input and output of a system. None of the literature supports the idea of using the coherence function to compare two output signals such as the electromyograms (EMG) from two different muscles. Finally, as in all types of scientific tools, all assumptions must be met if the results are to be interpreted correctly.

18.7 REFERENCES

1. Bendat, J. S. and Piersol, A. G. *Random Data: Analysis and Measurement Procedures.* Wiley Interscience, 1971.
2. Bendat, J. S. "Statistical Errors in Measurement of the Coherence Function and Input–Output Quantities," *Journal of Sound and Vibration*, September 1978.
3. Lessard, C. S. "Analysis of Nystagmus Response to Pseudorandom Velocity Input," *Compute Methods and Programs in Biomedicine*, Vol. 23, 1986.
4. Neter, J., Wasserman, W., and Hunter M.H., *Applied Linear Regression Models.* Irwin, 1983.

5. Roth, P. R. "Measurements Using Digital Signal Analysis," *IEEE Spectrum*, April 1971.
6. Roth, P. R. "How to Use the Spectrum and Coherence Function," *Sound and Vibration*, January 1971.
7. Winer, B. J. *Statistical Principles in Experimental Design*. McGraw-Hill, 1971.

CHAPTER 19

Error in Random Data Estimate Analysis (Information Is Paraphrased from Bendat & Piersol)

This chapter presents the various statistical errors in random data analysis. Emphasis is on frequency domain properties of joint sample records from two different stationary (ergodic) random processes. The advanced parameter estimates discussed in this chapter include magnitude and phase estimates of autospectral density functions, cross-spectral density functions, transfer functions, and coherence functions. In particular, statistical error formulas are given without development or proof for the following functions:

1. frequency response function estimates (gain and phase),
2. coherence function estimates,
3. cross-spectral density function estimates, and
4. autospectral density function estimates.

19.1 CROSS-SPECTRAL DENSITY FUNCTION ESTIMATES

Recall that the cross-spectral density function between two stationary (ergodic) Gaussian random processes $\{x(t)\}$ and $\{y(t)\}$ is defined by (19.1). Specifically, given a pair of sample records $x(t)$ and $y(t)$ of unlimited length T, the one-sided cross-spectrum is given by (19.1).

$$G_{xy}(f) = \lim_{T\to\infty} \frac{2}{T} E\left[\bar{X}(f, T) Y(f, T)\right] \tag{19.1}$$

where $X(f, T)$ is the conjugate of the input and $Y(f, T)$ are the finite Fourier transforms of $x(t)$ and $y(t)$, respectively.

It should be noted that the raw spectrum (no averages) of a finite length has the same equation as (19.1), without the limit as given by (19.2) and will have a resolution bandwidth given by (19.3).

$$\hat{G}_{xy}(f) = \frac{2}{T}[\bar{X}(f) Y(f)] \tag{19.2}$$

$$B_e = \Delta f = \frac{1}{T} \tag{19.3}$$

The ordinary coherence function is defined by (19.4).

$$\gamma_{xy}^2 = \frac{|G_{xy}|^2}{G_{xx} G_{yy}} \tag{19.4}$$

The corresponding variance errors for "smooth" estimates of all of the "raw" estimates in (19.4) will be reduced by a factor of n_d when averages are taken over n_d statistically independent "raw" quantities. To be specific, the variance expressions are given by (19.5), (19.6), and (19.7).

$$\mathrm{Var}[\tilde{G}_{xx}] = \frac{G_{xx}^2}{n_d} \tag{19.5}$$

$$\mathrm{Var}[\tilde{G}_{yy}] = \frac{G_{yy}^2}{n_d} \tag{19.6}$$

$$\mathrm{Var}[|\tilde{G}_{xy}|] = \frac{|G_{xy}^2|}{\gamma_{xy}^2 n_d} \tag{19.7}$$

Then, the normalized root-mean-square errors, which are the same as the normalized random errors, are given by (19.8), (19.9), and (19.10), respectively.

$$\varepsilon[\tilde{G}_{xx}] = \frac{1}{\sqrt{n_d}} \tag{19.8}$$

$$\varepsilon[\tilde{G}_{yy}] = \frac{1}{\sqrt{n_d}} \tag{19.9}$$

$$\varepsilon[|\tilde{G}_{xy}|] = \frac{1}{|\gamma_{xy}|\sqrt{n_d}} \tag{19.10}$$

TABLE 19.1: Normalized Random Errors for Spectral Estimates

ESTIMATE	NORMALIZED RANDOM ERROR, ε
$\hat{G}_{xx}(f)$, $\hat{G}_{yy}(f)$	$\frac{1}{\sqrt{n_d}}$
$\lvert\hat{G}_{xy}(f)\rvert$	$\frac{1}{\lvert\gamma_{xy}(f)\rvert\sqrt{n_d}}$
$\hat{C}_{xy}(f)$	$\frac{[G_{xx}(f)G_{yy}(f)+C_{xy}^2(f)-Q_{xy}^2(f)]^{1/2}}{C_{xy}(f)\sqrt{2n_d}}$
$\hat{Q}_{xy}(f)$	$\frac{[G_{xx}(f)G_{yy}(f)+Q_{xy}^2(f)-C_{xy}^2(f)]^{1/2}}{Q_{xy}(f)\sqrt{2n_d}}$

The quantity $|\gamma_{xy}|$ is the positive square root of γ_{xy}^2. Note that ε for the cross-spectrum magnitude estimate $|\hat{G}_{xy}|$varies inversely with $\left|\gamma_{xy}\right|$ and approaches $\frac{1}{\sqrt{n_d}}$ as γ_{xy}^2 approaches 1.

A summary is given in Table 19.1 on the main normalized random error formulas for various spectral density estimates. The number of averages n_d represents n_d distinct (nonoverlapping) records, which are assumed to contain statistically different information from record to record. These records may occur by dividing a long stationary ergodic record into n_d parts, or they may occur by repeating an experiment n_d times under similar conditions. With the exception of autospectrum error estimation equation (19.8) and (19.9), all other error formulas are functions of frequency. Unknown true values of desired quantities are replaced by measured values when one applies these results to evaluate the random errors in actual measured data.

Let us work through an example to illustrate calculating the random errors in Frequency Response Transfer Function Estimate. Suppose the frequency response function between two random signals $x(t)$ and $y(t)$ is estimated using $n_d = 50$ averages. Assume that the coherence function at one frequency of interest is $\gamma_{xy}^2(f_1) = 0.10$ and at a second frequency of interest is $\gamma_{xy}^2(f_2) = 0.90$, the problem is to determine the random errors in the frequency response function gain and phase estimates at the two frequencies of interest.

TABLE 19.2: Multiple-Input/Output Model Estimates

FUNCTION BEING ESTIMATED	NORMALIZED RANDOM ERROR, ε
$\hat{\gamma}^2_{xy}(f)$	$\dfrac{\sqrt{2}[1-\gamma^2_{xy}(f)]}{\lvert\gamma_{xy}(f)\rvert\sqrt{n_d}}$
$\hat{G}_{vv}(f)=\hat{\gamma}^2_{xy}(f)\hat{G}_{yy}(f)$	$\dfrac{[2-\gamma^2_{xy}(f)]^{1/2}}{\lvert\gamma_{xy}(f)\rvert\sqrt{n_d}}$
$\lvert\hat{H}_{xy}(f)\rvert$	$\dfrac{[1-\gamma^2_{xy}(f)]^{1/2}}{\lvert\gamma_{xy}(f)\rvert\sqrt{2n_d}}$
$\hat{\phi}_{xy}(f)$	$\text{S.D.}\,[\hat{\phi}_{xy}(f)]\approx\dfrac{[1-\gamma^2_{xy}(f)]^{1/2}}{\lvert\gamma_{xy}(f)\rvert\sqrt{2n_d}}$

Equations (19.11) and (19.12) are used to calculate the variance and the normalized random error of the transfer function, respectively.

$$\text{Var}[\lvert\tilde{H}_{xy}\rvert]\approx\frac{\left(1-\gamma^2_{xy}\right)\lvert H_{xy}\rvert^2}{2\gamma^2_{xy}n_d}\tag{19.11}$$

$$\varepsilon[\lvert\tilde{H}_{xy}\rvert]=\frac{\text{SD}\,[\lvert\tilde{H}_{xy}\rvert]}{\lvert H_{xy}\rvert}\approx\frac{\left(1-\gamma^2_{xy}\right)^{1/2}}{\lvert\gamma_{xy}\rvert\sqrt{2n_d}}\tag{19.12}$$

Then from (19.12), the normalized random error in the gain factor estimates at frequencies $f_1$and f_2 will be given by (19.13) and (19.14).

$$\varepsilon[\lvert\tilde{H}_{xy}(f_1)\rvert]=\frac{(1-0.10)^{1/2}}{(0.32)(10)}=0.30\tag{19.13}$$

$$\varepsilon[\lvert\tilde{H}_{xy}(f_2)\rvert]=\frac{(1-0.90)^{1/2}}{(0.95)(10)}=0.033\tag{19.14}$$

The phase angle estimate, from the standard deviation (SD), is given by (19.15).

$$\text{SD}\,[\tilde{\phi}_{xy}]\approx\frac{\left(1-\gamma^2_{xy}\right)^{1/2}}{\lvert\gamma_{xy}\rvert\sqrt{2n_d}}\approx\varepsilon\,\lvert\tilde{H}_{xy}\rvert\tag{19.15}$$

Applying (19.15), the results also constitute approximations for the standard deviation (not normalized) in the phase factor estimate calculation as follows.

$$\begin{aligned} \text{SD}\,[\tilde{\phi}_{xy}(f_1)] &= 0.30 \text{ radians} \qquad & \text{SD}\,[\tilde{\phi}_{xy}(f_2)] &= 0.033 \text{ radians} \\ &= 17 \text{ degrees} & &= 1.9 \text{ degrees} \end{aligned}$$

Hence, the estimates made with a coherence of $\gamma_{xy}^2(f_2) = 0.90$ are almost ten times as accurate as those made with a coherence of $\gamma_{xy}^2(f_1) = 0.10$.

19.2 SUMMARY

A summary is given in Table 19.2 on the main normalized random error formulas for single-input/output model estimates.

Biography

CHARLES S. LESSARD

Charles S. Lessard, Ph.D., Lt Colonel, United States Air Force (Retired), is an Associate Professor in the Department of Biomedical Engineering at Texas A&M University. His areas of specialization include *Physiological Signal Processing, Design of Virtual Medical Instrumentation, Noninvasive Physiological Measurements, Vital Signs, Nystagmus, Sleep & Performance Decrement, Spatial Disorientation, G-induced Loss of Consciousness (G-LOC)*, and *Neural Network Analysis.* Dr. Lessard received a B.S. in Electrical Engineering from Texas A&M (1958) a M.S. from the U.S. Air Force Institute of Technology (1965), and his Ph.D. from Marquette University (1972).

As an officer in the U.S. Air Force, Lessard was a pilot of F86L Interceptors and B52G Strategic Bombers. He also served as Research Scientist and Chief of Biomedical Engineering Research for the Aerospace Medical Division of the School of Aerospace Medicine, at Brooks Air Force Base, Texas. In this capacity he planned and directed research efforts in biomedical projects associated with the Manned Orbiting Laboratory Program (MOL), developed medical instrumentation (EEG Analyzer), conducted research on computer on the analysis of sleep brainwaves and cardiac signals, and the effects of zero-gravity (0-G) on the cardiac response during valsalva maneuvers. At the U.S. Air Force Medical Research Laboratories, Wright-Patterson AFB, Lessard served with Biocybernetics Wing Engineering and worked on neural networks, self-organizing controls (SOC), and remotely piloted vehicles. He was the Field Office Director, Program Manager, with the Electronics Systems Division of the Air Force Systems Command during the installation and testing of Spain's Automated Air Defense System as a part of the Treaty of Friendship and Cooperation between the US and Spain. Dr. Lessard retired from the U.S. Air Force in 1981 after serving as the Deputy Director of the Bioengineering and Biodynamic Division at AMRL- Wright-Patterson Air Force Base. He began his academic career with Texas A&M University in 1981.

Charles Lessard was a Senior Scientist for Veridian Inc. at Brooks Air Force Base and lead scientist for KRUG Life Sciences, Inc. in the psychological and neuropsychological manifestations of spatial orientation, mechanisms of spatial orientation in flight, and countermeasures against spatial disorientation. Additionally, he was responsible for developing and conducting research in spatial disorientation and high acceleration (Gz-forces) induced loss of consciousness (G-LOC). He was a science and engineering expert for the USAF, Air Force Research Laboratories and Wyle Laboratories, Inc. on joint US military (Air Force and Navy) G-LOC studies performing analysis of physiological data, i.e., Auditory Evoked Responses (AER), electroencephalograms (EEG), electrocardiograms (ECG), electro-oculograms (EOG), Oxygen saturation (SaO_2), and Tracking Performance data.

Image and Signal Processing for Networked E-Health Applications

Ilias G. Maglogiannis
Department of Information and Communication System Engineering
University of the Aegean

Kostas Karpouzis
Institute of Communication and Computer Systems (ICCS)
University of Piraeus

Manolis Wallace
Department of Computer Science
University of Indianapolis, Athens Campus (UIA)

MORGAN & CLAYPOOL PUBLISHERS

ABSTRACT

E-health is closely related with networks and telecommunications when dealing with applications of collecting or transferring medical data from distant locations for performing remote medical collaborations and diagnosis. In this book we provide an overview of the fields of image and signal processing for networked and distributed e-health applications and their supporting technologies. The book is structured in 10 chapters, starting the discussion from the lower end, that of acquisition and processing of biosignals and medical images and ending in complex virtual reality systems and techniques providing more intuitive interaction in a networked medical environment. The book also discusses networked clinical decision support systems and corresponding medical standards, WWW-based applications, medical collaborative platforms, wireless networking, and the concepts of ambient intelligence and pervasive computing in electronic healthcare systems.

KEYWORDS

Telemedicine, Biosignals, Clinical Support Systems, Coding, Standards, Web technologies, Medical Collaboration, Mobile health, Ambient Intelligence, Pervasive Computing, Virtual Reality

Contents

Preface

In all societies and in all eras, health has been an issue of major importance. Our times are, of course, no different in this, except for one thing: the kind and amount of resources available to us are augmented. During the last couple of centuries, technology has certainly been the most important new resource made available to mankind. Thus, we have witnessed the generation of a large series of new tools and devices that have helped the field of medicine progress and enhance the services it is able to offer.

In our time, it is computers and communication technology that have this role in all aspects of human activities, including medicine. As the computer-based patient record system expands to support more clinical activities, healthcare organizations are asking physicians and nurses to interact increasingly with computer systems to perform their duties. We have already seen the development of a number of medical information systems and, a few years back, we witnessed the birth of telemedicine with the utilization of networking in the field of medicine. As work in this direction progresses, medicine, medical information systems, and networking become more closely integrated, forming the area of e-Health and more specifically the field of networked and distributed e-Health.

Although e-Health has already provided a number of useful tools and applications, clearly showing its potential, it is still a field in its very early stages. As is often the case when a field has not yet matured, its literature is distributed in a wide range of publications of related fields. This makes it hard for researchers to monitor the state of the art and for new researchers to enter the field, both of which have the negative effect of slowing down the growth of the field.

In this book, we review the field of networked and distributed e-Health, hoping to provide a solid starting point for anyone interested in this marvelous science. Readers will find the chapters to be short and concise, briefly covering main achievements and estimated future directions in each important component of networked e-health applications, while at the same time offering a considerable number of current references for further reading.

We hope that the material included in this book will assist the potential reader to learn about the state of the art and future trends in the field of networked e-health applications and telemedicine and its supporting technologies.

The authors would also like to thank the publisher Joel Claypool for his endless patience and support, as well as the series editor, John Enderle, for his trust. Ilias Maglogiannis dedicates this book to Vicky, Kostas Karpouzis dedicates this book to Eleni and their two newborn daughters, and Manolis Wallace dedicates this book to Coula, Lambrini, and Nicoletta.

Ilias Maglogiannis,
Kostas Karpouzis,
Manolis Wallace
December 2005, Athens

CHAPTER 1

Introduction

For years, research in medicine was related closely only to natural sciences, as for example with chemistry in the research for new drugs. In the last decades we have seen a closer link to engineering as well, with mechanical parts being developed to substitute damaged or missing human parts, with microdevices being used as implants, with miniature cameras being utilized to avoid pervasive diagnostic examinations, and so on. With the rapid growth of networking technology and the enhancement of the communications capabilities it offers, it was only a matter of time before interest was also turned to the benefits that may be obtained from the integration of networking capabilities in medical procedures. Thus, the field of Telemedicine was born.

With computers continuing to further their penetration into everyday life, and information systems being utilized in practically every area where organizational tasks are relevant, it was only a matter of time before it became obvious that the way to go was to combine networked health applications with existing and emerging medical information systems. The new term coined just before the turn of the century in order to represent this new hybrid type of systems is e-Health.

E-Health applications quickly showed its potential, facilitating exchange of information between clinicians or between institutions, reducing costs, extending the scope and reach of medical facilities, enhancing the quality of service offered on- and off-site, providing for new means of medical supervision and preemptive medicine, and so on. Currently, the integration of medical networking and medical information systems is treated as an obvious and irrefragable rule; thus, standalone medical networking environments are no longer a reality and term telemedicine is in practice used interchangeably with e-Health.

In this book we provide an overview of the field of Networked e-Health applications and Telemedicine and its supporting technologies. We start from what is typically referred to as the lower end—that of signals. Thus, in Chapter 2 we review basic related concepts from the fields of signal processing. We focus more on the case of the electroencephalogram (EEG) and the electrocardiogram (ECG or EKG) and also look at medical signal processing and classification from the point of view of urgent medical support, where not every possible type of medical equipment is readily available. In addition to this medical imaging is discussed: building upon and extending the general purpose techniques developed in the parent field of image processing, medical image processing has provided a wealth of algorithms for the enhancement, analysis, and segmentation of medical images in particular as well as for the registration of different images to each other; we review all these issues in this chapter.

Chapter 3 brings us closer to the pure field of Telemedicine, discussing encoding for transmission of medical data. With compression being of central importance, the loss of information, as well as ways to minimize it, is discussed. Special attention is given to the utilization of wavelets, using which compression is also combined with the extremely useful and wanted capability of multiresolution presentation, which provides for great levels of flexibility in the design and application of medical communications.

Chapter 4, probably the lengthiest of the book, concludes the signal and data processing section of the book and discusses clinical decision support systems. We review their role and structure, as well as the most common approaches from the field of data mining that are used in order to automatically process medical data and provide support for medical decisions. We also discuss which types of e-Health applications each approach is most suitable for, given its characteristics.

In the last section of the book we move the discussion to topics related to the design, implementation, and operation of real life e-Health systems. Starting with Chapter 5 we review the standards that have been developed for the representation of medical data, so that their exchange between different systems may be facilitated.

Chapter 6 explains how the existing infrastructure of the Internet and the World Wide Web is exploited in the scope of e-Health, describing the way different general-purpose protocols are utilized in a purely medical scope. The important issue of data security is also discussed.

With Chapter 7 we focus specifically on collaborative platforms. These allow for clinicians to utilize the Internet in order to share files, medical data, and information in general, as well as to communicate with each other in an intuitive manner, even when they are geographically separated, thus facilitating their cooperation.

Wireless and *ad hoc* networks have received a lot of attention lately and are expected to have a greater role in everyday life in the future. In Chapter 8 we review their role in telemedicine together with their supporting protocols and standards.

In Chapter 9 we present the concepts of ambient intelligence and pervasive computing in electronic healthcare systems. The concepts refer to the availability of software applications and medical information anywhere and anytime, and the invisibility of computing modules when they are hidden in multimedia information appliances, which are used in everyday life.

Chapter 10 concludes the book with a reference to virtual reality systems and techniques and their application toward more intuitive interaction in a networked medical environment.

Overall, we believe we have provided a complete and balanced view of the field; we hope our readers will think so too, and that they find this book to be a good point to start and a useful reference to come back to.

CHAPTER 2

Signal Processing for Telemedicine Applications

2.1 INTRODUCTION TO SIGNAL PROCESSING

A very general definition of a signal is a "measurable indication or representation of an actual phenomenon," which in the field of medical signals refers to observable facts or stimuli of biological systems or life forms. In order to extract and document the meaning or the cause of a signal, a medical practitioner may utilize simple examination procedures, such as measuring the temperature of a human body or have to resort to highly specialized and sometimes intrusive equipment, such as an endoscope. Following signal acquisition practitioners go on to a second step, that of interpreting its meaning, usually after some kind of signal enhancement or "preprocessing" that separates the captured information from noise and prepares it for specialized processing, classification, and recognition algorithms. It is only then that the result of the acquisition process reveals the true meaning of the physical phenomenon that produced the signal under investigation.

As a general rule, the particular techniques used in all the above mentioned steps depend on the actual nature of the signal and the information it may convey. Signals in medicine-related applications are found in many forms, such as bioelectric signals that are usually generated by nerve and muscle cells. Besides these, tissue bioimpedance signals may contain important information such as tissue composition, blood volume and distribution, or endocrine activity, while bioacoustic and biomechanical signals may be produced by movement or flow within the human body (blood flow in the heart or veins, blood pressure, joint rotation or displacement, etc.). While acquiring these kinds of signals is usually performed with unintrusive examinations, biomagnetic signals, produced at the brain, heart, and lungs, and biochemical signals can be collected only with specialized instruments and in controlled hospital or research center environments.

While medical signal processing techniques can sometimes be performed on raw, analog signals, advanced, frequency-domain methods typically require a digitization step to convert the signal into a digital form. Besides catering for the deployment of techniques that would be otherwise impossible, digital signals are much more efficient when it comes to storing and transmitting them over networks or utilizing automated feature-extraction and recognition techniques. This process begins with acquiring the raw signal in its analog form, which is then fed into an analog-to-digital (A/D) converter. Since computers cannot handle or store continuous data, the first step of the conversion procedure is to produce a discrete time series from the analog form of the raw signal. This step is known as "sampling" and is meant to create a sequence of values sampled from the original analog signals at predefined intervals, which can faithfully reconstruct the initial signal waveform. In order for this to happen, the sampling frequency must be at least double the signal bandwidth; this requirement is known as the Shannon theorem and, in theory, it is the deciding requirement for the sampling process to produce a faithful representation of the captured signal. In practice, however, picking the sampling frequency with only Shannon's theorem in mind may lead to other equally important problems such as aliasing or noise replication.

The second step of the digitization process is quantization, which works on the temporally sampled values of the initial signal and produces a signal that is both temporally and quantitatively discrete. This means that the initial values are converted and encoded according to properties such as bit allocation and value range. Essentially, quantization maps the sampled signal into a range of values that is both compact and efficient for algorithms to work with. In the case of capturing a digital image or scanning a printed one the process of sampling produces a set of values, each of which represents the magnitude of visible light falling on each single sensor cell integrated over the area of the sensor. Depending on the number of bits allocated for the representation of the digital image, each single value is then quantized over the relevant range of values; for example, if 8 bits per pixel are available, 256 discrete levels of luminosity can be represented and, thus, the sampled value is quantized to fit into the interval [0, 255].

2.2 SIGNAL PROCESSING APPLICATIONS

Although there exist a number of algorithms that operate on the sampled and quantized digital signals as is, most of the techniques that extract temporal information from and recognize patterns in a signal are performed in the frequency domain. The foundation of these algorithms is the fact that any analog or digital signal can be represented as an integral or sum of fundamental sine functions with varying amplitudes and phases. The reversible transformation between the initial representation and that of the frequency domain is given by the Fourier transform (FT) [10], which is defined in both continuous and discrete forms. An interesting alternative to the frequency representation is to apply the FT to the correlation function, resulting in the power spectral density function (PSD) [17], which finds a prominent use in recognizing pathological states in electroencephalograms (EEGs).

2.2.1 Electrocardiography

The electrocardiogram (ECG) signal represents the electrical activity produced by the heart and acquired at skin surface level and has been available from nearly the beginning of the 20th century [1]. Actually some of the techniques used to capture the signal (usually at the arms and legs) and to label salient features originate from the pioneering work of Einthoven [7]; ECGs were also among the earliest acquired signals to be transferred to remote locations over phone lines. In recent capture stations dedicated hardware is used to perform the A/D conversion of the differential signals captured at as many as 12 sites of the human body. These 12 "leads" are then combined to produce the conclusive measurements, which are in the following fed to classification algorithms. ECGs are typically used to determine the rate and rhythm for the atria and ventricles and pinpoint problems between or within the heart chambers. In addition to this, this signal can illustrate evidence of dysfunction of myocardial blood perfusion, or ischemia, or chronic alteration of the mechanical structure of the heart, such as enlarged chamber size [19].

Emergency ECG processing caters for early detection of possibly severe pathologies, since the permanent or temporary effects of even minor cardiac dysfunctions on the waveform can be detected and classified. Regarding acquisition, bulky Holter [9] devices built in the 1960s are currently substituted with compact devices that cater for easy deployment at mobile sites. Initially, Holter probes were combined with 24-h recording and playback equipment in order to help identify periods of abnormal heart rate, possibly attributed to a developing heart block. These devices possessed minimal heart beat segmentation capabilities and could also illustrate results visually and audibly; in addition to this, identification of premature ventricular complexes (PVCs) and relevant treatment is also possible and is recently enhanced by high-speed, hardware-assisted playback. Despite the decline in its use, emergency ECG is still a very useful early detection tool, based on devices with dedicated memory and storing capabilities, capable to provide automated signal processing and classification without the need for a host computer. High-resolution ECGs are the most recent signal acquisition and processing improvement tools [2], capable to record signals of relatively low magnitude that occur after the QRS complex (the deflection in the ECG that represents ventricular activity of the heart) but are not evident on the standard ECG and can be related to abnormally rapid heart rate (ventricular tachycardia) [20].

ECG signal processing lends well to knowledge-based techniques for classification. As a result the best known and most widely used such methods have been employed in this framework, including Bayesian [26] algorithms and Markov models [4]. Neural networks (NNs) [6, 21] have also been put to use here with often better results, since NNs can be designed to operate with and handle incomplete data, which is quite often the case with ECG data sets.

2.2.2 Electroencephalography

Electroencephalograms capture the very low electrical activity produced by the brain [3, 8, 16]. Initially, EEGs were used to detect epilepsy-related conditions, but in the process, their use was extended to nonpathological, behavioral situations, such as detecting and studying the ground truth of emotion-related stimuli in the human brain [23]. Another factor of improvement in the study of EEGs is related to the means of analysis, which developed from being merely visual and subjective to semiautomated techniques working on signal waveform properties such as amplitude and phase. Acquisition is performed using recording systems consisting of conductive electrodes, A/D converters, and an attached recording device. Classification is one of the major signal processing tasks performed on the EEG signal: usually this is carried out by segmenting the signal into "chunks," and then calculating the closest "bin" in which the particular chunk is classified. Statistical analysis follows, and the distribution of chunks along the different classes is checked against predefined patterns. This last stage of control is not performed on the distribution curve *per se*, but more often on its quantitative control parameters, such as standard deviation or moments.

Regarding frequency domain signal processing, the most usual algorithms put to use are the fast Fourier transform (FFT) and analysis of the PSD. Both techniques are best suited for the extraction of specific features, but perform badly and essentially destroy some of the signal parameters. For example, FFT is typically computed over a predefined, *albeit* short time window, and thereby does not take into account variation within that time frame; nevertheless, it provides a reliable estimation of the PSD, which in turn is related to changing psychological states of the subject. In order to overcome the windowing effects, one may opt to move to autoregressive models.

2.3 MEDICAL IMAGE PROCESSING

Medical image processing can be thought of as a specialization of signal processing techniques and their application to the two-dimensional (2D) domain, in the case of still images, or three-dimensional (3D) domain for volumetric images or video sequences. A wide number of issues exist here such as image acquisition and coding, image reconstruction (e.g., from slices), image enhancement and denoising, and image analysis. On a higher level, automatic or user-assisted techniques may be used to segment and detect specific objects in an image, e.g., organs in a computed axial tomography (CAT) slice or process and visualize volumetric data [5].

Progress in medical imaging coincides not only with the advent of maturing processing techniques, but also with the introduction of better acquisition equipment and the integration of knowledge-based strategies. Regarding complexity and tackling of novel concepts, the arrival of magnetic resonance imaging (MRI) and other related acquisition techniques that produce

3D or higher resolution images paved the way for visualization applications and model-driven approaches.

In the field of 2D imaging image registration, segmentation, and pattern recognition and classification are still in the forefront of interest. In the case of segmentation, knowledge-based approaches [25] that are constantly integrated even into specialized equipment cater for tackling problems such as edge detection and region growing [12], and subsequent object boundary detection. CAT and MRI scans are eminent examples of such techniques, since lower level segmentation of image intensity can be combined with knowledge of tissue structure and positioning to cater for organ detection and segmentation. In addition to this, deformable models [13, 14] can be employed to track the deformation of organs such as the heart and model several functional organ parameters, while successive scan slices can be used to reconstruct the 3D volume of an internal organ [22]. It has to be noted here that the initial thickness of these slices would not cater for faithful, usable reconstruction; however, modern equipment can produce multiple successive slices in the order of a few millimeters, hence only minute volume details are lost.

Image texture is another useful tool, especially in cases where tissue density is mapped to pixel intensity (e.g., ultrasound [15]), but can also be used for motion analysis (e.g., in echocardiograms). Several issues still exist here, such as correct calibration of the instrument and provision of contrast enhancement tools that cater for faithful mapping of light intensity to tissue density; result of inconsistencies between different acquisition devices and setups result in texture information being put to better use in segmentation problems than in, e.g., modeling organ tissue using knowledge-based techniques.

A very important concept of medical imaging arises from the fact that information related to the same organ or tissue can be collected from a multitude of sources; as a result, there is the need to productively integrate these inputs, while still retaining the interrelation of what is provided from each source. With the arrival of multiresolution data, the problem of image matching and registration [11] became more apparent, since there could be cases where the same observation could be contained in both 2D slices and 3D volumetric scans, taken from different points of view, during different examinations and with different resolution. One approach [18], which tackled matching manually segmented surfaces from the brain, worked on CAT and MRI scans, as well as positron emission tomography (PET) data. Here, since temporal and spatial resolutions were diverse, the best solution would come from simple geometrical distance metrics in order to calculate the best matching transformation between the different sets of data [24].

REFERENCES

[1] J. Bailey, A. Berson, and A. Garson , "Recommendations for standardization and specifications in automated electrocardiography: bandwidth and digital signal processing. A report for health professionals by an ad hoc writing group of the committee on

electrocardiography and cardiac electrophysiology of the Council on Clinical Cardiology, American Heart Association," *Circulation*, vol. 81, p. 730, 1990.

[2] E. Berbari, "High-resolution electrocardiography," *CRC Crit. Rev. Bioeng.*, vol. 16, p. 67, 1988.

[3] J. Bronzino, "Quantitative analysis of the EEG: general concepts and animal studies," *IEEE Trans. Biomed. Eng.*, vol. 31, no. 12, p. 850, 1984.

[4] A. Coast, R. Stern, G. Cano, and S. Briller, "An approach to cardiac arrhythmia analysis using hidden Markov models," *IEEE Trans. Biomed. Eng.*, vol. 37, pp. 826–835, 1990.

[5] J. Duncan, "Medical image analysis: progress over two decades and the challenges ahead," *IEEE Trans. PAMI*, vol. 22, no. 1, pp. 85–106, 2000.

[6] L. Edenbrandt, B. Devine, and P. Macfarlane, "Neural networks for classification of ECG ST-T segments," *J. Electrocardiol.*, vol. 25, no. 3, pp. 167–173, 1992.

[7] W. Einthoven, "Die galvanometrische Registrirung des menschlichen Elektrokardiogramms, zugleich eine Beurtheilung der Anwendung des Capillar-Elecktrometers in der Physiologie," *Pflugers Arch. Ges. Physiol.*, vol. 99, p. 472, 1903.

[8] A. Givens and A. Remond, Eds., "Methods of analysis of brain electrical and magnetic signals," in *EEG Handbook*, vol. 1. Amsterdam: Elsevier, 1987.

[9] N. Holter, "New method for heart studies: continuous electrocardiography of active subjects over long periods is now practical," *Science*, vol. 134, p. 1214, 1961.

[10] S. Kay, *Modern Spectral Estimation: Theory and Application*. Englewood Cliffs, NJ: Prentice-Hall, 1988.

[11] J. Maintz and M. Viergever, "A survey of medical image registration," *Med. Image Anal.*, vol. 2, no. 1, pp. 1–16, 1998.

[12] A. Martelli, "An application of heuristic search methods to edge and contour detection," *Comm. ACM*, vol. 19, pp. 73–83, 1976.

[13] T. McInerney and D. Terzopoulos, "Deformable models in medical image analysis: a survey," *Med. Image Anal.*, vol. 1, no. 2, pp. 91–108, 1996.

[14] D. Metaxas and D. Terzopoulos, "Constrained deformable superquadrics and nonrigid motion tracking," in *Computer Vision and Pattern Recognition*, G. Medioni and B. Horn, Eds. IEEE Computer Society Press, 1991 pp. 337–343.

[15] D. Morris, "An evaluation of the use of texture measures for tissue characterization of ultrasound images of in vivo human placenta," *Ultrasound Med. Biol.*, vol. 14, no. 1, pp. 387–395, 1988.

[16] E. Niedermeyer and F. Lopes da Silva, *Electroencephalography: Basic Principles, Clinical Applications and Related Fields*, 3rd ed. Philadelphia: Lippincott, Williams & Wilkins, 1993.

[17] T. Ning and J. Bronzino, "Autoregressive and bispectral analysis techniques: EEG applications," *IEEE Eng. Med. Biol. Mag.*, vol. 9, Special Issue, p. 47, 1990.

[18] C. Pelizzari, G. Chen, D. Spelbring, R. Weichselbaum, and C. Chen, "Accurate three-dimensional registration of CT, PET, and/or MR images of the brain," *J. Comput. Assist. Tomogr.*, vol. 13, pp. 20–26, 1989.

[19] T. Pryor, E. Drazen, and M. Laks, Eds., *Computer Systems for the Processing of Diagnostic Electrocardiograms*. Los Alamitos, CA: IEEE Computer Society Press, 1980.

[20] M. Simson, "Use of signals in the terminal QRS complex to identify patients with ventricular tachycardia after myocardial infarction," *Circulation*, vol. 64, p. 235, 1981.

[21] R. Silipo and C. Marchesi, "Artificial neural networks for automatic ECG analysis," *IEEE Trans. Signal. Proc.*, vol. 46, no. 5, pp. 1417–1425, 1998.

[22] L. Staib and J. Duncan, "Model-based deformable surface finding for medical images," *IEEE Trans. Med. Imag.*, vol. 78, no. 5, pp. 720–731, 1996.

[23] J. Taylor and C. Mannion, Eds., *New Developments in Neural Computing*. Bristol, England: Adam Hilger, 1989.

[24] J. P. Thirion, "New feature points based on geometric invariants for 3D image registration," *Int. J. Comput. Vis.*, vol. 18, no. 2, pp. 121–137, 1996.

[25] J. Tsotsos, "Knowledge organization and its role in representation and interpretation for time-varying data: the ALVEN system," *Comput. Intell.*, vol. 1, no. 1, pp. 16–32, 1985.

[26] J. Willems and E. Lesaffre, "Comparison of multigroup logistic and linear discriminant ECG and VCG classification," *J. Electrocardiol.*, vol. 20, pp. 83–92, 1987.

CHAPTER 3

Medical Data Encoding for Transmission

3.1 INTRODUCTION

Medical information, especially in time-varying or multidimensional and multiresolution forms, typically creates large databases and requires relevant storage media and strategies. High quality stereo sound, for example, is sampled in 44 kHz, with 16 bits allocated per sample and two channels combined to produce the final file; a simple multiplication shows that this acquisition scenario produces data at 1.4 Mb/s. In the same framework, uncompressed NTSC TV signal at 30 frames/s utilizes almost 300 Mb/s, while for quarter-frame sizes, e.g. webcams, this drops to almost 73 Mb/s. Different applications may impose additional requirements: for example, machine-learning techniques may well operate offline and thus alleviate the need for real time availability, while emergency processing for early detection needs as much as data as possible on demand. In the case of transmission, the choice of medium is another critical factor, since the difference in bandwidth capacity between T3 networks and normal telephony networks (PSTN) is in the order of 1000 to 1. Data compression may be an obvious option here, but this also presents a number of interesting questions, especially when lossy compression techniques are put to use, thereby removing possibly critical information from the data set. In addition to this, the huge amounts of data produced by the acquisition equipment may also be a serious obstacle [2].

Compression of digital information comes in two forms: lossy, where data is encoded reversibly and, as a result, decoding will faithfully reproduce the initial data set, and lossless, in which facets of the data set are sacrificed to achieve bigger gains in file size and bandwidth for transmission. There are apparent advantages in either category, but the most prominent of them are the trade-off between compression reversibility and complexity, on one hand and resulting size of information and required bandwidth, on the other. In the case of lossless encoding, typical compression rates are in the order of 2 to 1, meaning that the size of the file produced by the compression algorithm is half of the initial file or that the size gain is only 50%; however, all information contained in the original data set is retained and can be retrieved with minimal decoding effort. When lossy algorithms are employed, a portion of the data is deemed either unnecessary to retain or "almost" correlated with another portion and thus can be dismissed. In the case of audio compression using the MPEG-1 standard, for example, besides the other compression schemes proposed by the standard itself, the more advanced encoders exploit what is known as the "psychoacoustic phenomenon." Here the existence of particular audible frequencies in sounds is thought to prevent the human ear from receiving neighboring frequencies, which as a result can be dropped from the original sound spectrum. While this process effectively alters the sound without the possibility to reproduce faithfully, the difference between the original waveform and the one reconstructed from the encoded file is thought to be unnoticeable for practical purposes. In the case of images, the human eye is thought to be more sensitive to the brightness (luminance) value of a pixel than to the color (chrominance) information; as a result, a portion of color information can be discarded (subsampled) without introducing perceptible deterioration. Here lossless techniques can provide compression ratios in the order of 20 to 1, with minimal visible artifacts in the image; as a result, for most applications, lossless image compressions techniques with measurable and small introduction of error are preferred [5] [11].

3.2 DATA COMPRESSION

Besides the lossy/lossless division, one may categorize compression efforts with respect to their analogy to signal representation and processing, i.e., distinguish between algorithms operating on temporal representations and the frequency domain. Usually the particular technique that is used depends on the actual signal to be compressed. In the case of electrocardiograms (ECGs) a compression ratio of 7.8 has been reported [9] using differential pulse code modulation (DPCM), an algorithm that uses linear prediction in order to decorrelate the samples of the input signal. The rationale behind this is that data samples with higher probabilities contribute less to what is already known and, since they occur more often than the rest, are assigned to smaller "codewords," i.e., are represented with less bits in a look-up table. Reversely, sample values that do not arise as frequent and, hence, have small probabilities can be assigned to larger representations, a fact that does not harm file size since the values in question are infrequent. This lossless process, which is called Huffman encoding, can be performed in both 1D and 2D signals, making it a very important tool for data compression. It has to be noted though that the remaining modules of the DPCM compression algorithm, and especially quantization of the values of the error sequence that is produced by the prediction/decorrelation step, introduce irreversible loss of information that, however, permits higher compression rates.

DPCM is a typical example of algorithms used to prepare the signal for either storage or transmission; signal encoding may also be performed to segment the signal or detect prominent features from it. In this framework the amplitude zone time epoch coding (AZTEC) algorithm [3] can be used to assist in analyzing ECGs, producing QRS detection automatically.

One may describe this method as "adaptive downsampling," where temporally static or almost static information within a specific time window is replaced by a constant value. To reconstruct the waveform, linear interpolation between the remaining samples is sufficient. However, this produces a lower resolution waveform that has an inherently "jagged" look.

When frequency-domain data compression techniques are put to use, regardless of the dimensionality of the data, input signals are first segmented to blocks or "chunks" of predefined size and then each of these chunks is transformed into the frequency domain. The Discrete Cosine Transform (DCT) [1] is the algorithm most often employed, especially in the case of images and extended to videos in the sense that they are approached as a sequence of frames. A number of popular video compression standards (MPEG-1, −2, and −4, as well as H.26x [8]) utilize DCT at some stage of encoding. In the case of images, JPEG is the most usual case of lossy compression and is based on the energy-packing characteristics of DCT, performing processes where amounts of information are irreversibly lost, resulting in huge compression rates. Its modes of operation are further divided into baseline, extended progressive, and extended hierarchical. There also exists a lossless variant of JPEG that operates using predictive algorithms, but fails to produce similar compression ratios as in the lossy case. In the beginning of the JPEG algorithm, the input image is segmented into two sets of blocks of 8×8 pixels, one for the luminance and one for the chrominance channel of the image, starting from the top left of the image; these blocks are in the following transformed to produce the 2D DCT, resulting in 64 DCT coefficients. Irreversible loss of information occurs in the next step, where the DCT coefficient values are quantized and thus a part of the variation among them is lost. This step hurts the larger values of the DCT coefficients in the block in question, resulting in loss of higher frequencies, which means that edges are blurrier and object boundaries are not so well separated. The amount of rounding off in each quantization is the deciding factor for the amount of compression and the decrease of the signal-to-noise ratio, which is a metric of image quality versus any introduced noise. In the final step of the algorithm, DCT coefficients within a block are arranged in a zigzag fashion and fed into a Huffman encoder to produce the final encoded image.

In the case of video sequences, the MPEG standard exploits both spatial and temporal redundancy. When handling a single frame, the encoding scheme can be very close to JPEG compression; for subsequent frames, the standard adopts the intra- and intercoding schemes. Here an MPEG GOP (group of pictures) is made of I- (intra), P- (predictive), and B- (bidirectional) coded frames. An I-frame has the role of a "key frame," in the sense that all processing is independent of following and previous information, and is uncompressed or compressed in a lossless fashion. P- and B-frames are intercoded reference frames, where temporal redundancy on reference frames is removed by performing motion estimation: a P-frame is coded based on its preceding I- or P-frame, while a B-frame is coded using block-based information from both of the preceding and the following I- and/or P-frames. Combination of the two methods is performed in MJPEG compression, where each frame of the video sequence is compressed individually using the JPEG compression algorithm.

3.3 WAVELET COMPRESSION

Most recent trends in the field of medical image compression are focused on the wavelet transformation and compression technology [12], leading to the introduction of the new image compression standard, JPEG 2000 [10], that, however, is yet to enjoy wide adoption. Wavelets cater for a multiresolution representation of the data set [6] [7], irrespective of dimensionality and without loss of information, as is the case with DCT-based techniques. This very important characteristic enables users to browse through a general view of the complete data set, e.g., a volumetric 3D scan, and focus on a specific portion which is decoded and presented in greater detail. Scalable rendering is of utmost importance in cases of large data sets, which can be in the order of hundred of millions of volumetric pixels, since it caters for visualization in computers with everyday capabilities as well as higher level observation [12].

Signal-wise, the wavelet transform is a form of subband coding, where the transformed signal information contained in both high-pass and low-pass parts can be repeatedly divided in a similar fashion. This process can be repeated, effectively producing a hierarchical tree structure in which each segment can be separately encoded; in fact, the wavelet transform concept can be thought as a windowed-preparation technique, with variable-sized regions, that allows the use of longer intervals where more precise low-frequency information is required and shorter regions where we want high-frequency information. In addition to this, wavelets provide means for local processing of the signal, without the need to decode the complete waveform and can also easily reveal unique higher level information like discontinuities in higher derivatives or self-similarity [4].

REFERENCES

[1] N. Ahmed, T. Natarajan, and K. Rao "Discrete cosine transform," *IEEE Trans.Comput.*, pp. 90–93, 1974.

[2] A. Cetin and H. Koymen, "Compression of digital biomedical signals," in *The Biomedical Engineering Handbook*, J. Bronzino, Ed., 2nd ed. Boca Raton: CRC Press LLC, 2000.

[3] J. Cox, "AZTEC: a preprocessing program for real-time ECG rhythm analysis," *IEEE Trans. Biomed. Eng.*, vol. 15, pp. 128–129, 1968.

[4] J. Crowe, "Wavelet transform as a potential tool for ECG analysis and compression," *J. Biomed. Eng.*, vol. 14, pp. 268–272, 1992.

[5] M. Karczewicz and M. Gabbouj, "ECG data compression by spline approximation," *Signal Process.*, vol. 59, pp. 43–59, 1997.

[6] S. Mallat, "A theory for multiresolution signal decomposition: the wavelet representation," *IEEE Trans. PAMI*, vol. 11, pp. 674–693, 1989.

[7] P. Morettin, "From Fourier to wavelet analysis of time series," in *Proc. Comput. Stat.*, New York: Physica, 1996, pp. 111–122.

[8] I. Richardson, *H.264 and MPEG-4 Video Compression: Video Coding for Next-Generation Multimedia*. England: Wiley, 2003.

[9] U. Ruttiman and H. Pipberger, "Compression of the ECG by prediction or interpolation and entropy coding," *IEEE Trans. Biomed. Eng.*, vol. 26, no. 4, pp. 613–623, 1979.

[10] D. Taubman, and M. Marcellin, *JPEG2000: Compression Fundamentals, Standards and Practice*. Norwell, MA: Kluwer Academic, 2001.

[11] J. Willems, "Common standards for quantitative electrocardiography," *J. Med. Eng. Techn.*, vol. 9, pp. 209–217, 1985.

[12] Z. Xiong, X. Wu, S. Cheng, and J. Hua, "Lossy-to-lossless compression of medical volumetric data using 3-D integer wavelet transforms," *IEEE Trans. Med. Imag.*, vol. 22, pp. 459–470, Mar. 2003.

CHAPTER 4

Clinical Decision Support Systems for Remote and Commuting Clinicians

4.1 INTRODUCTION

This chapter presents a special branch of expert systems called clinical decision support systems (CDSSs). In general, defining what is and what is not a CDSS is not an easy task [1]. Still, one can generally say that the term generally refers to automated systems that process medical data such as medical examinations and provide estimated diagnoses [2]. The estimations are often based on the analysis of details that elute the human eye as well as large amounts of medical history that humans cannot possibly consider. Although such systems typically do not reach 100% success, which means that they cannot substitute the clinician, the input they provide is extremely helpful as an independent source of evidence concerning the correct diagnosis [3].

CDSSs have an augmented role in the framework of networked e-Health [4]. Clinicians in small-scale facilities located at remote and isolated locations, as well as clinicians providing their services at the site of events that require medical attention before transportation can be allowed, have to face an important handicap: they do not have access to the large medical knowledge bases, such as medical libraries, warehouses of medical history, or the benefit of having other clinicians around to consult for a second opinion. In this case, the support of an automated system that can refer to such medical knowledge bases and provide a second opinion is certainly an important help.

Although a CDSS is a very complex software system, its operation in the framework of e-Health, as well as the steps typically followed in its design and development, can be generally broken down as shown in Fig. 4.1. As shown, information extracted from a patient running various medical tests—this could be from something as simple as body temperature to detailed blood analysis results or f-MRI images—is processed by the clinician in order to reach a conclusion regarding the diagnosis. In this process the clinician may also utilize the networking capabilities offered by the system to retrieve data from distributed electronic health records and/or consult other clinicians who may be located at different locations. The CDSS attempts to simulate the same process. Thus, the extracted information is processed by the

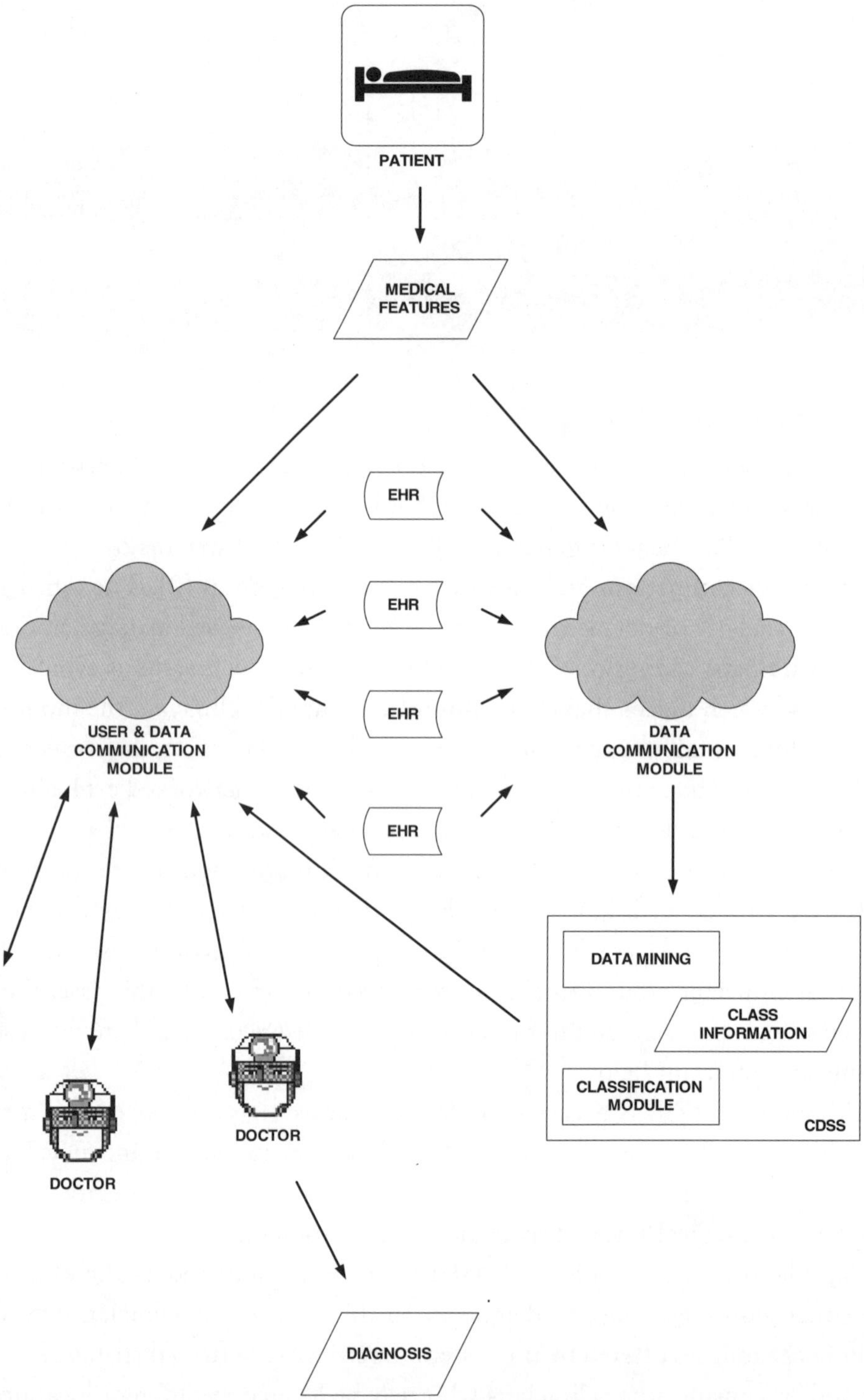

FIGURE 4.1: General architecture of a clinical decision support system

CDSS in order to reach a diagnosis. This diagnosis is provided by the classification module, which maps available information to one of the known classes of diagnoses. This information about the known classes is either provided manually by human experts or, most often, extracted automatically through processing of the medical history contained in the distributed electronic health records by a data mining module. The estimate produced by the CDSS is yet another input made available to the clinicians toward their diagnosis.

The most important processes in the development and operation of a CDSS are i) the acquisition of the information regarding the diagnosis classes and ii) the actual classification of a given case to a diagnosis. The two steps are actually closely related to each other, as the type of classifier chosen in most cases also indicates the training methodology to use. Although there is an extremely wide variety of classification methodologies that one may choose to apply, the most well known and widely applied genres of approaches can be briefly summarized and categorized as follows:

1. The examined case is compared directly to other cases in the medical history, and similarities are used in order to provide a most probable diagnosis.
2. Different types of artificial neural networks (ANNs) are trained based on available medical history, so that the underlying data patterns are automatically identified and utilized in order to provide more reliable classification of future data.
3. Information extracted automatically from medical history, or provided manually by human experts, is organized so that diagnosis estimation is provided in a dialogical manner through a series of question/answer sessions.
4. Multiple simple classifiers are combined to minimize the error margin.
5. Information provided by human experts in the form of simple rules is utilized by a fuzzy system in order to evaluate the case in hand.

Each one of these approaches has its characteristics, benefits, and drawbacks; these are discussed in detail in the following section. Especially for the case of remote or commuting clinicians, in addition to issues related to the performance of the CDSS, another important issue to consider is that of the location of the different components of the system. Depending on whether the classification information is located at a centralized system or stored on a portable device as well as depending on whether the classification system itself is portable or operates centrally and exchanges information with the clinician over a network, the types of classifiers, classification information structures, and data mining methodologies that can be applied in practice differ, based mainly on the networking resources that they require.

4.2 CLASSIFICATION SYSTEMS

Classifiers are a breed of software that has been studied for long in the field of artificial intelligence. Depending on the application, input data format, and computing resources available, a wide variety of methodologies can be selected in the literature [5, 6]. Given the diverse problems and circumstances that distributed medical applications have to deal with, many of these come in handy in different situations. In the following we review the main categories of classification systems that find application in a CDSS framework.

4.2.1 *k*-Nearest Neighbors

The *k*-nearest neighbors methodology, often referred to as *k*-NN, constitutes a breed of classifiers that attempt to classify the given patient data by identifying other similar cases in the database of medical history. The simplest, as well as most common, case is when all considered medical features are scalars. In that case, assuming that n features are available in total, one can consider all cases to be positioned perfectly and accurately in an n-dimensional space. Using an Euclidean distance it is possible to identify the historical medical case(s) that is closest to the one in question; the classification (diagnosis) for this medical case is then considered as the classification for the case in question as well. Parameter k is the number of similar cases to consider in order to reach a diagnosis. This simplistic approach has a number of drawbacks, the most important of which are described below.

The n-dimensional space described above is rarely partitioned perfectly in regions belonging to a single class, i.e., to regions corresponding to medical cases all diagnosed in the same way. Quite the contrary, in most regions of this space a number of classes coexist. In the quite common situation that not all cases in the considered neighborhood correspond to the same diagnosis, the diagnosis appearing the most times is the one selected by the system.

An extension to this approach is the weighted *k*-NN approach. This approach additionally considers the distance (or similarity) between the different cases in the neighborhood and the one in question, thus allowing most similar cases to affect the final decision to a greater extend.

Of course, in all the above we assume that the medical features measured are all scalars, and thus that it is possible to directly map the problem in the n-dimensional space. Of course, this is not always the case in medicine. Electrical signals such as electrocardiograms (ECGs) and images such as computed tomography (CT) scans are also medical features that often need to be considered but cannot be equally easily mapped to a geometrical space. Still, this does not mean that *k*-NN based approaches cannot be applied. In such cases, one needs to define a metric for each measured feature. In the above example, all one needs to define is a function providing the difference between two ECGs and another one providing the difference between two CT scan images. Typically, although differences are not that easy to define, it is quite simpler to define similarities. Thus, usually one identifies what it is that makes two medical signals or images similar and then defines a metric quantifying the inverse of this.

The k-NN methodology has found many applications in the field of CDSSs. For example, the following works are based on k-NN based approaches [7–10].

The applicability of k-NN and its variations in networked health applications are governed mainly by the need to compare a sample with all samples in the medical history in order to reach a conclusion. This makes it practically impossible to have the classification module located separately from the class information, as it would require the transmission of large amounts of data in every run. Therefore, it is the medical information about the current case that needs to be transmitted to a centralized system. This makes k-NN an excellent choice when these medical features are small in size, but not that practical when large signals and high resolution medical images are involved; in these cases one of the methodologies presented in the following sections needs to be utilized.

4.2.2 Artificial Neural Networks

Artificial Neural Networks are the main representatives of a more robust approach to classification; these classifiers, prior to being put to use, process available medical history in order to extract useful information concerning the underlying data patterns and structure, thus also acquiring the information required in order to optimize the classification process. Following a massively parallel architecture, quite similar to that of neurons in the human brain, ANNs construct powerful processing and decision-making engines through the combination of quite trivial processing units—also named neurons (Fig. 4.2).

In order to provide for some kind of control of the network operation and to facilitate training and fine-tuning, contrary to the case of networks of real neurons, connectivity in ANNs is usually constrained, in the sense that neurons are organized in layers and each neuron can

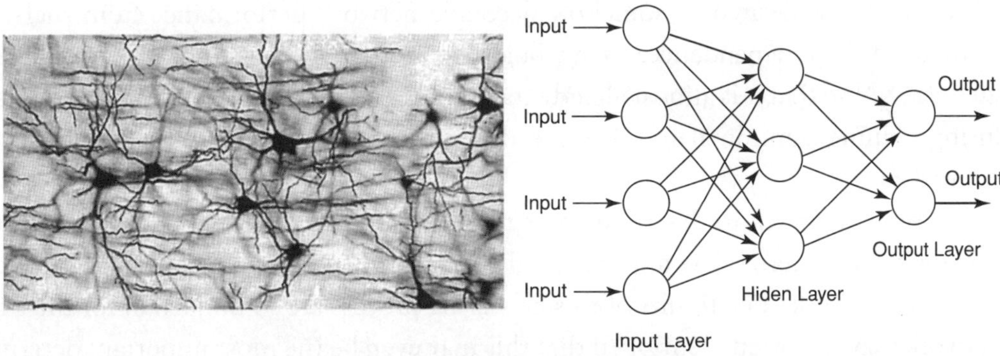

FIGURE 4.2: A network of real neurons compared to an artificial neural network

only accept inputs from the directly preceding layer and provide output to the directly following one.

Although ANNs are typically blamed by the scientific community for their nonsymbolic nature—i.e., the fact that their exact operation and the real meaning of their output is not easily revealed to the human inspector when their size grows—they are widely used in numerous classification problems, including CDSSs, due to their remarkable performance related characteristics. For example, the following works are based on applications of ANNs [11–15] while [16] reviews the benefits of ANN application in medicine in general.

The characteristic that makes ANNs so popular is the fact that given a set of labeled data (the medical history), an ANN will tune itself in an automated manner as to match these data in the best possible way. Unfortunately, this training process is not a trivial one. Numerous methodologies are presented in the literature for training ANNs, each one focusing on a different feature or situation. Thus, different training methodologies (as well as network structure) can be used when the number of training data is small, computing resources are limited, some or all of the historical medical data are unlabeled, there are missing data, the number of distinct patterns in the data is not known beforehand, and so on. For information on choosing, training, and using an ANN in a theoretical, research, or application environment, [17] is one of the best textbooks for both the novice and expert reader.

4.2.2.1 **Clustering**

One of the first things one finds out when trying to apply ANNs in a practical situation is the fact that their training is far from trivial. Even when having chosen the type of neurons, count of layers and training methodology to follow—neither of which is a trivial decision—the result of the training process is neither guaranteed nor set. Quite the contrary, depending on delicate issues such as the initial parameters of the network or even the order in which the training data are presented, the training procedure may differ in both time required and quality of results acquired. Clearly, it is desired to somehow decouple network performance from such issues. In order to avoid these dependences, one would need to have adequate prior knowledge as to initialize the ANN so that it is almost already trained correctly, in this case the adaptation taking place during training is minimal and does not depend that much on issues such as the order of the training data.

A common approach to this problem is to utilize a data-clustering approach in order to roughly extract the patterns that underlie in the data, and then map this information on the structure of the network. The results of this clustering process are so important for the training of the network that it is often mentioned that this may even be the most important determinant of the later operation of the CDSS as well [18].

Although a wide variety of clustering methodologies exists and has been used in the framework of ANN initialization, most methods fall under one of two categories: partitioning and agglomerative approaches. Partitioning approaches, much similarly to ANNs, start from some initial parameters and recursively optimize them. Specifically, given some initial estimation of the cluster's positions (centers, shapes, spreads, etc.) partitioning methods will classify all available data points to these clusters, and then use the results in order to update cluster parameters. Continuing recursively until equilibrium is found, changes are below a given threshold or enough repetitions have been made, the results of the process are refined. Variations, as for example having clusters with multiple centers, have also been proposed. Of course, in order for such a method to be applied, it is a requirement that available data can be positioned in an n-dimensional space.

Although partitioning approaches are the most robust and error resilient, the fact that they require an initial estimation as initialization makes them poor candidates for the initialization of ANNs, as they do not provide for complete automation of the process. Specifically, it is the fact that one needs to know the count of clusters, i.e., the count of distinct patterns in the data, beforehand that makes their application difficult.

On the other hand, agglomerative approaches start by considering each data sample as a distinct cluster and recursively merge the two most similar clusters until no similar clusters remain. The exact definition of the termination criterion as well as the definition of similarity between clusters are not trivial problems with unique answers; quite the contrary, they provide for a very wide variety of agglomerative clustering flavors.

The following works are based on the utilization of clustering approaches toward the initial exploration of the data patterns underlying in the available data [19–21], while [22] focuses specifically on the effect this has on the performance of the resulting ANN.

4.2.2.2 Self-Organizing Maps

Clearly, the fact that one needs to have a clear idea concerning the structure of the network (count of layers, count of neurons) before even training and testing can start is a very important limitation for ANNs. As one might expect, there have been efforts to produce ANN structures that overcome this. Resource-allocating neural networks were the first attempt in this direction. These networks have the ability to alter their structure during training by adding neurons when needed—and some times by even removing neurons no longer needed. Unfortunately, these networks remain sensitive to initial parameters, as incorrect initialization leads to the generation of too many neurons, thus resulting in over-fitting to the training data and poor classification performance when it comes to future data [22].

Kohonen's self-organizing maps (SOMs) constitute a more interesting and robust approach to training a network with no prior knowledge of the underlying data structures.

A number of details about the selection of the parameters, variants of the map, and many other aspects have been covered in the monograph [23]. SOMs, following an unsupervised learning approach, also achieve excellent data reduction and visualization, making it easier for humans to grasp the detected structures.

Due to their excellent properties and characteristics, SOMs have found numerous applications in CDSSs, such as [19, 24–27].

4.2.2.3 Support Vector Machines

Support vector machines (SVMs) are a relatively new type of training methodology for ANNs. Their characteristic is that they transform the input space, as to project training data from the original input space to a new one, where classes are linearly separable and thus linear learning algorithms can be applied. SVMs guarantee the maximal margin between different classes through global optimization of the decision boundary.

An important characteristic for SVM learning is that the margin between two classes depends solely on those data samples that are next to it. These training data samples—marked as gray in Fig. 4.3—are named support vectors and are the only ones to be considered when training the network (optimizing the margin between the classes).

Of course, once data have been projected to a space where they are linearly separable, training is both trivial and fast. Unfortunately, the choice of this mapping function, the kernel function, is not equally trivial. Prior knowledge of the characteristics of the data may contribute to a proper choice, but in practice selecting the optimal kernel is the main challenge in the application of SVMs [28].

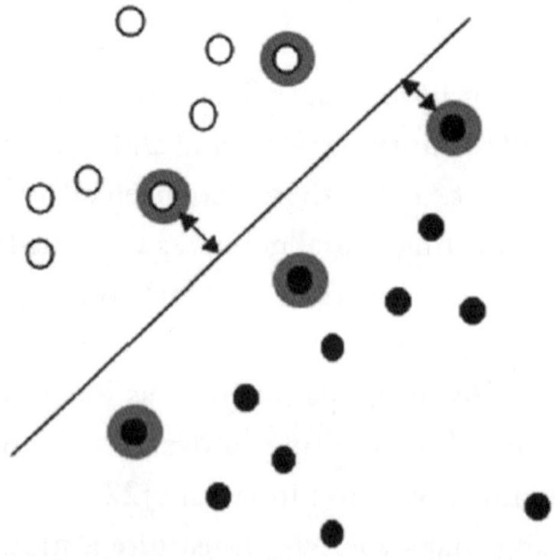

FIGURE 4.3: Operation of the support vector machine

Due to the fact that, unlike other ANNs, SVMs focus on maximizing the margin between classes, thus minimizing the probability of misclassification, they are extremely popular in CDSSs, where the cost of a misclassification may have a direct impact on human life. The following works are just a few examples of works in the medical field that are based on SVM learning [29–32].

4.2.3 Decision Trees

Humans using CDSSs, even those experts who are active in the research that develops them, are often reluctant to leave important medical decisions up to a subsymbolic, and thus generally incomprehensible, automated engine. A main drawback of ANNs in this framework is the fact that in order for a single output to be computed, numerous inputs are considered simultaneously and in different ways, which makes the process practically impossible for a human to follow. Decision trees offer an alternative computing methodology that reaches a decision through consecutive, simple question and answer sessions.

In the learning phase (when the decision tree is constructed) exactly one of the available features needs to be selected as the root feature; i.e. the most important feature in determining the diagnosis. Then data are split according to the value they have for this feature, and each group of data is used in order to create the corresponding child (subtree) of the root. If all of the data in a group belong to the same diagnosis, then that child becomes a leaf to the tree and is assigned that diagnosis. Otherwise, another feature is selected for that group, and data are again split leading to new groups and new children for this node. The choice of feature to use in order to split the data at any given step turns out to be a crucial one, as it can make the difference between a compact and a huge decision structure. Once a tree has been constructed in this manner, using the tree in order to classify a future case is quite similar to answering a small set of questions, each one regarding only one feature.

The decision tree in Fig. 4.4, for example, indicates that successful patients have more than one embryo transferred (NUM_TRANS > 1), have no additional hormonal stimulation (FLARE $= 0$), are of age less than or equal to 40 (AGE ≤ 40), and have no trauma during implantation of the embryos (TRAUMA $= 0$). Actually, in order to cope with errors in data, overlapping classes, and so on, real life decision trees also incorporate chance nodes, where instead of a clear decision, probabilities for different outcomes are presented.

Overall, decision trees are known for being robust to noisy data and capable of learning disjunctive expressions. Decision tree learning is one of the most widely used and practical methods for inductive inference. On the low side, the discrete structure of decision trees does not allow them to be applied in cases where the input variables are neither discrete nor easily partitioned in a small number of meaningful distinct ranges.

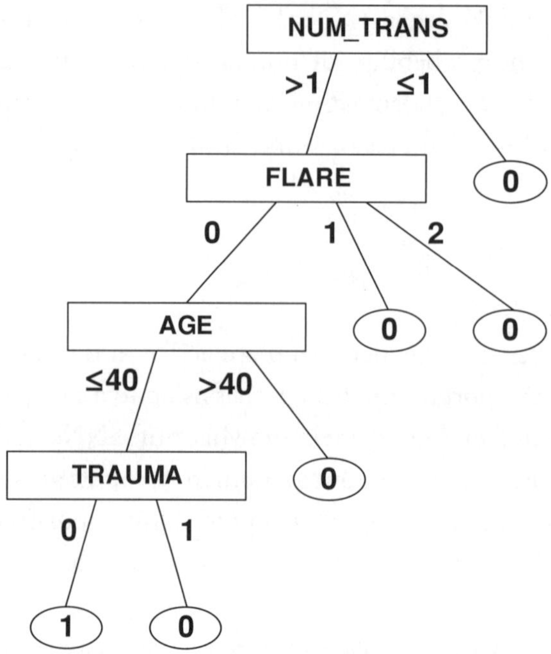

FIGURE 4.4: A sample decision tree

When it comes to CDSSs, decision trees were for long the most popular approach to decision making, and apparently, even though other data mining and machine learning approaches progress, still remain close to the top when it comes to popularity [33–37]. For a review of decision tree applications in medicine one may also consult [38]. It seems that decision trees will not fade out soon, as there is still work in progress toward further improving their performance, and their fields of application are constantly expanding.

4.2.4 Ensemble Machines

Classifiers such as the ones mentioned above, as well as many others, have found numerous applications in medicine and elsewhere. Each one of these classifiers finds a situation in which it performs its best, typically outperforming every other known approach. On the other hand, it is almost never possible to know beforehand which classifier will perform best on future data, which is the actual goal of designing and building a CDSS. In other words, although we have developed a powerful arsenal comprising a wide range of methodologies that one may use at any given time, we are quite unable to select the perfect tool for the job when it comes to developing a specific decision support system.

The ensemble methodology is the approach that helps overcome this obstacle. According to it, much similarly to having a group of doctors provide their independent opinions and

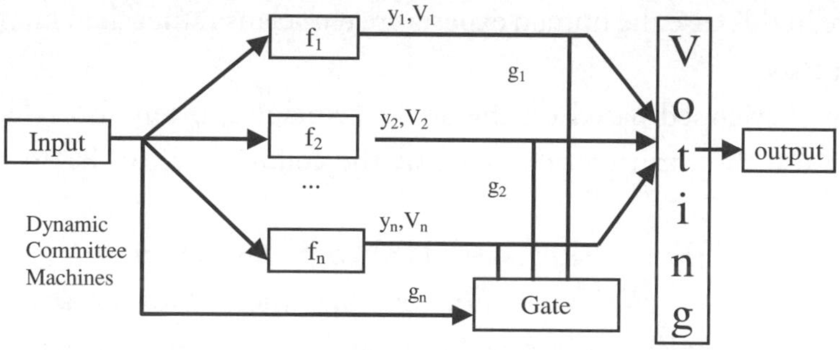

FIGURE 4.5: A dynamic committee machine architecture

then combining them to come up with a more reliable diagnosis, multiple simple classifiers are applied and their results are combined in order to acquire the output of the decision support system [39]. This combination can range from a simple averaging, to weighted averaging, or even to dynamic committee machines that consider the certainty in the output of each classifier, as estimated by the margin with which the classifier separates the classes for the specific inputs (Fig. 4.5).

As new information always reduces uncertainty, the consideration of more than one classifiers that have been trained and applied independently can only enhance the results of the classification process. Thus, ensemble approaches to classification allow for the generation of CDSSs with very high classification rates. For example, the following works have enhanced results due to the fact that they are based on ensemble approaches [40–42]. In general, it is now generally accepted that best results that one can acquire are those that come from the combination of multiple classifiers.

4.2.5 Fuzzy Logic and Fuzzy Rule Systems

In all classification schemes presented in the previous sections, automated analysis of medical history provided for the acquisition of the knowledge required for the operation of the CDSS, while the role of the human expert was limited to the labeling of historical medical data, i.e., to providing correct diagnoses for past cases. Alternatively to examining medical history in order to extract information, a CDSS may operate based on information already available to humans and provided directly to the system; in that case the role of the CDSS is to make up for human experts that cannot be available at all times when they may be needed.

There is typically no learning process in the development of a knowledge-based system. Instead, human experts provide directly the knowledge that the system needs to operate. Of course, this cannot be done at a subsymbolic level. The format of the knowledge needs to be one that facilitates the human experts; more so in knowledge-based systems developed for medical applications, since in this case the human experts are clinicians rather than knowledge engineers or computer scientists.

The systems developed based on the above principles, often also referred to as expert systems, typically use the form of rules such as the following in order to represent human knowledge:

IF *high_heart_rate* THEN *hyperthyroidism*

Although simple in definition, it turns out that these rules are not equally simple in evaluation. First of all, as can clearly be seen from the above example, measured medical features cannot be fed directly to the decision-making component; they first have to be transformed to lexical variables such as *has_euphoria*. One soon realizes that this cannot be done in a binary sense; for example there are different degrees of euphoria, as there are different degrees to which one may display manic symptoms. As a consequence, rule-based systems are closely related to the theory that best supports degrees and lexical variable, the theory of fuzzy sets and fuzzy logic.

In a fuzzy logic system, measured inputs are mapped to linguistic variables, as indicated in Fig. 4.6. A given measurement can be mapped to contradicting variables at the same time and to different degrees. As a result multiple, and often contradicting rules may fire, again each one to its degree. In order to acquire a classification, the result of this process is de-fuzzified; i.e. the activation levels of all rules are examined so that a crisp (not fuzzy) decision can be made. It is worth noting that, although the term "fuzzy rule" is quite common, the rules themselves are anything but fuzzy; it is the way they are being evaluated that associates them to degrees.

Due to the human perceivable nature of the knowledge representation format that rule-base systems use, it has always been a goal to design a rule-base system that is trained automatically from medical history. That would mean that the system, additionally to decision support, could offer the means to automatically explore data and propose possible associations to researchers. Neurofuzzy systems are one breed of classifiers that possess this property [17]. They have the structure and training characteristics of traditional ANNs, but their structure can be directly generated from, or be directly mapped to fuzzy rules. Unfortunately, the proper

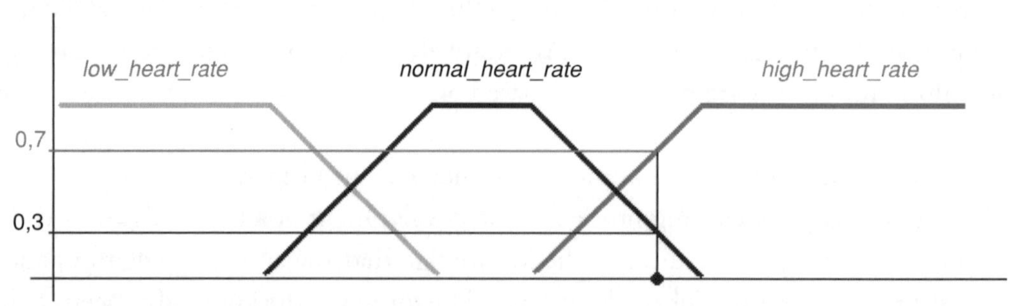

FIGURE 4.6: Three linguistic variables for the same measured feature

initialization and training of a neurofuzzy system turns out to be overly complex, as one needs to know beforehand how measure variables are mapped best to linguistic variables.

What seems to be more promising is the less direct combination of ANN operation with rule-base systems. As ANN-learning focuses on detecting underlying data patterns, if one is able to somehow extract the information stored in the structure and parameters of the network, then one can automatically generate the fuzzy rules that characterize the observed biological system; this process is known as rule extraction from ANNs. For some types of ANNs, as are, for example radial basis function (RBF) networks, the definition of the linguistic variable intervals, as well as the definition of the rules is not overly complex.

Still, the rules typically extracted in this process differ from what is desired, due to the fact that they are overly complex as they combine almost all available inputs. In order for both meaningful and human perceivable rules to be extracted, rule antecedents (inputs) that are not as important need to be trimmed; the ideal is to get rules that are as simple as possible and yet provide for a classification rate as high as that of the untrimmed rules. Various approaches have been proposed in this direction, with evolutionary ones being the ones that are currently gaining in momentum [43].

The fact that rule-base systems can i) accept information provided by human experts in a natural way and ii) provide the knowledge generated during their training in a human perceivable way has made them a very popular choice for CDSSs. As far as information extraction is concerned, there are numerous works focusing on rule extraction from ANNs that have been trained on medical data [44]. As far as the application of fuzzy logic and rule-base systems in a medical context, the following are only few examples [45, 46]. For a review of the field, readers may refer to [47, 48]. When it comes to actual implemented systems, they are far too many to mention. For an indicative list of more than 60 real life systems, one may refer to Federhofer's page [49].

4.3 EVALUATION

Although all of the methodologies and approaches presented in this chapter have found applications in the medical field with important results, none of them is a panacea that is suitable for utilization in all cases. Quite the contrary, any of these methodologies will perform at a level much below what one might consider as acceptable when not properly initialized or fine-tuned to the problem at hand. As a result, before a CDSS is put into actual use, it needs to be carefully evaluated, so that its error margins and weaknesses in general are identified and clinicians are informed about them; obviously, a clinician should not weigh the diagnosis of an automated system with 99% accuracy similarly to that of a system with 85% accuracy.

In classification research in general, a system is both trained and evaluated using available medical history. When the data suffice in number, a part of it (typically the smaller part) is used for training, while remaining data are used for testing. When less data are available, techniques such as leave one out may be used, where all but one available data points are used to train the classifier and the remaining point is used for testing; all points recursively play the role of the testing data and overall results are evaluated statistically [5].

While both approaches are meaningful from the evaluator's point of view, they are not necessarily 100% reasonable for a CDSS. As far as the leave one out approach is concerned, it provides an estimation of the potential of the overall classification methodology applied; as multiple distinct classifiers are trained, one cannot tell at the end which classifier parameters the results refer to, in order to construct the CDSS that can achieve similar classification rates. As far as training data/testing data separation is concerned, using the smaller part of the data for training means that useful information is not considered in the construction of the CDSS in order to pursue maximum performance. One can see that the way to evaluate a CDSS is a challenge on its own, often exceeding in difficulty and importance the development of the system itself. Typically, a training data/testing data approach is followed, but with most of the available medical history used for training and a smaller part as test data.

A more subtle but equally important issue to consider is that of the suitability of the medical history for the evaluation of the system. Clinicians now accept that living in different regions of the world, as well as living in different climates and so on, affects one's body in a way that may make different medical feature measurements qualify as "normal" in each case. As a result, medical history collected from patients from a specific part of the world may not be suitable in order to evaluate a system that will be used in another. Additionally, medical history of previous years may not always be suitable for the evaluation of a system that will be put in use today. It is exactly this last remark that makes the evaluation of CDSSs an ongoing process; a CDSS that has been optimally configured may need to be reconfigured in order to adapt to an era where a difference in feeding habits or even simply in lifestyle has an effect on typically measured medical values, when clinicians empirically observe that error margins have augmented.

From a technical point of view, an interesting point to consider is the way to measure the performance of a CDSS. For classification systems in general, evaluation is typically based on a true/false and positive/negative scheme. When adopted in the medical case, true positive (TP) is a correct illness estimation, true negative (TN) a correct healthy estimation, false positive (FP) an illness estimation for a healthy case, and a false negative (FN) a healthy estimation for an ill case. Based on these, accuracy is defined as follows:

$$Accuracy = \frac{TP + TN}{TP + TN + FP + FN} \tag{4.1}$$

The simplistic approach of simply counting correct and incorrect classifications in order to estimate accuracy, although generally accepted in other expert systems and classifiers, is not sufficient for the case of medical systems, where one type of mistake may be much more important—as far as the possible consequences are concerned—compared to another. For example a false positive estimation has a result of a patient taking extra tests in order to verify one's health status, whereas a false negative diagnosis may deprive one from an early diagnosis and treatment. Finally, classes in a medical setting are rarely balanced; it is only typical that only a small percentage of people examined will actually be ill. As a result, a system that always provides a "healthy" diagnosis reaches high classification rates.

In order to compensate for this, a more flexible consideration of errors needs to be used, in order for class probabilities to be considered as well. A simple approach that is commonly followed in this direction is the utilization of specificity and sensitivity measures, defined as follows:

$$Specificity = \frac{TN}{TN + FP} \tag{4.2}$$

$$Sensitivity = \frac{TP}{TP + FN} \tag{4.3}$$

where specificity and sensitivity are actually measures of accuracy, when considering only healthy or only ill cases, respectively, thus decoupling the measures from class probabilities.

4.4 CONCLUSIONS

CDSSs originate from data mining and generic decision support systems, both of which are fields that have matured over the years and are now at a position to provide reliable and robust practical solutions. As a result, high-quality CDSSs have been developed. Still, one cannot forget that an automated decision support system cannot become perfect, when the data it relies its operation on is not.

It is perhaps not surprising that numerous clinicians hesitate to utilize decision support systems in their line of work [50]. This is of course disappointing, given the fact that CDSSs when properly used have been found to make a clear difference in the quality of treatment patients receive [51]. Lately, the opposite, and quite possible more dangerous, extent is also observed at times, with clinicians often trusting CDSSs more than their own judgment [52].

Clearly, the goal is to reach a stage where CDSSs are integrated in the process of everyday clinical work, but without being assigned roles they are not made for, such as the role of the actual clinician. It seems that a number of parameters will have an effect in this process, ranging from purely financial issues to degree of automation, from availability at the time and location support is needed to the ease of the user interface, and from the quality of the data acquisition components to the success of the system development and integration procedures [53]. The areas in which a CDSS could make a difference are, of course, countless. Table 4.1 provides only a summary of the most important ones.

TABLE 4.1: Potential Users and Uses for a CDSS

USER	APPLICATION
Pharmacists	Drug levels, drug/drug interactions, culture and sensitivity results, adverse drug events
Physicians	Critical lab results, culture and sensitivities
Nurses	Critical lab results, drug/drug interactions
Dietary	Patient transfers, lab support for tube feedings
Epidemiology/infection control	VRE and MRSA results, reportable organisms
Housekeeping	ADT functions
Billing	Excessively expensive tests and treatments
Administration	Patient chart administration
Chaplain	Counseling opportunity
Case manager	Premature readmission
Patient	Drug/drug interactions, drug dosing, missing tests

Regardless of their penetration degree in traditional clinical practice, CDSSs have not yet reached the extent of application one might hope for in the telemedicine case. In this framework, e-Health continues to refer mainly to distributed information systems for the organization and exchange of medical records and information in general; automated processing and decision support components are still considered by many as a luxury for e-Health, mainly due to their augmented networking and processing needs.

As networking bandwidth is now becoming cheaper and more easily available, and portable and handheld devices are becoming more powerful, it is only natural to expect a growth in the number and type of services e-Health will incorporate, with decision support being a primary candidate. Already a number of Internet-based or distributed medical decision support systems have been proposed and developed [54–56], while others are certainly on their way.

REFERENCES

[1] L. E. Perreault, and J. B. Metzger, "A pragmatic framework for understanding clinical decision support," *J. Healthc. Inf. Manage. Syst. Soc.*, vol. 13, no. 2, pp. 5–21, 1999.

[2] J. C. Wyatt and J. L. Liu, "Basic concepts in medical informatics," *J. Epidemiol. Community Health*, vol. 56, pp. 808–812, 2002.

[3] B. Kaplan, "Evaluating informatics applications—clinical decision support systems literature review," *Int. J. Med. Inf.*, vol. 64, no. 1, pp. 15–37, 2001.

[4] K. Zheng, R. Padman, M. P. Johnson, and H. S. Diamond, "Understanding technology adoption in clinical care: Clinician adoption behavior of a point-of-care reminder system," *Int. J. Med. Inf.*, vol. 74, no. 7–8, pp. 535–543, 2005.

[5] T. S. Lim, W. Y. Loh, and Y. S. Shih, "A Comparison of prediction accuracy, complexity, and training time of thirty-three old and new classification algorithms," *Mach. Learn.*, vol. 40, pp. 203–229, 2000.

[6] L. Bull, *Applications of Learning Classifier Systems*. New York: Springer-Verlag, 2004.

[7] M. Ohlsson, "WeAidU—A decision support system for myocardial perfusion images using artificial neural networks," *Artif. Intell. Med.*, vol. 30, no. 1, pp. 49–60, 2004.

[8] M. Hilario, A. Kalousis, M. Muller, and C. Pellegrini, "Machine learning approaches to lung cancer prediction from mass spectra," *Proteomics*, vol. 3, pp. 1716–1719, 2003.

[9] J. Prados, A. Kalousis, J. C. Sanchez, L. Allard, O. Carrette, and M. Hilario, "Mining mass spectra for diagnosis and biomarker discovery of cerebral accidents," *Proteomics*, vol. 4, pp. 2320–2332, 2004.

[10] M. Wagner, D. Naik, A. Pothen, S. Kasukurti, R. Devineni, B. L. Adam, O. J. Semmes, and G. L. Wright Jr., "Computational protein biomarker prediction: A case study for prostate cancer," *BMC Bioinform.*, vol. 5, p. 26, 2004.

[11] A. E. Smith, C. D. Nugent, and S. I. McClean, "Evaluation of inherent performance of intelligent medical decision support systems: Utilising neural networks as an example," *Artif. Intell. Med.*, vol. 27, no. 1, pp. 1–27, 2003.

[12] M. E. Futschik, M. Sullivan, A. Reeve, and N. Kasabov, "Prediction of clinical behaviour and treatment for cancers," *OMJ Appl. Bioinform.*, vol. 2, no. 3, pp. 53–58, 2003.

[13] G. Ball, S. Mian, F. Holding, R. O. Allibone, J. Lowe, S. Ali, G. Li, S. McCardle, I. O. Ellis, C. Creaser, and R. C. Rees, "An integrated approach utilizing artificial neural networks and SELDI mass spectrometry for the classification of human tumours and rapid identification of potential biomarkers," *Bioinformatics*, vol. 18, no. 3, pp. 395–404, 2002.

[14] L. J. Lancashire, S. Mian, I. O. Ellis, R. C. Rees, and G. R. Ball, "Current developments in the analysis of proteomic data: Artificial neural network data mining techniques for the identification of proteomic biomarkers related to breast cancer," *Curr. Proteom.*, vol. 2, no. 1, pp. 15–29, 2005.

[15] I. Maglogiannis and E. Zafiropoulos, "Utilizing support vector machines for the characterization of digital medical images," *BMC Med. Inf. Decis. Making*, vol. 4, no. 4, 2004, http://www.biomedcentral.com/1472-6947/4/4/

[16] P. J. Lisboa, "A review of evidence of health benefit from artificial neural networks in medical intervention," *Neural Netw.*, vol. 15, no. 1, pp. 11–39, 2002.

[17] S. Haykin, *Neural Networks: A Comprehensive Foundation*, 2nd ed. Prentice Hall, 1999.

[18] J. H. Chuang, G. Hripcsak, and R. A. Jenders, "Considering clustering: A methodological review of clinical decision support system studies," in *Proc. AMIA Symp.*, 2000, pp. 146–150.

[19] H. Wang, F. Azuaje, and N. Black, "An integrative and interactive framework for improving biomedical pattern discovery and visualization," *IEEE Trans. Inf. Technol. Biomed.*, vol. 8, no. 1, pp. 16–27, 2004.

[20] H. M. Kuerer, K. R. Coombes, J. N. Chen, L. Xiao, C. Clarke, H. Fritsche, S. Krishnamurthy, S. Marcy, M. C. Hung, and K. K. Hunt, "Association between ductal fluid proteomic expression profiles and the presence of lymph node metastases in women with breast cancer," *Surgery*, vol. 136, no. 5, pp. 1061–1069, 2004.

[21] P. V. Purohit and D. M. Rocke, "Discriminant models for high-throughput proteomics mass spectrometer data," *Proteomics*, vol. 3, pp. 1699–1703, 2003.

[22] M. Wallace, N. Tsapatsoulis, and S. Kollias, "Intelligent initialization of resource allocating RBF networks," *Neural Netw.*, vol. 18, no. 2, pp. 117–122, 2005.

[23] T. Kohonen, *Self-Organizing Maps*, 2nd ed. Berlin: Springer-Verlag, 1997.

[24] T. P. Conrads, V. A. Fusaro, S. Ross, D. Johann, V. Rajapakse, B. A. Hitt, S. M. Steinberg, E. C. Kohn, D. A. Fishman, G. Whitely, J. C. Barrett, L. A. Liotta, E. F. Petricoin, and T. D. Veenstra, "High-resolution serum proteomic features for ovarian cancer detection," *Endocr. Relat. Cancer*, vol. 11, no. 2, pp. 163–178, 2004.

[25] D. J. Johann Jr., M. D. McGuigan, S. Tomov, V. A. Fusaro, S. Ross, T. P. Conrads, T. D. Veenstra, D. A. Fishman, G. R. Whiteley, E. F. Petricoin, and L. A. Liotta, "Novel approaches to visualization and data mining reveals diagnostic information in the low amplitude region of serum mass spectra from ovarian cancer patients," *Dis. Markers*, vol. 19, pp. 197–207, 2004.

[26] D. Ornstein, W. Rayford, V. Fusaro, T. Conrads, S. Ross, B. Hitt, W. Wiggins, T. Veenstra, L. Liotta, and E. Petricoin, "Serum proteomic profiling can discriminate prostate cancer from benign prostates in men with total prostate specific antigen levels between 2.5 and 15.0 NG/ML," *J. Urol.*, vol. 172, no. 4, pp. 1302–1305, 2004.

[27] J. H. Stone, V. N. Rajapakse, G. S. Hoffman, U. Specks, P. A. Merkel, R. F. Spiera, J. C. Davis, E. W. St.Clair, J. McCune, S. Ross, B. A. Hitt, T. D. Veenstra, T. P. Conrads, L. A. Liotta, and E. F. Petricoin, "A serum proteomic approach to gauging the state of remission in Wegener's granulomatosis," *Arthritis Rheum.*, vol. 52, pp. 902–910, 2005.

[28] H. Shi and M. K. Markey, "A machine learning perspective on the development of clinical decision support systems utilizing mass spectra of blood samples," *J. Biomed. Inf.*, to be published.

[29] A Statnikov, C. F. Aliferis, and I. Tsamardinos, "Methods for multi-category cancer diagnosis from gene expression data: A comprehensive evaluation to inform decision support system development," *Medinfo*, vol. 11, pp. 813–817, 2004.

[30] L. Li, H. Tang, Z. Wu, J. Gong, M. Gruidl, J. Zou, M. Tockman, and R. A. Clark, "Data mining techniques for cancer detection using serum proteomic profiling," *Artif. Intell. Med.*, vol. 32, pp. 71–83, 2004.

[31] B. Wu, T. Abbott, D. Fishman, W. McMurray, G. Mor, K. Stone, D. Ward, K. Williams, and H. Zhao, "Comparison of statistical methods for classification of ovarian cancer using mass spectrometry data," *Bioinformatics*, vol. 19, no. 13, pp. 1636–1643, 2003.

[32] I. Maglogiannis, S. Pavlopoulos, and D. Koutsouris, "An integrated computer supported acquisition, handling and characterization system for pigmented skin lesions in dermatological images," *IEEE Trans. Inf. Technol. Biomed.*, vol. 9, no. 1, pp. 86–98, 2005.

[33] J. R. Trimarchi, J. Goodside, L. Passmore, T. Silberstein, L. Hamel, and L. Gonzalez, "Assessing decision tree models for clinical in-vitro fertilization data," Dept. of Computer Science and Statistics, Univ. Rhode Island, Tech. Rep. TR03-296, 2003.

[34] R. D. Niederkohr and L. A. Levin, "Management of the patient with suspected temporal arteritis: A decision-analytic approach," *Ophthalmology*, vol. 112, no. 5, pp. 744–756, 2005.

[35] N. Ghinea and J. M. Van Gelder, "A probabilistic and interactive decision-analysis system for unruptured intracranial aneurysms," *Neurosurg. Focus*, vol. 17, no. 5, p. E9, 2004.

[36] M. K. Markey, G. D. Tourassi, and C. E. J. Floyd, "Decision tree classification of proteins identified by mass spectrometry of blood serum samples from people with and without lung cancer," *Proteomics*, vol. 3, no. 9, pp. 1678–1679, 2003.

[37] H. Zhu, C. Y. Yu, and H. Zhang, "Tree-based disease classification using protein data," *Proteomics*, vol. 3, no. 9, pp. 1673–1677, 2003.

[38] V. Podgorelec, P. Kokol, B. Stiglic, and I. Rozman, "Decision trees: An overview and their use in medicine," *J. Med. Syst.*, vol. 26, no. 5, pp. 445–463, 2002.

[39] G. Fumera and F. Roli, "A theoretical and experimental analysis of linear combiners for multiple classifier systems," *IEEE Trans. Pattern Anal. Mach. Intell.*, vol. 27, no. 6, pp. 942–956, 2005.

[40] P. Mangiameli, D. West, and R. Rampal, "Model selection for medical diagnosis decision support systems," *Decis. Support Syst.*, vol. 36, no. 3, pp. 247–259, 2004.

[41] A. C. Tan and D. Gilbert, "Ensemble machine learning on gene expression data for cancer classification," *Appl. Bioinform.*, vol. 2, no. 3, pp. 75–83, 2003.

[42] P. L. Martelli, P. Fariselli, and R. Casadio, "An ENSEMBLE machine learning approach for the prediction of all-alpha membrane proteins," *Bioinformatics*, vol. 19, pp. 205–211, 2003.

[43] M. Wallace and N. Tsapatsoulis, "Combining GAs and RBF neural networks for fuzzy rule extraction from numerical data," in *Int. Conf. Artificial Neural Networks*, Warsaw, Poland, Sep. 2005.

[44] J. R. Rabunal, J. Dorado, A. Pazos, J. Pereira, and D. Rivero, "A new approach to the extraction of ANN rules and to their generalization capacity through GP," *Neural Comput.*, vol. 16, no. 7, pp. 1483–1523, 2004.

[45] J. H. T. Bates and M. P. Young, "Applying fuzzy logic to medical decision making in the intensive care unit," *Am. J. Respir. Crit. Care Med.*, vol. 167, no. 7, pp. 948–952, 2003.

[46] P. R. Innocent, R. I. John, and J. M. Garibaldi, "Fuzzy methods for medical diagnosis," *Appl. Artif. Intell.*, vol. 19, no. 1, pp. 69–98, 2005.

[47] S. Barro and R. Marin, *Fuzzy Logic in Medicine*. Physica-Verlag (Heidelberg), 2002.

[48] M. Mahfouf, M. F. Abbod, and D. A. Linkens, "A survey of fuzzy logic monitoring and control utilisation in medicine," *Artif. Intell. Med.*, vol. 21, no. 1–3, pp. 27–42, 2001.

[49] http://www.computer.privateweb.at/judith/special_field3.htm

[50] N. Rousseau, E. McColl, J. Newton, J. Grimshaw, and M. Eccles, "Practice based, longitudinal, qualitative interview study of computerized evidence-based guidelines in primary care," *Br. Med. J.*, vol. 326, pp. 1–8, 2003.

[51] A. X. Garg, N. K. Adhikari, H. McDonald, M. P. Rosas-Arellano, P. J. Devereaux, J. Beyene, J. Sam, and R. B. Haynes, "Effects of computerized clinical decision support systems on practitioner performance and patient outcomes: A systematic review," *J. Am. Med. Assoc.*, vol. 293, no. 10, pp. 1223–1238, 2005.

[52] S. Dreiseitl and M. Binder, "Do physicians value decision support? A look at the effect of decision support systems on physician opinion," *Artif. Intell. Med.*, vol. 33, no. 1, pp. 25–30, 2005.

[53] K. Kawamoto, C. A. Houlihan, E. A. Balas, and D. F. Lobach, "Improving clinical practice using clinical decision support systems: A systematic review of trials to identify features critical to success," *Br. Med. J.*, vol. 330, no. 7494, pp. 740–741, 2005.

[54] D. Jegelevicius, V. Marozas, A. Lukosevicius, and M. Patasius, "Web-based health services and clinical decision support," *Stud. Health Technol. Inform.*, vol. 105, pp. 27–37, 2004.

[55] P. Fortier, S. Jagannathan, H. Michel, N. Dluhy, and E. Oneil, "Development of a hand-held real-time decision support aid for critical care nursing," *HICSS'03*, vol. 6, no. 6, p. 160, 2003.

[56] G. Eysenbach, J. Powell, M. Englesakis, C. Rizo, and A. Stern, "Health related virtual communities and electronic support groups: Systematic review of the effects of online peer to peer interactions," *Br. Med. J.*, vol. 328, no. 7449, pp. 1–6, 2004.

CHAPTER 5

Medical Data Coding and Standards

5.1 INTRODUCTION

Standards are generally required when excessive diversity creates inefficiencies or impedes effectiveness. Such a case exists in the electronic healthcare environment. In healthcare, standards are needed for encoding data about the patient that are collected by one system and used by another. One obvious need is the development of standardized identifiers for individuals, healthcare providers, health plans, and employers so that such individuals can be recognized across systems. Because of the diversity of patient data creation and storage points and the variety of medical IT-based equipment, healthcare is one area that can ultimately greatly benefit from implementation of standards [1].

Patients receive care across primary, secondary, and tertiary care settings with little bidirectional communication and coordination among the services. Patients are cared for by one or more primary physicians as well as by specialists. There is little coordination and sharing of data between inpatient and outpatient care. Within the inpatient setting, the clinical environment is divided into clinical specialties that frequently treat the patient without regard to what other specialties have done [1].

The lack of "standards" for electronic coding of medical data has been a major obstacle to establishing an integrated electronic medical record. However, during the last years significant efforts by major standard bodies (CEN, IEEE, HL7, ANSI, ISO) have produced several standards in medical informatics; some of them (i.e., HL7 and DICOM) are well established in the e-health sector. This chapter provides short descriptions of the dominant standards, indicating how they may be used for the development of distributed networked e-health applications.

5.2 THE MAJOR MEDICAL INFORMATICS CODING STANDARDS

5.2.1 Health Level 7 (HL7)

HL7 stands for Health Level 7 and is the highest level of the ISO communication model. It is a standard for data interchange. The model's purpose is to archive OSI (open systems interconnections), a way to get different systems to work together. The OSI model is not specific to medical informatics; however, HL7 is specific to the healthcare domain.

The current version of HL7 is version 3 with basic features being the object-oriented approach, the definition of a *Reference Information Model* (*RIM*) in UML defining all messages, objects and their relationships, the inclusion of specific conformance to the standard statements, and the use of *extensive markup language* (*XML*) as a structured communication language.

The use of the XML and the RIM has enabled the exchange of medical data among different healthcare organizations and devices and consequently the development of distributed networked e-health applications. For instance external systems with XML-aware browsers can communicate with a hospital information system (HIS) through the HL7 standard.

The structure of an HL7 message is as follows:

- Each message is constituted by specific **segments** in a concrete line.
- Each message has a **message type** that determines his scope.
- Each type of message corresponds in one or more actual incidents (**trigger event**). Actual incidents can be, for example, a patient admission or a medical examination order.
- Each segment has specific **data fields**.
- Segments in an HL7 message are either mandatory or optional while it is possible they are repeated in a message. Each segment has a concrete name. For example a message related to patient admission discharge and transfer functions (Type: ADT) can include the following segments: Message Header (MSH), Event Type (EVN), Patient ID (PID), and Patient Visit (PV1)

Such a message produced to notify the admission to surgery of a patient is as follows:

- MSH|**ADT1**|MCM|**LABADT**|MCM|**199808181126**|SECURITY|ADT^A01|MSG00001|**P**|**3.0**|
- EVN|A01|**199808181123**||
- PID|||PATID1234^5^M11||**JONES^WILLIAM^A^III**||19610615|M||C|1200 N ELM STREET^^**GREENSBORO**^NC^27401-1020|GL|(919)379-1212|(919)271434||S||PATID**12345001**^2^M10|123456789|987654^NC|
- NK1|**JONES^BARBARA^K**|**WIFE**||||||NK^NEXT OF KIN
- PV1|1|I|**2000^2012^01**||||**004777^LEBAUER^SIDNEY**^J.|||**SUR**|||| ADM| A0|

The message explanation is as follows: Patient William A. Jones, III was admitted for surgery (SUR) on 18th July 1998, 11.23 p.m. from the doctor Sidney J. Lebauer (# 004777).

He entered the surgical room 2012, bed 01 in the nursing unit 2000. This message was sent by the system ADT 1 to the system LABADT, 3 minutes afterward admission.

5.2.2 International Classification of Diseases– (ICD)

ICD is a specific standard for medical terminologies. It was first published in 1893 and has been revised at roughly every 10 year. The revisions are underlaid the World Health Organization (WHO). The tenth edition was published in 1992. The coding consists of a core classification of three digits. A fourth digit (the decimal) is used for further detail. The terms are arranged in a hierarchy based on the digits.

The ICD has become the international standard diagnostic classification for all general epidemiological and many health management purposes. These include the analysis of the general health situation of population groups, and monitoring of the incidence and prevalence of diseases, and other health problems in relation to other variables such as the characteristics and circumstances of the individuals affected.

It is used to classify diseases and other health problems recorded on many types of health and vital records including death certificates and hospital records. In addition to enabling the storage and retrieval of diagnostic information for clinical and epidemiological purposes, these records also provide the basis for the compilation of national mortality and morbidity statistics by WHO member states.

ICD-10 uses alphanumeric codes, with an alphabetic character in the first position. It includes the letters "I" and "O," which are usually avoided in alphanumeric coding because of the likelihood of confusion with the numerals "1" and "0". Somewhat surprising is the fact that chapters in *ICD-10* do not necessarily start with new alphabetic characters. For example see Table 5.1.

ICD is incorporated in most of electronic health record (EHR) implementations; therefore, it is quite powerful tool for the development of integrated distributed health records, concentrating the total medical history of a patient. Further information regarding the ICD standard is out of the scope of this textbook and may be found in [8].

5.2.3 Digital Imaging and Communications in Medicine (DICOM)

DICOM (DICOM homepage URL: http://medical.nema.org/) is a standard for exchanging medical images. Medical images comply with the DICOM format [4]. A single DICOM file contains both a header (which stores information about the patient's name, the type of scan, image dimensions, etc.) and all of the image data. The header contains information regarding its size, the size of the whole image, and usually information that has to do with the image type and characteristics (resolution, possible compression, number of frames, etc). Table 5.2 shows a DICOM compliant image file representation, where after the header section the image data follow.

The standard also specifies the following:

TABLE 5.1: An Example of an ICD-10 Classification Table

Chapter I	Certain infectious and parasitic diseases	A00–B99
Chapter II	Neoplasms	C00–D48
Chapter III	Diseases of the blood and blood-forming organs and certain disorders involving the immune mechanism	D50–D89
Chapter IV	Endocrine, nutritional, and metabolic diseases	E00–E90
...		
Chapter IX	Diseases of the circulatory system	I00–I99
...		
Chapter XV	Pregnancy, childbirth, and the puerperium	O00–O99
...		

1. A set of protocols for devices communicating over a network.
2. The syntax and semantics of commands and associated information that can be exchanged using these protocols.

TABLE 5.2: DICOM Compliant Image File Format Representation

<—— Header ——><—— Image ——>

Transfer Syntax UID	Definition	Value
0002,0000	File meta elements group length	150
0002,0001	File meta info version	256
0008,0008	Image type	SINGLE PLANEA
0008,0020	Study date	19941013
0008,0030	Study time	141917
0010,0030	Patient date of birth	19751025
0010,0040	Patient sex	M
0028,0004	Photometric interpretation	MONOCHROME2
0010,0020	Patient ID	556342B
0008,0090	Referring physician's name	Dr. A
0008,0080	Institution name	Medical Center
0010,0010	Patient's name	Patient A

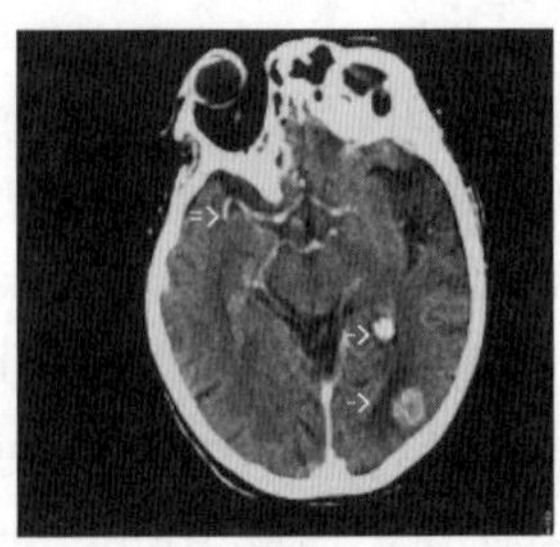

3. A set of media storage services and devices claiming conformance to the standard, as well as a file format, and a medical directory structure to facilitate access to the images and related information stored on media that share information.

The DICOM standard is used in almost all e-health applications handling medical images (i.e, picture-archiving and communicating systems, PACS, and telemedicine and teleradiology).

5.2.4 Other Standards

ASTM (ASTM homepage. URL: http://www.astm.org/) is a standard for message exchange about clinical observations, medical logic, and electrophysiologic signals.

ADA is a standard for data exchange and processing in the dental health care sector.

ANA is a standard for data exchange and processing for nursing services.

LOINC (The logical observation identifiers names and codes) provides a standard for a set of universal names and codes, clinical observations, and diagnostic study observations.

EDIFACT is a set of standards for interchange of data between independent computer-based systems.

5.3 BIOSIGNAL CODING STANDARDS

Biosignals in networked e-health applications are provided in numerical format (i.e., heart rate, respiratory rate, temperature of body, arterial blood pressure, etc.) and as waveforms (Electrocardiogram, ECG, and Electroencephalogram, EEG). Numerical data are very easy to handle and code in applications. Regarding the waveforms the acquisition and transmission of ECG is the most usual technique for the determination and the verification of a patient's condition in a remote position. Almost all modern ECG (mobile/wearable or not) are equipped with digital outputs and use digital techniques for communication. The computer-based coding and communication of ECG is regulated by the Standard Communications Protocol, SCP-ECG. SCP-ECG is a file format for ECG traces, annotations, and metadata. It is defined in the joint ANSI/AAMI standard EC71:2001 and in the CEN standard EN 1064:2005.

Practical experience during implementation and in the field confirmed its usability for telemetric applications as well as for data volume effective storage and retrieval (demonstrated in the OEDIPE project [9]).

5.3.1 SCP-ECG Standard Description

SCP-ECG is similar to the DICOM formatting concept. The ECG data are divided into different sections. Contents and format of each section are defined in the SCP document [10]. A global overview of the SCP-ECG data structure is presented in Table 5.3.

TABLE 5.3: Overview of the SCP-ECG Data Structure

CRC CHECKSUM
SIZE OF THE ENTIRE SCP-ECG RECORD
POINTERS TO DATA AREAS IN THE RECORD ("Table of content")
HEADER, e.g., PatID, Device ID, Recording ID (Time stamp, etc.)
ECG Data in optional formats without/with (selectable) compression methods
Various types of processing and over reading results

More specifically the information provided above is included in the SCP standard shown in Table 5.4 [10].

Each section is divided into two parts: The section ID Header and the section Data Part (see Fig. 5.1).

Although the section ID Header always has a length of 16 bytes, the section data part is variable. The SCP standard allows for a rather large number of options to store and format the ECG data. ECG data may be acquired at different sampling rates, with different quantization levels; they may be not compressed or be compressed by selectable methods and an SCP-ECG record may or may not contain analysis and over reading results. Also, the number of leads, the length of the recording interval and even the simultaneity of leads is left open to the manufacturers.

TABLE 5.4: SCP-ECG Data Fields

Mandatory	2 BYTES—CHECKSUM—CRC—CCITT OVER THE ENTIRE RECORD (EXCLUDING THIS WORD)
Mandatory	4 BYTES—(UNSIGNED) SIZE OF THE ENTIRE ECG RECORD (IN BYTES)
Mandatory	(Section 0)
	POINTERS TO DATA AREAS IN THE RECORD
Mandatory	(Section 1)
	HEADER INFORMATION—PATIENT DATA/ECG ACQUISITION DATA
Optional	(Section 2)
	HUFFMAN TABLES USED IN ENCODING OF ECG DATA (IF USED)

TABLE 5.4: (*Continued*)

Optional	(Section 3) ECG LEAD DEFINITION
Optional	(Section 4) QRS LOCATIONS (IF REFERENCE BEATS ARE ENCODED)
Optional	(Section 5) ENCODED REFERENCE BEAT DATA IF REFERENCE BEATS ARE STORED
Optional	(Section 6) "RESIDUAL SIGNAL" AFTER REFERENCE BEAT SUBTRACTION IF REFERENCE BEATS ARE STORED, OTHERWISE ENCODED RHYTHM DATA
Optional	(Section 7) GLOBAL MEASUREMENTS
Optional	(Section 8) TEXTUAL DIAGNOSIS FROM THE "INTERPRETIVE" DEVICE
Optional	(Section 9) MANUFACTURER SPECIFIC DIAGNOSTIC AND OVERREADING DATA FROM THE "INTERPRETIVE" DEVICE
Optional	(Section 10) LEAD MEASUREMENT RESULTS
Optional	(Section 11) UNIVERSAL STATEMENT CODES RESULTING FROM THE INTERPRETATION

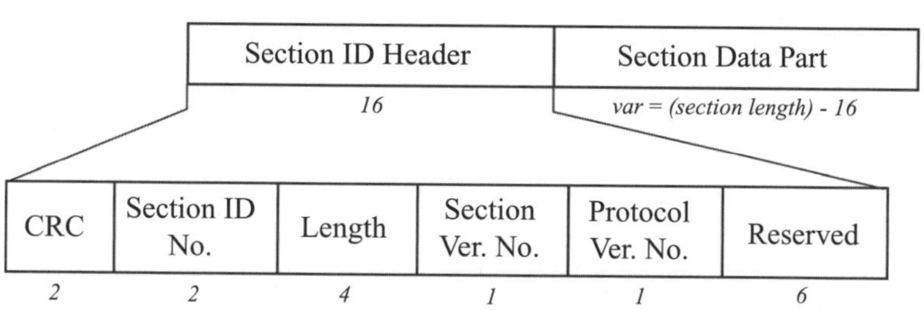

FIGURE 5.1: SCP-ECG section structure.

REFERENCES

[1] W. E. Hammond, J. J. Cimino, "Standards in medical informatics (Chapter 6)," in *Medical Informatics: Computer Applications in Health Care and Biomedicine*, E. H. Shortliffe and L.E. Perreault, Eds. New York: Springer-Verlag, 2000.

[2] http://www.journeyofhearts.org/jofh/jofh_old/minf_528/ intro.htm

[3] K. W. Goodman, *Ethics, Computing, and Medicine. Information and the Transformation of Health Care.* Cambridge: Cambridge University Press, 1998.

[4] http://www.deepthought.com.au/health/HIS_manifesto/Output/his_manifesto.html, Thomas Beale.

[5] http://www.diffuse.org/medical.html

[6] www.dicom.org

[7] www.hl7.org

[8] http://www.who.int/classifications/icd/en/

[9] www.ehto.org/aim/volume2/oedipe.html

[10] http://www.openecg.net/

CHAPTER 6

Web Technologies for Networked E-Health Applications

6.1 INTRODUCTION

When networking was first introduced in the field of medical applications, its role was limited to facilitating the connectivity and exchange of data between a central data center and various terminals in the scope of a clinical facility. In those days the resources provided by any trivial and proprietary local area network were more than sufficient for the development of the desired applications. Nowadays that networked e-Health applications of much larger scale and more importantly much broader reach are produced, the scope of medical applications has by far surpassed that of local area networks [1]. As one might expect, rather than reinventing the wheel, for reasons related not only to cost but mainly to interoperability, current and emerging networked e-Health applications rely on the existing networking infrastructure and standards, i.e., the Internet and its protocols, in order to reach its users.

There have been at times, and still are, barriers of various natures for the utilization of the Web and its technologies in a medical framework. For example, parameters such as computer literacy and Internet access greatly affect the degree to which the existence of Web-based applications provides for augmented accessibility to health services [2]. Ethical issues also have a major role in the development of Web-based medical applications [3]. Still the utilization of Web technologies in telemedicine and e-Health has already reached remarkable extents and continues to grow. In this chapter we review the most important Web technologies that commonly find application in a medical framework. We start with a brief overview of the utilization of simple Web sites and the World Wide Web and continue to discuss XML data format, mobile agents, Web Services, and security related issues.

6.2 THE WORLD WIDE WEB

Health related sites in the World Wide Web have reached an impressive population. It is estimated that as many as 2% of all Web sites focus on health-related information. The degree to which these resources are used is similarly high, with an estimated 50–75% of Internet users having looked online for such information at least once [4, 5]. Clearly, rather than such a wealth of uncontrolled information, what is needed by individuals and clinicians alike is clear, accurate, and well-presented information that one may rely on [6]. In this direction a number of attempts have been made to filter the reliable from the not so reliable medical information portals [7, 8].

Of course, the application of Web in e-Health is not limited to health-related portals; much more sophisticated applications have been developed, which use the Web as their supporting technology. As simple examples, [9] presents a Web-based tool for nutrition counselling of patients at high cardiovascular risk, while [10] also combines the e-mail service, and [11] develops customized browser plug-ins to suit both the expert and the novice Internet user patient. Decision support, more details on which are presented in Chapter 5, is also found in a Web-based setting [12, 13].

Asynchronous communication is another area in which the Web provides a solution to networked e-Health; as clinicians cannot be available online 100% of the time, communication can be facilitated by tools that allow for asynchronous exchange of messages and medical data. There are today a number of Web sites that have been developed specifically for this purpose. Through a store and forward approach, asynchronous communication can be achieved between patient and doctor [14]; this finds application mainly in the case of remote and chronic patients who are not closely and directly supervised by a clinician. The asynchronous option is not only supported for the communication between patients and doctors, but also for the communication and interaction between different doctors [15].

A situation when networked e-Health proves its worth the most is disaster. As disasters come across with no warning and at any place and time, it is practically impossible to have the required resources, this including experienced medical personnel, in place in order to cope with them; this is a gap that Web-based medical applications have been proven able to cover, in both military and civilian circumstances. Medical networks set up specifically to provide the infrastructure needed to support networked e-Health in a disaster framework include Global Health Network (GHNet) [16], Global Health Disaster Network (GHDNet) [17], Relief Web [18], and Virtual Information Center (VIC) [19], while the Rapid Health Assessment Module also provides for quick sentence translation so that disaster relief can be provided in a multilingual setting [20].

6.3 XML DATA REPRESENTATION

Similarly to any other system increasing in scale, networked e-Health systems are often formed from the integration of multiple parts, often ones designed and developed by different people, in the framework of different projects and sometimes even at different times. In order for the integration of such components to be feasible, and the compatibility of future components

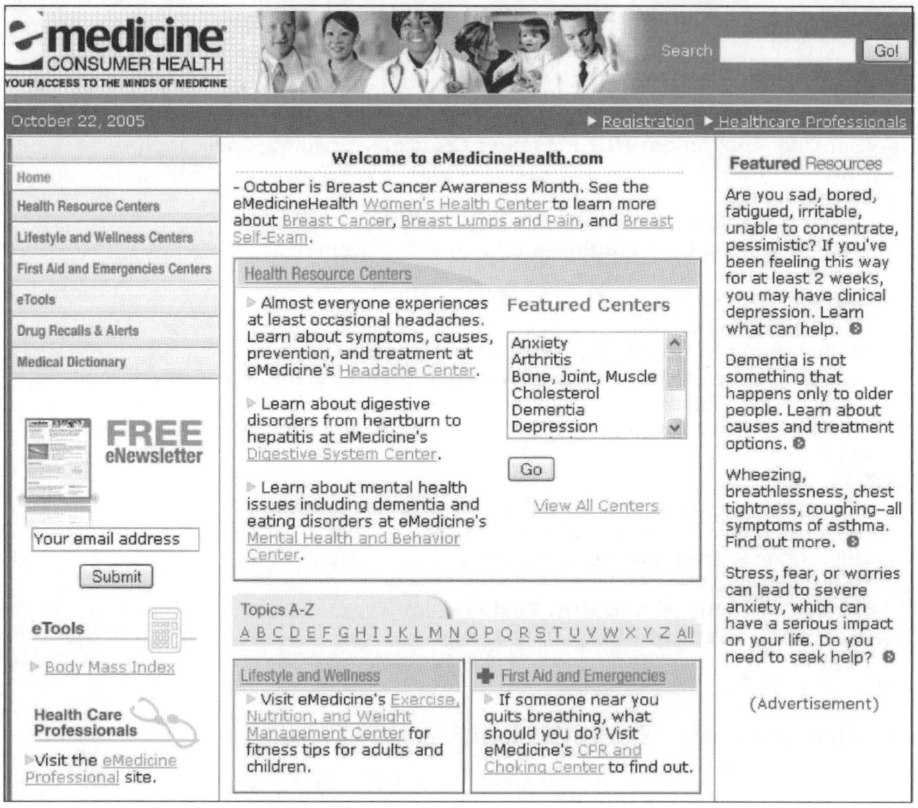

FIGURE 6.1: http://www.emedicinehealth.com/ portal

to be assured, data representation issues need to be addressed in a formal manner. Ideally, single representation standard would exist and all applications would adhere to it, making the integration of any e-Health components a much simpler task.

As seen in Chapter 6, a number of standards have been proposed in the medical field, and as a result useful medical information is already coded in a number of different formats in digital information systems around the world. Certainly, telemedicine and e-Health applications can benefit from the consideration of this information; rather than moving toward a totally novel standard, the research community is focusing on ways to facilitate the exchange of data stored in these existing formats over different types of Internet connections.

The problem of uniform data exchange over the Internet has been addressed by W3C with the extensible mark-up language (XML) standard [21]. Using XML Schema (which has rendered the use of Document Type Definitions obsolete) one may define the structure and permitted contents of an XML file. Once the XML Schema has been shared, applications developed independently by different people at different times and using different software development platforms can safely exchange data using any networking channel; safely, of course,

```xml
<?xml version="1.0" encoding="ISO-8859-1" ?>
<HL7_Message>
  <MSH>
    <Field_Separator>.</Field_Separator>
    <Encoding_Characters>ASCI</Encoding_Characters>
    <Sending_Application>HL7 Message Creator</Sending_Application>
  </MSH>
  <PID>
    <Patient_ID>001</Patient_ID>
    <Patient_Name>Ilias Maglogiannis</Patient_Name>
    <Maiden_Name />
    <Date_of_Birth>1/1/1970</Date_of_Birth>
    <Sex>Male;</Sex>
    <Phone_Number>22730-82239</Phone_Number>
    <Patient_Address>University of Aegean Karlovasi Samos</Patient_Address>
    <Nationality>Hellenic</Nationality>
    <Marital_Status>Single;</Marital_Status>
  </PID>
  <PV1>
    <Current_Date>2005-10-20 15:34:40</Current_Date>
    <Medical_Supervisor>Dr. Manolis Wallace</Medical_Supervisor>
    <Institution>Athens General Hospital</Institution>
    <Institution_Address />
    <Exam_Type>Blood and Urin Test</Exam_Type>
    <Comments>Sent to LIS</Comments>
  </PV1>
</HL7_Message>
```

FIGURE 6.2: An example of an HL7 XML document

referring to the certainty of their ability to correctly connect to each other rather than to the security of any established connection.

It is already clear that XML is the way to go for networked e-Health applications too. Efforts have been made to combine XML with existing medical standards such as Health Level Seven (HL7) [22] and Digital Imaging and Communications in Medicine (DICOM) [23], some new XML-based standards have been proposed [24], while XML in general is being used in numerous European and international systems to facilitate data exchange between end-users often located in different countries.

6.4 WEB SERVICES AND MOBILE AGENTS

It is not only data format that one needs to consider on the Internet; the way to establish communication is an issue itself at times. In this section we discuss Web services and mobile agents, two technologies that aim to address issues related to connectivity and the way to achieve it.

Web servers using HTTP to format pages and display information offer for a very intuitive data presentation format that users can easily relate to. On the other hand, without strong natural language processing and understanding tools, it is not possible for software components to automatically extract useful information from Web sites. Similarly, although search engines offer for very intuitive searches by providing both simple and complex input options, their proprietary form makes it impossible to develop generic software modules to perform searches using them.

In order to allow for the Web present information and services to become available to software components in an automated manner, W3C has proposed using Web Services to share services and the XML-based Web services description language (WSDL) to describe them [21]. Similarly to XML and XML Schema, once the query and result parameters for a Web Service have been described using WSDL and the file has been shared, any informed application is able to successfully query the service and parse the provided results.

In Fig. 6.3 we present a typical setup for a Web service-based application. Web Services formalize the structure of the exchanged data, so that server and client components can be developed independently and in possibly quite different platforms, while the SOAP protocol formalized the format of the data, thus making the application independent of the underlying data communication channel.

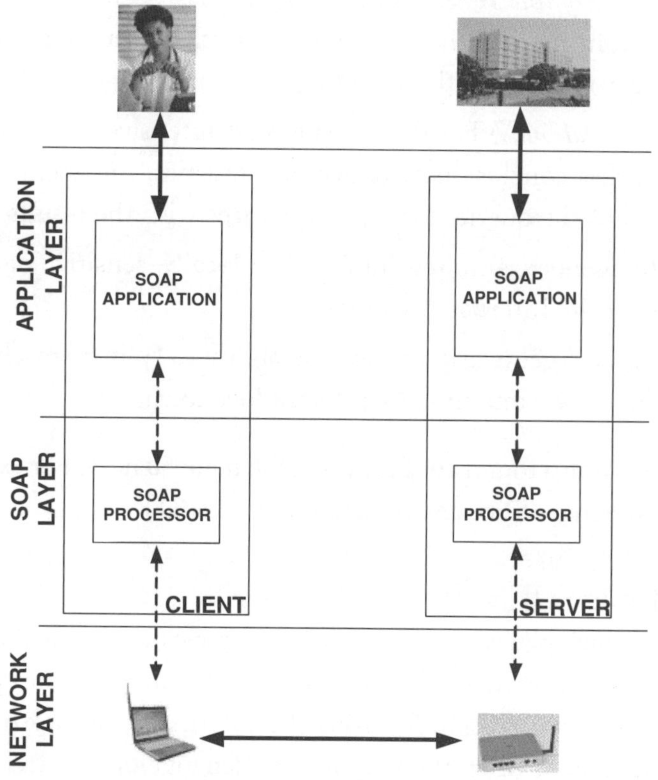

FIGURE 6.3: A sample architecture for Web Service application

Although Web Service technology is quite new, its excellent characteristics have allowed it to have an early adoption in various fields, networked e-Health being one of them. More and more medical systems are developed using Web Services and WSDL to describe the developed features, rather than simply using XML and sharing XML Schemas [25–28].

Other issues one might choose to consider include that of services that might be unavailable at a given time, as well as that of large amounts of data needing to be processed for a simple conclusion to be reached. In both cases the traditional client–server approach proves to be suboptimal, with a lot of time being spent online before the required information is acquired. As an example of online time being too expensive to maintain for long, one can see, for example, the case of Antarctica. There, with INMARSAT having to be used in order to support for exchange of medical data, a different approach needs to be followed [29].

Mobile agents have been proposed in order to tackle such problems. They are agents that can physically travel across a network and perform tasks on machines that provide agent-hosting capability. This allows processes to migrate from computer to computer, for processes to split into multiple instances that execute on different machines, and to return to their point of origin. Their benefits include:

- *Task delegation at offline time.* Once the agent has been sent, the local host can be disconnected from the Internet and only connected at a later time again in order to receive the agent together with its findings.
- *Bandwidth conservation.* When data and network intensive queries need to be performed in order to reach a conclusion or to acquire statistics, the agent travels to the host of the database, runs the queries locally, and returns with the results.
- *Privacy.* With agents examining information locally, sensitive data do not have to be transmitted over the Internet any more.
- *Parallel processing.* Agents also possess the ability to split themselves, so that different copies are sent at the same time to different locations.

Mobile agents have also found application in the medical field. For example, the following networked e-Health systems utilize agent technology [29–31].

6.5 SECURITY ISSUES

Security-related issues have played a major role as obstacles to the further penetration of the Internet and its services in everyday life. For example, *Business Week* has provided evidence in the past that the main user reservation for using the Internet is the lack of privacy, rather than cost-related issues. Obviously, when it comes to the exchange of sensitive private information, especial medical information, user reservation is even higher. Additionally, the legal issues that need to be considered are equally severe, causing service providers, i.e. medical organizations such as hospitals, to also have reservations [32]. As the contribution made by telemedicine and e-Health is evident and immense, this is not an issue of whether the Internet and communications in general should be used in a medical framework. The question is rather how these communications shall be protected from the different types of risks that lurk in network environments.

First of all one needs to identify the risks, starting from what it is that may be in danger. In the case of networked e-Health, there are two main risks to consider: security and privacy violations. Although often mistakenly treated as synonyms, the two terms refer to distinct and equally important properties. Security of the data refers to the protection of their content, while privacy refers to the protection of the identity of their source or owner. For these to be protected, technical, procedural, and organizational measures need to be taken by organizations and/or individuals communicating sensitive data over insecure channels [33]. As far as the technical aspect is concerned, it is typically general purpose Information Security Technologies (ISTs) and Privacy-Enhancing Technologies (PETs) that are being ported in the medical framework in order to enhance protection levels.

An important effort is being made in order to standardize the way to pursue protection at an international level [34], and works are presented that attempt to set the guidelines for secure and private communications in the medical sector through the careful examination of real life scenarios [33]. Unfortunately, a common conclusion is that regardless of the technical solutions available, major attention needs to be given to procedural and organizational issues, as systems' security is often jeopardized by user actions far before malicious attacks of a technical nature take place.

When it comes to the specific technologies and methodologies used in order to maximize privacy and security, with the exception of some standardization bodies, such as CEN and HL7, that have proposed Electronic Healthcare Record architectures that address issues of authorization, access permissions, and data exchange between different applications [35], there are no networked e-Health specific tools; it is the general purpose security techniques that are applied in the medical field as well. And although there have been at times concerns that augmented security might imply reduced utility [36], it is now generally accepted that security measures can be transparent and have no obvious effect on the amount and quality of services available.

Given that communication channels are insecure by nature, the most common technical measure used to ensure security is that of encrypted data exchange. Using the Secure Sockets Layer (SSL) protocol, data can travel over insecure channels and third parties are able to intercept it, but only the intended recipient is able to decrypt the message in order to retrieve the original data. The process relies on the public key infrastructure (PKI) [37, 38].

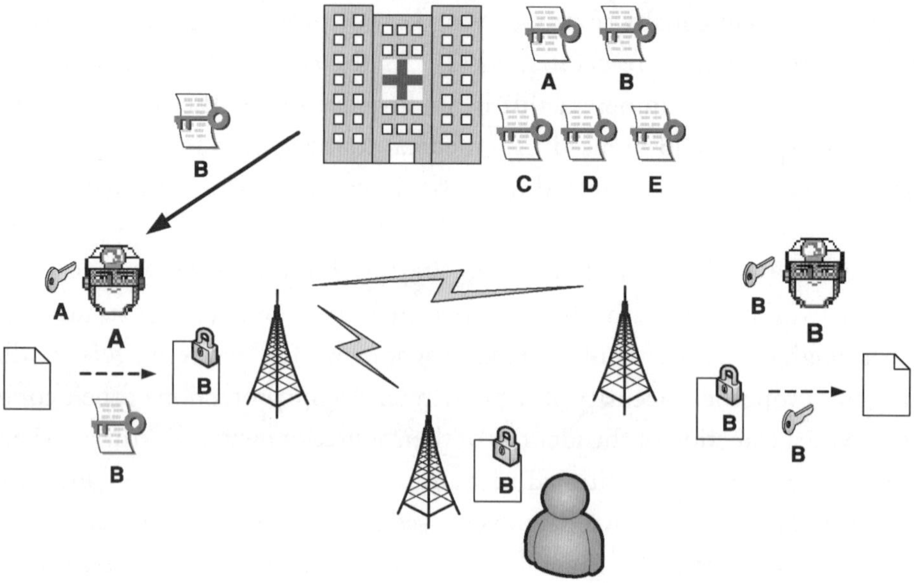

FIGURE 6.4: Using the PKI for data exchange

A key is a sequence of symbols which is utilized by the SSL algorithms in order to process data and produce an encrypted form of it. The complexity of the algorithm and the length of the key are such that a brute force attack on an encrypted message is computationally intractable, and thus encrypted data are generally considered as secure. The process that generates the public key also produces as a combined product the private key. This is another sequence of symbols of equal length that can be utilized by a similar decrypting algorithm in order to retrieve the original message. The first key is made public by publishing in the PKI, so that everyone can send secure messages to a given recipient, while the second is kept private at the recipient, and thus the names public and private key.

In Fig. 6.4 we outline the process of secure communication utilizing public and private keys. The PKI authority issues public/private key pairs for every authorized host and makes the public ones freely available to anyone that requests them. In order to securely transmit a document to host B, host A requests the public key that corresponds to B and utilizes it to encrypt the message. Thus, even if a third part intercepts the communication, the intercepted document is encrypted and cannot be accessed without the private key that only host B has.

Although this assures to a high degree the security of the transmitted data, it leaves another door open for fraud: third parties may be unable to intercept transmitted data, but they are able to impersonate the other end of the communication, thus intercepting sensitive data stored on the other end. In order to protect from this possibility, a reliable method for identity verification is required. IP address checks, MAC address checks, username password sessions, and even using encrypted transmission and frequent challenges as in the Challenge Handshake

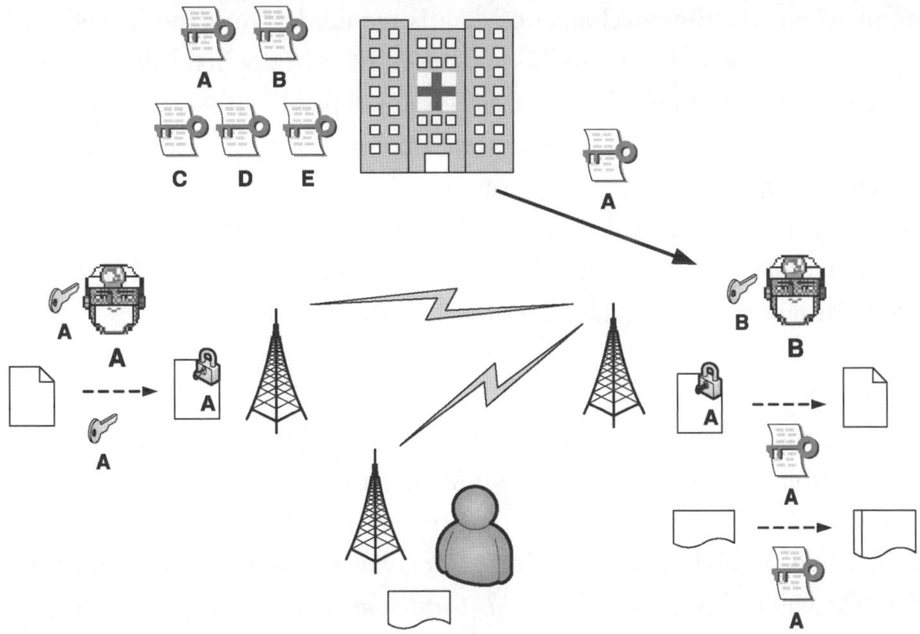

FIGURE 6.5: Using the PKI for identity verification

Authentication Protocol (CHAP) are all imperfect measures that do not guarantee absolute security.

The most reliable known approach to identity authentication is again provided by the PKI and relies on the commutative property of the keys: alternatively to using the public key for encryption and the private one for encryption, the private one may be used on the encrypting end and the public one on the decrypting end, as seen in Fig. 6.5. This process does not guarantee data security, as any third party is able to use the publicly available public key to perform the decryption. What it does provide for is reliable identity verification as only the real sender possesses the private key and is able to correctly encrypt the data; data transmitted by any third party cannot be meaningfully decrypted by the recipient, and thus are rejected as generated by an unauthorized sender.

In real life it is customary to maximize security by utilizing the sender's private key in order to reencrypt the message, thus providing for identity verification, and then using the recipient's public key to encrypt the data, thus providing for data security. At the recipient end the reverse process is followed, with the recipient's private key used first and the sender's public key second.

Although the SSL and the PKI provide a reliable solution for secure communications over the Internet, when wireless *ad hoc* networks are utilized encryption can only provide for reduced security levels. The reason behind this is mainly that portable networking devices typically have reduced computational capabilities and are unable to process adequately complex keys and algorithms when real-time exchange of data is required. Thus, specialized (and typically reduced in performance) encryption methodologies need to be applied [39, 40]. Actually, security concerns and corresponding solutions exist for any type of connectivity pursued, as well as for any type of protocol or methodology utilized; one can consider security in Web servers, in XML documents, in mobile agents, and so on.

What one typically finds in real life installations is the combination of different security measures, each one focusing at different levels and stages of the communication process, thus protecting from different types of vulnerabilities [32, 41].

REFERENCES

[1] G. Kormentzas, I. Maglogiannis, D. Vasis, D. Vergados, and A. Rouskas, "A modeling and simulation framework for compound medical applications in regional healthcare networks," *Int. J. Electron. Healthc.*, vol. 1, no. 4, pp. 427–441, 2005.

[2] M. S. Cashen, P. Dykes, and B. Gerber, "eHealth technology and Internet resources: barriers for vulnerable populations," *J. Cardiovas. Nurs.*, vol. 19, no. 3, pp. 209–214, 2004.

[3] G. R. Eysenbach, "Towards ethical guidelines for e-health: JMIR theme issue on ehealth ethics," *J. Med. Internet Res.*, vol. 2, no. 1, p. e7, 2000.

[4] H. Tatsumi, H. Mitani, Y. Haruki, and Y. Ogushi, "Internet medical usage in Japan: current situation and issues," *J. Med. Internet Res.*, vol. 3, no. 1, p. e12, 2001.

[5] M. Murero, G. D'ancona, and H. Karamanoukian, "Use of the Internet by patients before and after cardiac surgery: telephone survey," *J. Med. Internet Res.*, vol. 3, no. 3, p. e27, 2001.

[6] J. Powell and A. Clarke, "The WWW of the World Wide Web: who, what, and why?" *J. Med. Internet Res.*, vol. 4, no. 1, p. e4, 2002.

[7] S. Lapinsky and S. Mehta, "Web reports: critically appraising online resources," *Crit. Care*, vol. 6, no. 6, p. 462, 2002.

[8] F. R. Jelovsek, "Assessing the credibility of medical web sites," OBG Manage., vol. 12, no. 9, 2000.

[9] M. Verheijden, P. van Genuchten, M. Godwin, C. van Weel, and W. van Staveren, "Heartweb: a web-based tool for tailoring nutrition counselling in patients at elevated cardiovascular risk," *Technol. Health Care*, vol. 10, no. 6, pp. 555–557, 2002.

[10] J. Castrén, "Ask doctor per email—a modern health service among Finnish university students, carried out by general practitioners," *Technol. Health Care*, vol. 10, no. 6, pp. 458–459, 2002.

[11] C. Lau, R. S. Churchill, J. Kim, F. A. Matsen, and Y. Kim, "Asynchronous web-based patient-centered home telemedicine system," *IEEE Trans. Biomed. Eng.*, vol. 49, no. 12, pp. 1452–1462, 2002.

[12] D. Jegelevicius, V. Marozas, A. Lukosevicius, and M. Patasius, "Web-based health services and clinical decision support," *Stud. Health Technol. Inform.*, vol. 105, pp. 27–37, 2004.

[13] G. Eysenbach, J. Powell, M. Englesakis, C. S. Rizo, and A. Stern, "Health related virtual communities and electronic support groups: systematic review of the effects of online peer to peer interactions," *Br. Med. J.*, vol. 328, no. 7449, pp. 1–6, 2004.

[14] H. M. Kim, J. C. Lowery, J. B. Hamill, and E. G. Wilkins, "Accuracy of a web-based system for monitoring chronic wounds," *Telemed. J. e-Health*, vol. 9, no. 2, pp. 129–140, 2003.

[15] C. E. Chronaki, F. Chiarugi, E. Mavrogiannaki, C. Demou, P. Lelis, D. Trypakis, M. Spanakis, M. Tsiknakis, and S. C. Orphanoudakis, "An eHealth platform for instant interaction among health professionals," *Comput. Cardiol.*, vol. 30, pp. 101–104, 2003.

[16] F. Sauer, R. E. LaPorte, S. Akazawa, E. Boostrom, E. Ferguson, C. Gamboa, *et al.*, "Global health network task force. Towards a global health network," *Curr. Issues Public Health*, vol. 1, pp. 160–164, 1995.

[17] I. M. Libman, R. E. LaPorte, S. Akazawa, E. Boostrom, C. Glosser, E. Marler, E. Pretto, F. Sauer, A. Villasenor, F. Young, and G. Ochi, "The need for a global health disaster network," *Prehospital Disaster Med.*, vol. 12, no. 1, pp. 11–12, 1997.

[18] http://www.reliefweb.int/

[19] R. E. LaPorte, "Improving public health via the information superhighway," *The Scientist*, vol. 11, p. 10, 1997.

[20] Dragon Systems, *Rapid Health Assessment Module for Disaster Management and Humanitarian Assistance: Operations Guide*, Newton, MA: Dragon Systems, 1997.

[21] http://www.w3.org/

[22] M. Tobman, C. Nätscher, H. Sussmann, and A. Horsch, "A new approach for integration of telemedicine applications into existing information systems in healthcare," *Stud. Health Technol. Inform.*, vol. 90, pp. 152–155, 2002.

[23] A. Anagnostaki, S. Pavlopoulos, E. Kyriakou, and D. Koutsouris, "A novel codification scheme based on the 'VITAL' and 'DICOM' standards for telemedicine applications Anagnostaki," *IEEE Trans. Biomed. Eng.*, vol. 49, no. 12, pp. 1399–1411, 2002.

[24] A. Anagnostaki, S. Pavlopoulos, and D. Koutsouris, "XML and the VITAL standard: the document-oriented approach for open telemedicine applications," *MEDINFO*, vol. 10, no. 1, pp. 77–81, 2001.

[25] V. Bicer, G. Laleci, A. Dogac, and Y. Kabak, "Artemis message exchange framework: semantic interoperability of exchanged messages in the healthcare domain," *ACM SIGMOD Rec.*, vol. 34, no. 3, pp. 71–76, 2005.

[26] R. J. Rodrigues and A. Risk, "ehealth in Latin America and the Caribbean: development and policy issues," *J. Med. Internet Res.*, vol. 5, no. 1, p. e4, 2003.

[27] G. Demiris, "Disease management and the Internet," *J. Med. Internet Res.*, vol. 6, no. 3, p. e33, 2004.

[28] J. Anhøj and L. Nielsen, "Quantitative and qualitative usage data of an Internet-based asthma monitoring tool," *J. Med. Internet Res.*, vol. 6, no. 3, p. e23, 2004.

[29] S. Pillon and A. R. Todini, "eHealth in Antarctica: a model ready to be transferred to every-day life," *Int. J. Circumpolar Health*, vol. 63, no. 4, pp. 436–442, 2004.

[30] S. Kirn, C. Heine, R. Herrler, and K.-H. Krempels, "Agent.hospital—a framework for clinical applications," in *Applications of Software Agent Technology in the Health Care Domain*, A. Moreno and J. Nealon, Eds. Basel: Birkhäuser Publisher, 2003, pp. 67–85.

[31] M. Rodriguez, J. Favela, V. M. Gonzalez, and M. Munoz, "Agent based mobile collaboration and information access in a healthcare environment," *e-Health: Appl. Comput. Sci. Med. Health Care*, vol. 5, pp. 133–148, 2003.

[32] B. Stanberry, *The Legal and Ethical Aspects of Telemedicine*. London: Royal Society of Medicine Press, 1998.

[33] A. Gritzalis, C. Lambrinoudakis, D. Lekkas, and S. Deftereos, "Technical guidelines for enhancing privacy and data protection in modern electronic medical environments," *IEEE Trans. Inf. Technol. Biomed.*, vol. 9, no. 3, pp. 413–423, 2005.

[34] P. Ruotsalainen and H. Pohjonen, "European security framework for healthcare," *Stud. Health Technol. Informat.*, vol. 96, pp. 128–134, 2003.

[35] KONA Editorial Group, "Patient record architecture, HL7 Document (Online)," http://www.hl7.org/special/committees/sgml/PRA/PRA_ballot_April00.zip, 2000.

[36] N. P. Terry, "An ehealth diptych: the impact of privacy regulation on medical error and malpractice litigation," *Am. J. Law Med.*, vol. 27, no. 4, pp. 361–419, 2001.

[37] Sax *et al.*, "Wireless technology infrastructures for authentication of patients: PKI that rings," *J. Am. Med. Informat. Assoc.*, vol. 12, pp. 263–268, 2005.

[38] G. Kambourakis, I. Maglogiannis, and A. Rouskas, "PKI-Based Secure Mobile Access to Electronic Health Services and Data," *Technol. Healthc.*, vol. 13, no. 6, pp. 511–526, 2005.

[39] P. K. Dutta "Security considerations in wireless sensor networks," in *Sensors Expo*, San Jose, CA, 2004.

[40] F. Hu and N. K. Sharma, "Security considerations in wireless sensor networks," *Ad hoc Netw.*, vol. 3, no. 1, pp. 69–89, 2005.

[41] V. Traver, C. Fernández, E. Montón, J. C. Naranjo, S. Guillén, and B. Valdivieso, "IST project Integrated Distributed Environment for Application Services in e-Health (IDeAS)," Jan. 31, 2006, http://www.ideas-ehealth.upv.es.

CHAPTER 7

Distributed Collaborative Platforms for Medical Diagnosis over the Internet

7.1 INTRODUCTION

Hospitals are nowadays sufficiently rich in their infrastructure to handle the internal administrative and clinical processes for their inpatients. However, this kind of information exchange is usually reduced at regional level, and the need to integrate the processes in geographically distributed and organizationally independent organizations and medical personnel has arisen. This challenge leads the design of health information applications to combine the principles of heterogeneous Workflow Management Systems (WfMSs), the peer-to-peer networking architecture, the World Wide Web, and the Visual Integration concepts to facilitate medical information exchange among geographically distributed personnel.

Under this scope, the goal of this chapter is to present the concept of Web-based collaborative systems for the facilitation of physician's communication at remote locations. Such systems enable communication via audiovisual means through a public network. The Health Level Seven (HL7), the Clinical Data Architecture (CDA), and the Digital Imaging and Communications in Medicine (DICOM) standards [1, 2] are used for the transfer and storing of the medical data.

7.2 STATE OF THE ART REVIEW

The development of a distributed collaborative system has received the attention of several research groups. For instance, Collaboratory for Research on Electronic Work (CREW) [8] is a typical example of a collaborative system implemented in the University of Michigan. The specific computer-based environment is able to give multiple access to shared documents and provides CREW community with information about other CREW members, their activities, location, availability, and interests. Moreover, it supports collaboration tools such as postable notes and e-mail.

Healthcare information collaborative systems preoccupy the medical informatics research community as well. InterMed [9] is a system aiming at the development of a robust framework for medical collaboration, using the Internet as the basic network. The project TeleMed [10] supports both real-time and off-line sessions. Multiple physicians at remote locations are able to view, edit, and annotate patient data on real-time. This project focuses on security and confidentiality services, access to online knowledge, clinical data access and sharing, as well as collaborative policies, procedures, and guidelines. The ARTEMIS project aims at "advance cooperative activities of healthcare providers to promote the delivery of total and real-time care" [11]. A group of developers, physicians, and healthcare researchers are using prototypes and commercial off-the-shelf technologies to develop an open collaboration healthcare environment. In ARTEMIS, community care networks consist of a collection of primary and specialized care providers, who collaborate to meet the healthcare needs of the community. The communication infrastructure allows primary care physicians to consult with remote specialists with computer support for X-rays, ultrasound, voice-annotations, and other multimedia information.

Another project focusing on the healthcare collaboration domain is the virtual medical office [12], which is an integrated environment that encourages users to take active participation in the management of their health. It provides access to digital medical libraries, yellow pages, and regional healthcare resources.

WebOnCOLL [13] has been designed in the context of the regional healthcare network of Crete and employs the infrastructure of regional healthcare networks to provide integrated services for virtual workspaces, annotations, e-mail, and on-line collaboration. Virtual workspaces support collaborative concepts like personal Web pages, bulletin boards, discussion lists, shared workspaces, and medical case folders. Annotations provide a natural way for people to interact with multimedia content, while e-mail is one of the most popular forms of communication today. Finally, another type of health collaborative application is described in [14]. The presented application uses a multiuser environment for collaboration with multimodal human machine interaction. Audio–video teleconferencing is also included to help the radiologists to communicate with each other simultaneously while they are working on the mammograms.

7.3 COLLABORATIVE ARCHITECTURAL ASPECTS

Examining networked applications from an architectural aspect, we find that the most common architectures used today are the client–server, the 3-tier or *n*-tier, and the peer-to-peer networking schemas. Client–Server scheme is the "network architecture in which each computer or process on the network is either a *client* or a *server*. Servers are powerful computers or processes dedicated to managing disk drives (*file servers*), printers (*print servers*), or network traffic (*network servers*). Clients are PCs or workstations on which users run applications. Clients rely on servers for resources, such as files, devices, and even processing power." N-tier application architecture [7] provides a model for developers to create a flexible and reusable application. By breaking up an application into tiers, developers only have to modify or add a specific layer, rather than having to rewrite the entire application over, if they decide to change technologies or scale up. In the term "N-tier," "N" implies any number—like 2-tier or 4-tier, basically any number of distinct tiers used in a specific architecture. Peer to Peer [15, 16] on the other hand is the "type of network in which each workstation has equivalent capabilities and responsibilities. This differs from client–server architectures, in which some computers are dedicated to serving the others."

The upside of the peer-to-peer architecture is that it is more dependable, reliable, and available, as there is no single point of failure; instead information is usually replicated at several peers. In addition it is relatively inexpensive, it causes low network workload by skipping extra reroutes, and it is fairly simple to set up and manage. Furthermore, it scales better because there is no single point of access since there is no machine called server, and also no server farm. The flip side is that it is limited in extensibility, tends to overburden user workstations by having them play the role of server to other users, is largely insecure, and is typically unable to provide system-wide services since the typical workstation will run a standard desktop operating system incapable of hosting any major service (e.g., a post office). Table 7.1 summarizes the benefits and the fault backs of the peer-to-peer and the client–server architectures related to the needs of the medical collaboration platform [15, 16].

7.4 COLLABORATIVE PLATFORM—MAIN FUNCTIONS AND MODULES

The subsections below describe the main functions and modules available in most of medical collaborative platforms.

7.4.1 Users' Registration

Each physician wishing to join a collaboration session has to register him- or herself at a medical collaborative platform. During the registration procedure, the physician provides the system with information about his/her identity and specialty. In most cases the registration is a face-to-face process in order to ensure the integrity of the physician's personal information.

7.4.2 On-line Calls

In order to initiate a session, a physician (named physician A) selects from the "on-line medical personnel" list the remote physician (physician B) whom he/she wishes to contact. His/her module sends a communication request to the selected doctor. On the remote doctor's side, physician B receives the call and accepts it. Physician B's module returns to the initiator of the communication an acknowledgment. Such a procedure is depicted in the sequence diagram of Fig. 7.1.

TABLE 7.1: Client–Server vs Peer-to-Peer Networking Architectures

PLATFORM REQUIREMENTS	PEER-TO-PEER	CLIENT–SERVER
Collaboration sessions: Best quality of communication	Point-to-point communication, direct exchange of data, no extra reroutes,	Extra rerouting through the server module
Centralized physician management	Does not support centralized control	Centralized user and network management
Security schema	Distributed security schema, difficult to set up and manage	Centralized security schema, easy to control
Administrative control	No need of a central administrator, limited responsibilities	The existence of an administrator is crucial, various and essential responsibilities
Use of a centralized medical database	No centralized control over the database	Server can provide centralized control of database access list, easy data distribution
Expandability	Significantly easy to expand, low complexity, ad hoc infrastructure	High complexity on expanding architecture, Slow-moving on new adaptations
Availability	High level of availability due to distributed architecture	Low level of availability in case of traffic or server malfunction, i.e., system crash, denial of service

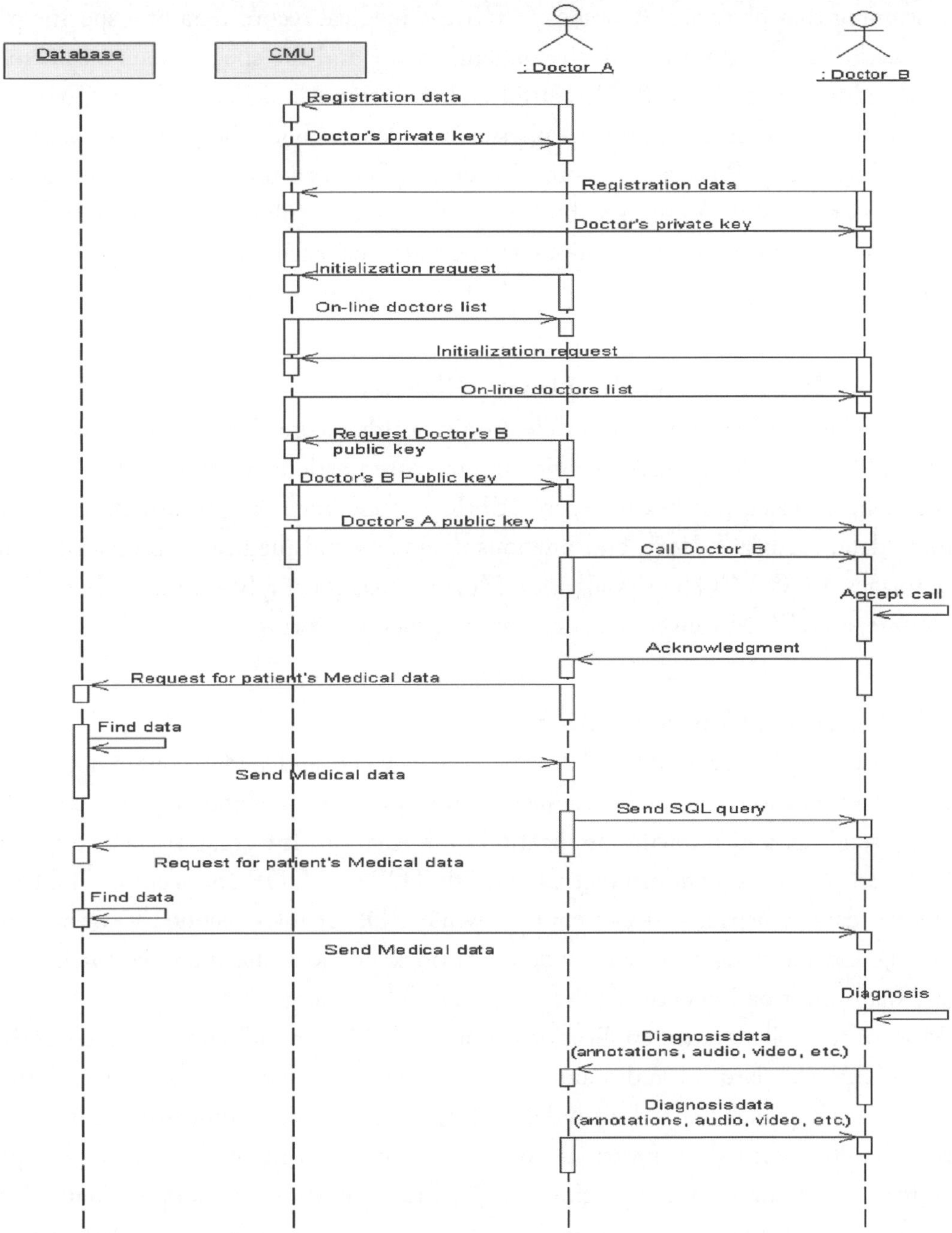

FIGURE 7.1: Communication sequence diagram of a collaborative session

Assuming that physician A wishes to retrieve medical record data of a specific patient, he/she fills up the related form with the patient's name and the specific data and commits a query to a middleware entity (called hereafter Collaboration Management Unit, CMU). Upon receiving the data, physician A informs physician-B device about the patient's name and the committed SQL query. Physician B from his/her part follows a similar procedure and receives the patient's data as well. As soon as both physicians successfully retrieve the requested part of the patient's record, they can manipulate the data and exchange opinion about patient's condition.

7.4.3 Patient Record and Medical Image Transfer

Patient's medical data are stored in an electronic health record database that connects with the CMU middleware database. Most of computer-based patient systems follow the standards for incident-based electronic health record (EHR) storage, including information such as demographic data, medical history, and previous diagnoses and medical examinations (such as medical images, ECG/EEG biosignals, etc). Medical images comply with the DICOM format [4]; therefore, a DICOM viewer is incorporated in such platforms.

7.4.4 Video–Audio Communication

Most of the medical collaborative platforms enable video and audio conference as well as message communication. On the lowest level of the applications, standards are responsible for translating, sending, and receiving application's information. Information can be transferred over the Internet or corporate intranet using both TCP and UDP connections. TCP is used primarily for data transport and call control, while UDP enables secondary connections for transmitting and receiving audio and video. Winsock provides the interface to the network stack and maps information between the programs and the network.

Most of the collaborative applications support H.323 for audio and video conferencing. The ITU H.323 standard for audio and video conferencing provides a set of standard audio and video codecs capable of encoding and decoding the input and output from audio and video sources that is transmitted over various connection rates and networks that do not provide a guaranteed quality of service. By using H.323, the application can support many different modes of Internet telephony and can send and receive audio and video information to H.323-compatible nodes. H.323 computers and equipment can carry real-time video, audio, and data, or any combination of these elements. This standard is based on the Internet Engineering Task Force (IETF) Real-Time Protocol (RTP) and Real-Time Control Protocol (RTCP), with additional protocols for framing, call signaling, call control, data and audiovisual communications, and for negotiating the information flow and sequencing.

7.4.5 Workspace Management

The workspace of the doctor's module consists of the basic image-processing tools offering the capability of interaction in real time with his/her colleague located on the other end of the session. Upon loading the medical images on the DICOM viewer, all editing taking place on his/her side is automatically displayed on the image of the other party. Note that one part at a time can have access to the editing tools.

The features of the tools mostly implemented are the following:

- *Image comments.* Doctors can comment the image by writing particular observations on specific points of interest.
- *Use of annotations.* This feature enables user to draw geometrical schemes with his/her mouse selection ROIs (region of interests) and relay this information to the other end.
- *Multiple color channels* Physicians are able to select different colors or spectrum channels for image viewing.
- *Contrast and brightness configuration.* Medical personnel can set via scrollbars the contrast and the brightness level of the medical image displayed.
- *Image scaling (zoom-in, zoom-out).* Medical personnel can increase and decrease the zoom level of a picture via special buttons.

7.4.6 Image Compression

The most amount of bandwidth during the on-line collaboration sessions is consumed by medical images. The data throughput is significantly enhanced if the images are compressed. Many compression algorithms have been presented in the past. These can be classified to lossless and lossy. In the former case the reconstructed image is identical to the uncompressed one while in the latter there is loss of information. All the reviewed platforms have the capability of transmitting lossless compressed images.

7.4.7 Security Issues

Specific attention in medical collaborative platforms is drawn toward the security requirements. The transmission of sensitive medical data related to the patient heath record through a public network arises the problem of communication privacy. Most of the approaches followed include the incorporation of encryption algorithms that achieve confidentiality, authenticity, and integrity of the data and the adoption of PKI infrastructure for consistent user authorization and authentication. The middleware CMU entity applies and manages the system's security policy. The policy is predetermined by the system administrator and authorized by the health organization that controls the system. The policy consists of specific rules assigning physicians to authorization groups. The group definition is based on the geographical regions; i.e., the group

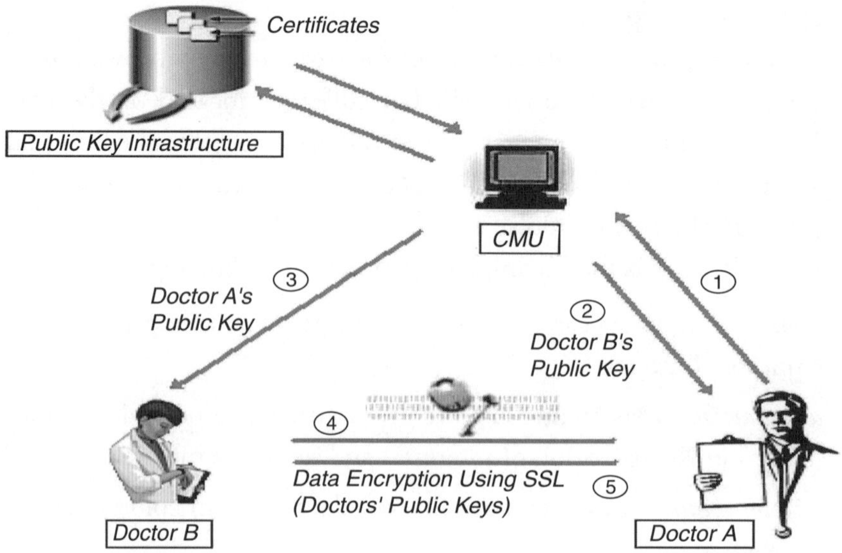

FIGURE 7.2: A PKI-based security scheme

of North Aegean consists of physicians whose hospitals are placed in the North Aegean region, group of South Aegean, etc. Each group has specific rights over patients' data; i.e., a doctor that belongs to the North Aegean group can access medical data stored on hospital databases placed in that specific region. Since patients' data are stored on databases, specific countermeasures have to be taken to ensure data confidentiality. In order to meet this requirement each group receives a username and password that provide read access to databases.

Note that during collaboration both physicians need to have access to data related to the patient cross-examined. What happens in case that only one part of the communication has access to that information? One possible solution is to allow the authorized doctor to forward the information directly to the nonauthorized doctor. However, this solution would be inefficient in terms of performance, since it would cause additional load to doctor's device. A better approach would include a daemon running on CMU machine dedicated to users' access management. The basic idea is to create a temporary user account for nonauthorized users with a specific validity period. By using this account users can have access to the patient's data during a specific session. As soon as the session ends the temporary account is removed from the database. A security scheme based on public key infrastructure (PKI) is depicted in Fig. 7.2.

7.5 CONCLUSIONS AND CHALLENGES FOR THE FUTURE

The chapter discusses the domain specific technological issues regarding medical collaboration platforms. Considering the fact that physicians are not computer experts, it is of high importance to build a user-friendly platform, adopting most of the desired functionality that users need during a collaboration session. Taking into account the great amount of data that a medical record has, the platform requires medium bandwidth resources; however, the data throughput may be enhanced by compression methods. Remaining future challenges are the enhancement of the main functionalities in terms of speed, user friendliness, HCI, ergonomics, data compression and security, and the transition to the mobile paradigm by implementing platforms to work on personal digital assistants (PDAs) and smart phones.

REFERENCES

[1] *Health Level Seven, Reference Information Model.* Ann Arbor, MI: Health Level Seven, Inc., 2002. http://www.hl7.org.

[2] *Digital Imaging and Communications in Medicine (DICOM),Part 1: Introduction and Overview.* Rosslyn, VA: National Electrical Manufacturers Association, 2001.

[3] http://www.rsna.org/practice/dicom/intro/index.html

[4] The ACR-NEMA DICOM Standard, http://medical.nema.org/

[5] J. F. Coyle, A. R. Mori, and S. M. Huff, "Standards for detailed clinical models as the basis for medical data exchange and decision support," *Int. J. Med. Inf.*, vol. 69, pp. 157–174, 2003.

[6] T. Penzel, B. Kemp, G. Klösch, A. Schlögl, and J. Hasan, "Acquisition of Biomedical Signals Databases," *IEEE Eng. Med. Biol.*, pp. 25–32, May–Jun. 2001.

[7] http://www.15seconds.com/issue/011023.htm

[8] A. Arbor, "Collaboratory for Research on Electronic Work," http://www.crew.umich.edu/CREW.home.html.

[9] InterMed Collaboratory, http://smi-web.stanford.edu/projects/intermedweb/Overview.html.

[10] D. G. Kilman and D. W. Forslund, "An international collaboratory based on virtual patient records," *Commun. ACM*, vol. 40, pp. 111–117, Aug. 1997.

[11] V. Jagannathan *et al.*, "An overview of the CERC Artemis project," Concurrent Eng. Res. Center, West Virginia Univ., Morgantown, Tech. Rep. CERC-TR-RN-95-002, Apr. 1995, http://www.cerc.wvu.edu/nlm/artemis/summary.html.

[12] Virtual telemedicine Office, http://www.telemedical.com

[13] C. E. Chronaki, D. G. Katehakis, X. C. Zabulis, M. Tsiknakis, and S. C. Orphanoudakis, "WebOnCOLL: medical collaboration in regional healthcare networks," *IEEE Trans. Inf. Technol. Biomed.*, vol. 1, no. 4, pp. 257–269, Dec. 1997.

[14] M. Akay, I. Marsic, A. Medl, and G. Bu, "A system for medical consultation and education using multimodal human/machine communication," *IEEE Trans. Inf. Technol. Biomed.*, vol. 2, no. 4, pp. 282–291, Dec. 1998.

[15] G. Cugola and G. P. Picco, "Peer-to-Peer for collaborative applications," in *Proc. Int. Workshop Mobile Teamwork Support, co-located with the 22nd International Conference on Distributed Computing Systems*, Vienna, H. Gall and G. P. Picco, Eds., Piscataway, NJ: IEEE Press, Jul. 2002, pp. 359–364.

[16] K. Albrecht, R. Arnold, and R. Wattenhofer, "Clippee: a large-scale client/peer system," in *Int. Workshop Large-Scale Group Communication, held in conjunction with the 22nd Symposium on Reliable Distributed Systems (SRDS)*, Florence, Italy, October 2003.

[17] S. Saha, "Image compression—from DCT to wavelets: a review," *ACM Crossroads*, http://www.acm.org/crossroads/2000.

[18] A. Cohen, I. Daubechies, and J. C. Feauveau, "Biorthogonal bases of compactly supported wavelets," *Comm. Pure Appl. Math.*, vol. 45, pp. 485–560, 1992.

[19] A. Shamir and E. Tromer, "On the cost of factoring RSA-1024," *RSA CryptoBytes*, vol. 6, no. 2, pp. 10–19, 2003.

[20] A. K. Lenstra and E. R. Verheul, "Selecting cryptographic key sizes," *J. Cryptol.*, vol. 14, no. 4, pp. 255–293, 2001.

[21] J. M. Shapiro, "Embedded image coding using zerotrees of wavelet coefficients," *IEEE Trans. Signal Process.*, vol. 41, no. 12, pp. 3445–3462, Dec. 1993.

[22] A. Said and W. A. Pearlman, "A new fast and efficient image codec based on set partitioning in hierarchical trees," *IEEE Trans. Circuits Syst. Video Technol.*, pp. 243–250, Jun. 1996.

[23] A. J¨arvi, J. Lehtinen, and O. Nevalainen, "Variable quality image compression system based on SPIHT," *Signal Process.: Image Commun.*, 1998.

[24] J. Lehtinen, "Limiting distortion of a wavelet image codec," *Acta Cybern.*, vol. 14, no. 2, pp. 341–356, 1999.

[25] M. Vettereli and J. Kova˘cevi´c, *Wavelets and Subband Coding*. Englewood Cliffs, NJ: Prentice Hall, 1995.

[26] W. Pennebaker and J. Mitchell. *JPEG : Still Image Data Compression Standard.* New York: Van Nostrand Reinhold, 1992.

CHAPTER 8

Distributed Telemedicine Based on Wireless and *Ad Hoc* Networking

8.1 INTRODUCTION

The introduction of small and portable personal digital assistants (PDAs), mobile phones, and wireless communication, in general, has already changed the way we communicate with each other for everyday routine interaction. In the healthcare sector this reality enabled the development of advanced wireless distributed telemedicine services for remote areas.

In recent years several telemedicine applications have been successfully implemented over wired communication technologies like POTS (Plain Old Telephone Services), ISDN, and xDSL. However, nowadays, modern wireless telecommunication means like the GSM, the GPRS, and the UMTS mobile telephony standards, as well as satellite communications and Wireless Local Area Network (WLAN) allow the operation of wireless telemedicine systems freeing the medical personnel and/or the subject-monitored bounded to fixed locations [1].

Telemedicine applications, including those based on wireless technologies, span the areas of emergency healthcare, telecardiology, teleradiology, telepathology, teledermatology, teleophtlalmology, and telepsychiatry. In addition, health telematics applications enabling the availability of prompt and expert medical care have been exploited for the provision of healthcare services at understaffed areas like rural health centers, ambulance vehicles, ships, trains, airplanes, as well as for home monitoring [1].

The scope of the present chapter is twofold. It provides, in its first part, a concise report in the enabling wireless technologies that are applied in the distributed and mobile telemedicine sector [2]. The second part of the paper presents the special features for digital signals, images, and videos transmission in mobile telemedicine systems and discusses the application of wireless *ad hoc* networks in telemedicine. A wireless *ad hoc* network is a collection of two or more wireless devices that have the possibility of wireless communication and networking without the help of any infrastructure. The chapter presents the particular characteristics of communication with *ad hoc* networks in mobile telemedicine applications and provides examples of such *ad hoc* networking utilization.

8.2 EMERGING MOBILE TELEMEDICINE TECHNOLOGIES

In this section we highlight the main wireless communications technologies that have been used and are about to be used in mobile telemedicine systems, namely GSM/GPRS, satellite, 3G (UMTS), DVB-H/T (Digital Video Broadcasting for Television/Handheld devices), WPAN (Wireless Personal Area Network), and WLAN (Wireless Personal Area Network) [1]. These systems are summarized in Table 8.1.

GSM is a system currently in use and is the second generation (2G) of the mobile communication networks. When it is in the standard mode of operation, it provides data transfer speeds of up to 9.6 kb/s. Through the years the HSCSD technique makes it possible to increase the data transmission to 14.4 kb/s or even triple at 43.3 kb/s [3, 4].

The evolution of mobile telecommunication systems from 2G to 2.5G (iDEN 64 kb/s, GPRS 171 kb/s, EDGE 384 kb/s) systems provides the possibility of faster data transfer rates, thus enabling the development of telemedicine systems that require high data transfer rates and are currently only feasible on wired communication networks [3].

The fast development of current 3G wireless communication and mobile network technologies will be the driving force for future developments in mobile telemedicine systems. 3G wireless technology represents the convergence of various 2G wireless systems. One of the most important aspects of 3G technology is its ability to unify existing cellular standards under one umbrella [3].

In recent years other mobile network technologies such as WLANs and WPANs have become popular. WLAN allows users to access a data network at high speeds of up to 54 Mb/s as long as users are located within a relatively short range (typically 30–50 m indoors and 100–500 m outdoors) of a WLAN base station (or antenna). In the US WLAN operates in two unlicensed bands: a) 802.11b and 802.11g operate in the 2.4 GHz band, together with many other devices including Bluetooth and cordless telephones and b) 802.11a (Wi-Fi 5.2 GHz) operates in the 5.2 GHz band, which at this point is relatively free of interference from other electrical devices operating in this band [5, 6].

WPANs are defined with IEEE standard 802.15. They are considered quite important for the e-health sector, and though they are included in Table 8.1 they will be discussed separately in the next chapter, which presents pervasive computing in e-health.

The recent surge research of mobile *ad hoc* networking will also trigger a parallel research activity in the application of these emerging technologies for mobile telemedicine. It is expected that 4G will integrate existing wireless technologies including UMTS, GSM, WLAN, WPAN, and other newly developed wireless technologies into a seamless system. Some expected key features of 4G networks are stated as: a) high usability b) support for telemedicine services at low transmission cost, c) 4G provides personalized services in order to meet the demands of different users for different services, and d) 4G systems provide facilities for integrating services.

TABLE 8.1: Main Wireless Communication Networks/Standards

TYPE	SUBTYPE	FREQUENCY BAND	DATA TRANSFER RATES	CORRESPONDING TELEMEDICINE APPLICATIONS
Wireless LAN/PAN	IEEE 802.11a IEEE 802.11b IEEE 802.11g HIPERLAN 1 HIPERLAN 2 Bluetooth HomeRF 1.0 HomeRF 2.0	5 GHz 2.4 GHz 2.4 GHz 5 GHz 5 GHz 2.4 GHz 2.4 GHz 2.4 GHz	54 Mb/s 11 Mb/s 54 Mb/s 23.5 Mb/s 54 Mb/s 1 Mb/s 2 Mb/s 10 Mb/s	• Access to medical data in hospital environments • In hospital patient telemonitoring • Home care and smart homes applications
GSM	GSM-900 GSM-1800 GSM-1900	900 MHz 1800 MHz 1900 MHz	9.6–43.3 kb/s 9.6–43.3 kb/s 9.6–43.3 kb/s	• Emergency telemedicine • Patient telemonitoring in random locations
GPRS	GPRS	900/1800/ 1900 MHz	171.2 kb/s	
3G/UMTS	UMTS		2 Mb/s	• High bandwidth mobile telemedicine
Satellite	ICO Globalstar Iridium Cyberstar Celestri Teledesic Skybridge	C, S band L, S, C band L, Ka band Ku, Ka band Ka band, 40–50 GHz Ka band Ku band	2.4 kb/s 7.2 kb/s 2.4 kb/s 400 kb/s–30 Mb/s 155 Mb/s 16 kb/s–64 Mb/s 16 kb/s–2 Mb/s	• Location based telemedicine • High bandwidth telemedicine • Regional healthcare networking infrastructure
DVB	DVB-T DVB-H	5/6/7/8 MHz 5/6/7/8 MHz	5–32 Mb/s 10 Mb/s	

The main technological characteristics of 4G systems are expected to be transmission speeds higher than in 3G (min 50–100 Mb/s, average 200 Mb/s), system capacity larger than in 3G by ten times, transmission costs per bit 1/10 to 1/100 of that of 3G, support for Internet protocols (IPv6), various quality of service (QoS) providing many kinds of best effort multimedia services corresponding to users demand, and user-friendly services where users can access many services in a short time span as compared with other wireless systems of longer waiting times for response.

4G advances will provide both mobile patients and citizens the choices that will fit their lifestyle and make easier for them to interactively get the medical attention and advice they need; when and where it is required and how they want it regardless of any geographical barriers or mobility constraints. The concept of including high-speed data and other services integrated with voice services is emerging as one of the main points of future telecommunication and multimedia priorities with the relevant benefits to citizen-centered healthcare systems [3]. These creative methodologies will support the development of new and effective medical care delivery systems into the 21st Century. The new wireless technologies will allow both physicians and patients to roam freely, while maintaining access to critical patient.

Finally, DVB-T/H stands for digital video broadcasting for television/handheld devices, and it is the standard for digital, terrestrial television in Europe and beyond. The MPEG-2 technology for data compression and multiplexing makes it possible to use the existing frequency channels for a multiplied number of TV programs or equivalents. Furthermore, IP data can be encapsulated directly into the MPEG-2 transport stream using the multiprotocol encapsulation method (MPE), standardized within the DVB system. By this means, a DVB-T or DVB-H network can be designed for mobile reception in order to deliver medical data and IP packet streams [10].

8.3 DISTRIBUTED MOBILE TELEMEDICINE REQUIREMENTS

The range and complexity of telecommunications technology requirements vary with specific medical or health applications. However, generically defined digital medical devices impose the telecommunications performance requirements. Table 8.2 illustrates a sampling of several of the more common digital medical devices that may be used in distributed telemedicine. Except for the last few items contained in the table (starting with ultrasounds and running through full motion video), the majority of vital sign medical devices require relatively low data transmission rates [11].

Regarding the transmission of medical images, there are essentially no theoretical bandwidth requirements, but lack of bandwidth needs longer transmission time. Yet, high-quality medical images such as a single chest radiograph may require from 40 to 50 Megabytes. In

TABLE 8.2: Data Rates of Typical Devices Used in Telemedicine

DIGITAL DEVICE	IMAGE RESOLUTION		DATA RATE REQUIRED
	SPATIAL	CONTRAST	
Digital blood pressure monitor (sphygmomanometer)	–	–	<10 kb/s
Digital thermometer	–	–	< 10 kb/s
Digital audio stethoscope and integrated electrocardiogram	–	–	<10 kb/s
Ultrasound, cardiology, radiology	512 × 512	×8	256 KB (image size)
Magnetic resonance image	512 × 512	×8	384 KB (image size)
Scanned x-ray	1024 × 1250	×12	1.8 MB (image size)
Digital radiography	2048 × 2048	×12	6 MB (image size)
Mammogram	4096 × 4096	×12	24 MB (image size)
Compressed and full motion video (e.g., ophthalmoscope, proctoscope, etc.)	–	–	384 kb/s to 1.544 Mb/s (speed)

practice, it is desirable to transmit medical images during a single patient visit, so as to at least avoid a follow-up visit.

Medical image compression techniques have primarily focused on lossless methods, where the image has to be reconstructed exactly from its compressed format due to the diagnostic use. The medical compression issue has been thoroughly discussed in Chapter 4.

About the digital video compression, the DICOM committee has not yet adopted any standard. The adoption of MPEG-2 is possible, but this is limited by the MPEG-2 requirement for constant delay method for frame synchronization. On the other hand, the transmission of off-line video is still possible. It is important to distinguish among the requirements for real-time video transmission, off-line video transmission, medical video and audio for diagnostic applications, and nondiagnostic video and audio. Real-time video transmission for diagnostic applications is clearly the most demanding. Off-line video transmission is essentially limited by the requirement to provide patient–doctor interaction. Real-time diagnostic audio applications include the transmission of stethoscope audio or the transmission of the audio stream that accompanies the diagnostic video. Good quality diagnostic audio at 38–128 kb/s using Dolby AC-2 has been achieved, while MPEG-1 Layer 2 audio (32–256 kb/s) or Dolby AC-3 (96–768 kb/s) may also be used. For nondiagnostic applications such as teleconferencing, H.261 (64 kpbs–1.92 Mb/s) and H.263 (15–64 kb/s) may be acceptable. A typical application will require a diagnostic audio and video bitstream, in addition to a standard teleconferencing bitstream [1, 13, 14].

Reviewing literature, from 150 papers (including both conference and journal papers) published under the mobile telemedicine from year 1979 until now, we selected 50 distributed applications. These systems covered a significant component of the whole spectrum of health telematics applications grouped under the wireless technologies of GSM/GPRS, satellite, radio, and WLAN and into the areas of emergency healthcare, telecardiology, teleradiology, telepsychology, teleopthalmology, and remote monitoring. The majority of these applications utilize the GSM network and the areas of emergency telemedicine, remote monitoring, and telecardiology. Satellite links were also used in many telemedicine applications and most of them were used for remote monitoring. Satellite systems have the advantage of worldwide coverage and offer a variety of data transfer speeds, even though satellite links have the disadvantage of high-operating cost. Finally, WLAN technology is an emerging technology already applied for emergency telematics and teleradiology [1, 13–17].

8.4 *AD HOC* NETWORKS IN DISTRIBUTED MOBILE TELEMEDICINE APPLICATIONS

Mobile *ad hoc* networks represent complex distributed systems that consist of wireless mobile nodes, which can freely and dynamically self-organize into arbitrary and temporary, "*ad hoc*" network topologies, allowing people and devices to seamlessly internetwork in areas with no preexisting communication infrastructure or centralized administration. In such an environment, provision of medical data communication is currently dependent on low-bandwidth point-to-point telemedicine systems. An easily field-deployable *ad hoc* wireless network or even a basic sensor network could provide a significant improvement in real-time patient monitoring and emergency communications capability in general [18, 19].

From the standards that have already presented for transmission in mobile telemedicine networks, the ones that have the possibility to support *ad hoc* communication or could do it through modifications and changes are: a) IEEE 802.11, b) HIPERLAN/2, c) Bluetooth, and d) HomeRF [6, 7, 20]. The Bluetooth and HomeRF standards could not be compared with IEEE 802.11 and HIPERLAN/2 because their possibilities are more for a WPAN than for a WLAN.

In a prehospital environment such as a patient's home it is important to capture his/her clinical data and transfer them to a diagnostic center (DC) for earlier diagnosis, with reasonable speed. A mobile *ad hoc* network with various devices such as PDAs, laptops, digital electrocardiogram machines, and cellular phones may be established in this case.

The advancement of mobile *ad hoc* network technology has the potential to allow a physician to get patient's information anywhere and even before the patient reaches the hospital. Timely access of patient's information may fundamentally improve patient care in both pre- and inhospital setting with doctors' earlier access of patient information and thus intervention. For example in prehospital environment, patient's data such as the electrocardiogram of a patient with chest pains, and other clinical data, are rarely send to hospital before the ambulance arrival [21].

In addition, inhospital environment the doctors and the nurses want sometimes to communicate based on not constant infrastructure. Similarly, in hospital the visiting doctors and the stretcher-bearers could communicate with the internal doctors and receive or transmit information through *ad hoc* networks [18, 22].

Another case that *ad hoc* networks use in mobile telemedicine systems is Wireless Sensor Networks (WSNs). WSNs consist of a large number of tiny devices capable of executing sensing, data processing, and communication tasks and as such can be applied in numerous user scenarios in healthcare monitoring. WSNs consist of a subcategory of WPANs, and they will be presented in the next chapter.

REFERENCES

[1] C. S. Pattichis, E. Kyriacou, S. Voskarides, M. S. Pattichis, R. Istepanian, and C. N. Schizas, "Wireless telemedicine systems: an overview," *IEEE Antennas Propag. Mag.*, vol. 44, no. 2, pp. 143–153, 2002.

[2] K. Shimizu, "Telemedicine by mobile communication," *IEEE Eng. Med. Biol. Mag.*, vol. 18, no. 4, pp. 32–44, Jul.–Aug. 1999.

[3] R. S. H. Istepanian, E. Jovanov, and Y. T. Zhang, "Introduction to the special section on m-health: beyond seamless mobility and global wireless health-care connectivity (Guest editorial)," *IEEE Trans. Inf. Technol. Biomed.*, vol. 8, no. 4, pp. 405–414, 2004.

[4] B. Woodward and M. F. A. Rasid, "Wireless telemedicine: the next step?" in *Proc. 4th Annu. IEEE Conf. Information Technology Applications in Biomedicine*, Birmingham, UK, 2003, pp. 43–46.

[5] K. Banitsas, R. S. H. Istepanian, and S. Tachkara, "Applications of medical wireless LAN (MedLAN) systems," *Int. J. Health Mark.*, vol. 2, no. 2, pp. 136–142, 2002.

[6] B. P. Crow, I. Widjaja, J. G. Kim, and P. T. Sakai, "IEEE 802.11 wireless local area networks," *IEEE Commun. Mag.*, vol. 35, no. 9, pp. 116–126, Sept. 1997.

[7] J. C. Haartsen, Ericsson Radio Systems B.V. "The bluetooth radio system," *IEEE Pers. Commun.*, vol. 7, no. 1, pp. 28–36, Feb. 2000.

[8] A. M. Hura, "Bluetooth—enabled teleradiology: applications and complications," *J. Digit. Imaging*, vol. 15, suppl. 1, pp. 221–223, 2002.

[9] ZigBee Alliance, "IEEE 802.15.4, ZigBee standard," http://www.zigbee.org

[10] DVB—Digital Video Broadcasting, http://www.dvb.org

[11] M. Ackerman, R. Craft, F. Ferrante, M. Kratz, S. Mandil, and H. Sapci, "Telemedicine technology," *Telemed. J. e-Health*, vol. 8, no. 1, pp. 71–78, 2002.

[12] NEMA—DICOM Standards Committee, (2004). "DICOM Strategic Document, Version 4.0, December 29 2004 http://medical.nema.org/dicom/geninfo/Strategy.htm

[13] International Telecommunication Union, Telecommunication Development Bureau, ITU-D Study Group 2, "Question 14-1/2: application of telecommunications in health care, technical information—a mobile medical image transmission system in Japan, PocketMIMAS," Rapporteurs Meeting Q14-1/2 Japan, 2004.

[14] C. LeRouge, M. J. Garfield, and A. R. Hevner, "Quality attributes in telemedicine video conferencing," in *Proc. 35th Hawaii Int. Conf. System Sciences*, 2002, p. 159.

[15] S. Tachakra, X. H. Wang, R. S. H. Istepanian, and Y. H. Song, "Mobile e-Health: the unwired evolution of telemedicine," *Telemed. J. e-Health*, vol. 9, no. 3, pp. 247–257, 2003.

[16] U. Engelmann, A. Schröter, E. Borälv, T. Schweitzer, and H.-P. Meinzer, "Mobile teleradiology: all images everywhere," *Int. Cong. Series 1230*, pp. 844–850, 2001.

[17] E Kyriacou, S Pavlopoulos, A Berler, M Neophytou, A Bourka, A Georgoulas, A Anagnostaki, D Karayiannis, C Schizas, C Pattichis, A Andreou, and D Koutsouris, "Multi-purpose healthcare telemedicine systems with mobile communication link support," *Biomed. Eng. Online*, vol. 2, p. 7, 2003.

[18] C.-K. Toh, *Ad Hoc Mobile Wireless Networks—Protocols and Systems*. New Jersey: Prentice Hall PTR, pp. 243–255, 266–283, 2002.

[19] R. Ramanathan and J. Redi, "A brief overview of ad hoc networks: challenges and directions," *IEEE Commun. Mag.*, vol. 40, 50th Anniversary Commemorative Issue, pp. 20–22, 2002.

[20] J. Khun-Jush, G. Malmgren, P. Schramm, and J. Torsner, "HIPERLAN type 2 for broadband wireless communication," *Ericson Review*, vol. 77, no. 2, pp. 108–119, 2000.

[21] V. Anantharaman and L. Swee Han, "Hospital and emergency ambulance link: using IT to enhance emergency pre-hospital care," *Int. J. Med. Inf.*, vol. 61, no. 2–3, pp. 147–161, 2001.

[22] E. Jovanov, A. O'Donnell Lords, D. Raskovic, P. G. Cox, R. Adhami, and F. Andrasik, "Stress monitoring using a distributed wireless intelligent sensor system," *IEEE Eng. Med. Biol. Mag.*, vol. 22, no. 3, pp. 49–55, May–Jun. 2003.

CHAPTER 9

Ambient Intelligence and Pervasive Computing for Distributed Networked E-Health Applications

9.1 INTRODUCTION

In this chapter we present the concepts of ambient intelligence and pervasive computing in electronic healthcare systems. Both concepts refer to the availability of software applications and medical information anywhere and anytime and the invisibility of computing; computing modules are hidden in multimedia information appliances, which are used in everyday life [1]. Pervasive healthcare systems refer mostly in patient remote monitoring or telemonitoring, which is an important sector of distributed telemedicine. Patient telemonitoring involves the sensing of a patient's physiological and physical parameters and transmitting them to a remote location, typically a medical center, where expert medical knowledge resides [11, 18].

A typical telemonitoring system has the ability to record physiological parameters and provide information to the doctor in real time through a wireless connection, while it requires sensors to measure parameters like arterial blood pressure, heart rate, electrocardiogram (ECG), skin temperature, respiration, and glucose or patient position and activity [19]–[23]. Filtered signals and medical data are either stored locally on a monitoring wearable device for later transmission or directly transmitted, e.g., over the public phone network, to a medical center. Such architecture is depicted in Fig. 9.1.

The vital signals that may be monitored by pervasive system and corresponding dependant variables are presented in Table 9.1. Each biosignal provides different and complementary information on a patient's status, and for each specific person the anticipated range of signal parameters is different. For example, heart rate may vary between 30 and 250 beats/min for normal people in different circumstances; likewise, breathing rate could be between 5 and 50 breaths/min. electroencephalogram (EEG) and ECG are considerably more complex biosignals with spectra up to 10 KHz. The voltage levels of the collected biosignals also vary from microvolts (EEG) to millivolts (ECG). The biosignals' characteristics are summed up in Table 9.2.

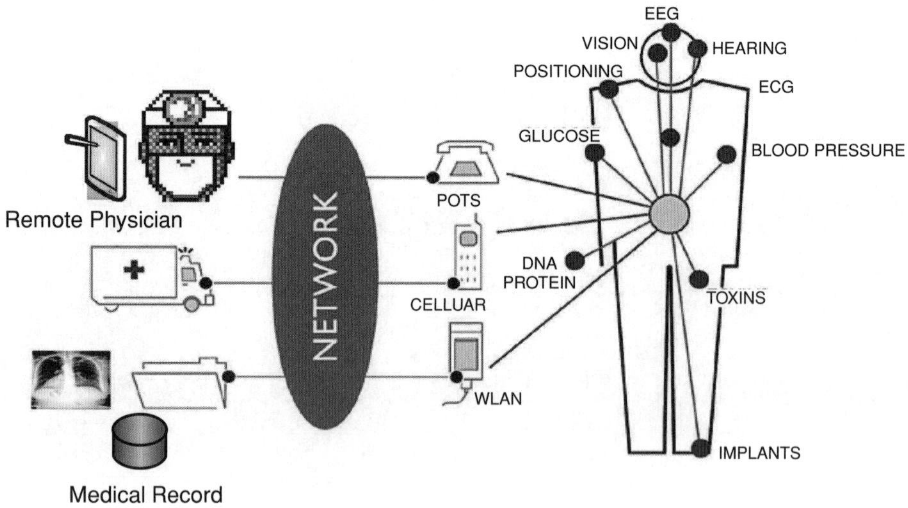

FIGURE 9.1: A typical pervasive healthcare system architecture

9.2 ENABLING TECHNOLOGIES IN PERVASIVE HEALTHCARE

The development and the standardization of the mobile IEEE 802.x family protocols [7], along with the, currently available, 2.5G and 3G networks and the positioning technologies provide sufficient tools for the development of such "intelligent medical environments" realizing location aware pervasive e-health services.

9.2.1 Networking Technologies

Regarding networking, there are two main enabling technologies according to their topology: *on-body* (wearable) and *off-body* networks. Recent technological advances have made possible a new generation of small, powerful, and mobile computing devices. A wearable computer must be small and light enough to fit inside clothing. Occasionally, it is attached to a belt or

TABLE 9.1: Physiological Signals and Dependent Variables

SIGNAL	DEPENDANT VARIABLES
ECG	Heart rate, ventricular beat, ST segment, QT time
Noninvasive blood pressure	Systolic, diastolic, mean, pulse rate
Respiration	Breath rate, expired CO_2
Pulse oximetry	Pulse rate, pulse volume, oxygen saturation

TABLE 9.2: Biosignals Characteristics

BIOMEDICAL MEASUREMENTS TS	VOLTAGE RANGE (V)	NUMBER OF USERS = K (SENSORS)	BANDWIDTH (HZ)	SAMPLE RATE (HZ)	RESOLUTION (B/SAMPLE)	INFORMATION RATE (B/S)
ECG	0.5–4 m	5–9	0.01–250	1250	12	15 000
Heart sound	Extremely small	2–4	5–2000	10 000	12	120 000
Heart rate	0.5–4 m	2	0.4—5	25	24	600
EEG	2–200 μ	20	0.5–70	350	12	4200
EMG	0.1–5 m	2+	0–10 000	50 000	12	600 000
Respiratory rate	Small	1	0.1–10	50	16	800
Temperature of body	0–100 m	1+	0–1	5	16	80

other accessory or is worn directly like a watch or glasses. An important factor in wearable computing systems is how the various independent devices interconnect and share data. An off-body network connects to other systems that the user does not wear or carry, and it is based on a wireless local area network (WLAN) infrastructure, while an on-body or wireless personal area network (WPAN) connects the devices themselves—the computers, peripherals, sensors, and other subsystems—and runs at *ad hoc* mode. Tables 9.3 and 9.4 present the characteristics of off-body and on-body networking technologies respectively.

WPANs are defined within the IEEE 802.15 standard. The most relevant protocols for pervasive e-health systems are Bluetooth and ZigBee (IEEE 802.15.4 standard). Bluetooth technology was originally proposed by Ericsson in 1994, as an alternative to cables that linked mobile phone accessories. It is a wireless technology that enables any electrical device to communicate in the 2.5-GHz ISM (license free) frequency band [6]. It allows devices such as mobile phones, headsets, personal digital assistants (PDAs) and portable computers to communicate and send data to each other without the need for wires or cables to link the devices together. It has been specifically designed as a low-cost, low-size, and low-power radio technology, which is particularly suited to the short range of a personal area network (PAN). The main features of Bluetooth are a) real-time data transfer usually possible between 10–100 m, b) support of point-to-point wireless connections without cables, as well as point-to-multipoint connections to enable *ad hoc* local wireless networks, and c) data speed of 400 kb/s symmetrically or 700–150 kb/s of data asymmetrically. On the other hand, ZigBee (IEEE 802.15.4 standard) has been developed as a low data rate solution with multimonth to multiyear battery life and very low complexity. It is intended to operate in an unlicensed international frequency band. The maximum data rates for each band are 250, 40, and 20 kb/s. The 2.4-GHz band operates worldwide while the sub-1-GHz band operates in North America, Europe, and Australia.

Pervasive healthcare systems set high demanding requirements regarding energy, size, cost, mobility, connectivity, and coverage. Varying size and cost constraints directly result in corresponding varying limits on the energy available, as well as on computing, storage, and communication resources. Low power requirements are necessary also from safety considerations since such systems run near or inside the body. Mobility is another major issue for pervasive e-health applications because of the nature of users and applications and the easiness of the connectivity to other available wireless networks. Both off-body and PANs must not have line-of-sight (LoS) requirements.

9.2.2 Positioning Technologies

Pervasive services often require specific infrastructure for the estimation of the user's location. Several techniques that provide such estimation are currently available. The most prominent are satellite-based (e.g., GPS) or terrestrial infrastructure-based (e.g., Cell-ID, TOA).

TABLE 9.3: Off-Body Networking Technologies [25]

TECHNOLOGY	OPERATIONAL SPECTRUM	PHYSICAL LAYER	CHANNEL ACCESS	MAXIMUM DATA RATE	COVERAGE
Wireless LANs					
IEEE 802.11b	2.4 GHz	DSSS with PBCC	CSMA-CA	22 Mb/s	100 m
HomeRF	2.4 GHz	FHSS with FSK	CSMA-CA and TDMA	10 Mb/s	>50 m
HiperLAN2	5 GHz	OFDM with QAM	Central resource control/ TDMA/TDD	32–54 Mb/s	30–150 m
Cellular telephony					
CDMA20 00 1 × Ev-DO	Any existing frequency; band with 2/1 × 3.75 MHz channels	DSSS with QPSK/BPSK; 1.2288 MHz or 3.6864 MHz	FDD or TDD	2 Mb/s	Area of a cell
UMTS (WCDMA)	1920–1980 MHz and 2110–2170 MHz; 2 × 5 MHz channels	DSSS with QPSK; 3.84 MHz (chip rate)	FDD	2.048 Mb/s	Area of a cell
ReFLEX (two-way paging)	150–900 MHz; 25- or 50-KHz channel	FSK	TDMA-based	9600 b/s in, 6400 b/s out	Typically a large city

(*Continued*)

TABLE 9.3: Off-Body Networking Technologies [25] (*Continued*)

TECHNOLOGY	OPERATIONAL SPECTRUM	PHYSICAL LAYER	CHANNEL ACCESS	MAXIMUM DATA RATE	COVERAGE
Wireless MANs					
Ricochet	900 MHz	FHSS	CSMA	176 Kb/s	As of Sept. 2003, only Denver and San Diego, USA
IEEE 802.16	10–66 GHz; line of sight required; 20/25/28 MHz channels	TDMA-based; uplink QPSK; 16- and 64-QAM	FDD and TDD variants	120/134.4 Mb/s for 25/28 MHz channel	Typically a large city

TABLE 9.4: On-Body Networking Technologies [25]

TECHNOLOGY	OPERATIONAL SPECTRUM	PHYSICAL LAYER	CHANNEL ACCESS	MAXIMUM DATA RATE	COVERAGE
IrDA	Infrared; 850 nm	Optical rays	Polling	4 Mb/s	<10 m
MIT/IBM personal area networks	0.1–1 MHz	DSSS with ASK/FSK	CSMA-CA/CD, TDMA, or CDMA	417 Kb/s (theoretical); 2400 b/s (IBM); 9600 b/s (MIT)	Typically the entire human body
BodyLAN	Not available	OOC (time-based spread spectrum)	Complex TDMA scheme	32 Kb/s	<10 m
Fabric area networks	125 KHz; RFID	RF fields	Polling	1–10 Kb/s (duplex); 100 Kb/s (simplex)	Multiple short-range antennas (<2 cm each)
IEEE 802.15.1 (Bluetooth)	2.4 GHz ISM band	FHSS;	Master-slave polling, TDD	Up to 1 Mb/s	<10 m
IEEE 802.15.3	2.402–2.480 GHz ISM band	Uncoded QPSK, trellis-coded QPSK, or 16/32/64-QAM scheme	CSMA-CA and GTS in superframe structure	11–55 Mb/s	<10 m
IEEE 802.15.4	2.4 GHz and 868/915 MHz	DSSS with BPSK or MSK (O-QPSK)	CSMA-CA and GTS in superframe structure	868 MHz– 20 Kb/s, 915 MHz–40 Kb/s, and 2.4 GHz– 250 Kb/s	<20 m

Satellite-based positioning does not operate properly in deep canyons and indoors where cellular coverage may be denser. Terrestrial-based positioning may be more imprecise with sparse deployment of base stations in rural environments, where satellite visibility is better. Position-fixing in indoor environments may also exploit other technologies such as WLAN. WLAN-based positioning requires access points (AP) to be installed on the building (e.g., hospital) structure and WLAN-equipped portable devices. In general, two types of WLAN positioning architectures exist. In the *centralized scheme* a server calculates the position of a client based, e.g., on triangulation/trilateration or scene analysis. Example of the centralized architecture is the Ekahau's Positioning Engine [3], which provides absolute information (i.e., depicted on a floor plan, building grid, or on a geographic coordinates system) and works for 802.11 and Bluetooth. The *distributed scheme* relies on specialized software that runs on clients (e.g., PDAs). Each client uses its own radio frequency (RF) measurements in order to calculate its position. Examples of such architecture are the Microsoft RADAR [1] and the Nibble [2] systems. RADAR uses both scene analysis and triangulation through the received signal's attenuation. It has been developed for 802.11 networks. It provides a positioning accuracy of 3–4 m, but with a precision of 50%. It uses the scene analysis technique to estimate location and provides symbolic (the position of a located object is determined in reference to the observer or another known object) and absolute positioning information. The accuracy of the system can reach 3 m. Other proprietary radio- or infrared-based positioning schemes also exist for indoor environments. Indicative examples of such systems are the Active Badge and Active Bat systems. An extensive survey of similar platforms can be found in [4].

9.3 PERVASIVE HEALTHCARE APPLICATIONS IN CONTROLLED ENVIRONMENTS

The use of pervasive systems in controlled environments, such as hospitals, may be divided into two broad categories. The first one relates to applications, enabling the mobile ubiquitous delivery of medical data and implementations of mobile electronic health records (EHR), accessible by PDA's or Tablet PC's in a hospital equipped with WLAN infrastructure [13]. Several research groups [11, 14, 15, 17] have experimented on the use of handheld computers of low cost and high portability, integrated through a wireless local computer network within the IEEE 802.11 or Bluetooth standards. Regarding the medical data exchange, DICOM (www.dicom.org) and HL7 (www.hl7.org) standards are used in the data coding and transmission via mobile client/server applications capable of managing health information.

On the other hand, pervasive systems are used for the monitoring and diagnosis of patients. A wide range of medical monitors and sensors may enable the mobile monitoring of a patient, who is able to walk freely without being tied to the bed. Pervasive systems in hospital environment are mostly based on Bluetooth communication technology. For example,

Khoor *et al.* [10] have used the Bluetooth system for short-distance (10–20 m) data transmission of digitized ECGs together with relevant clinical data. Hall *et al.* have demonstrated a Bluetooth-based platform for delivering critical health record information in emergency situations [11], while J. Andreasson *et al.* have developed a remote system for patient monitoring using Bluetooth enabled sensors [12]. The above examples show that the merging of mobile communications and the introduction of handhelds along with their associated technology has potential to make a big impact in emergency medicine. Moreover, many market projections indicate that mobile computer is both an emerging and enabling technology in healthcare [13].

9.4 PERVASIVE HEALTHCARE IN DISTRIBUTED NONHOSPITAL SETTINGS—HOME CARE APPLICATIONS

Facilities for medical practice in nonhospital settings are limited by the availability of medical devices suitable for producing biosignals and other medical data. There are a number of active research and commercial projects developing sensors and devices, which do not require local intervention to enable contact with a clinician remote from the care environment. These new systems provide automated connection with remote access and seamless transmission of biological and other data upon request. Pervasive systems in nonhospital systems aim at better managing of chronic care patients, controlling of health delivery costs, increasing quality of life and quality of health services, and the provision of distinct possibility of predicting and thus avoiding serious complications [8].

The patient or elder person mainly requires monitoring of his/her vital signals (i.e., ECG, blood pressure, heart rate, breath rate, oxygen saturation, and perspiration). Patients recently discharged from hospital after some form of intervention, for instance, after a cardiac incident, cardiac surgery, or a diabetic comma are less secure and require enhanced care. The most common forms of special home monitoring are ECG arrhythmia monitoring, post surgical monitoring, respiratory and blood oxygen levels monitoring, and sleep apnoea monitoring. In the case of diabetics, the monitoring of blood sugar levels resigns the patient to repeated blood sampling, which is undesirable and invasive. One possible solution is the development of implantable wireless sensor devices that would be able to give this information quickly, and in a continuous fashion. Current conditions where home monitoring might be provided include hypertension, diabetes (monitoring glucose), obesity and CHF (monitoring weight), asthma and COPD (monitoring spirometry/peak flow), and, in the near future, conditions utilizing oximetry monitoring. Other home-monitoring conditions might include preeclampsia, anorexia, low-birth-weight infants, growth abnormalities, and arrhythmias. Most chronic health conditions in children and adults could be managed and/or enhanced by home monitoring.

In most applications two monitoring modes are foreseen: the Batch and the Emergency Mode. Batch mode refers to the every day monitoring process, where vital signs are acquired and transmitted periodically to a health-monitoring center. The received data are monitored by the doctor on-duty and then stored into the patient's electronic health record maintained by a healthcare center. The emergency mode occurs because the patient does not feel well and, thus, decides to initiate an out-of-schedule session, or because the monitoring device detects a problem and automatically initiates the transfer of data to the corresponding center. Application of emergency episode detection and the corresponding alarm processing are important for the protection of the patient. An alarm represents a change in the status of a physiological condition or a sensor reading state outside of agreed limits.

9.5 CONCLUSIONS AND FUTURE CHALLENGES

The technological advances of the last few years in mobile communications, location-aware and context-aware computing have enabled the introduction of pervasive healthcare applications. These applications may assist medical personnel and people with health-related problems in two distinct ways. One concerns the remote monitoring of patients with chronic diseases, involving diagnosis and monitoring using biosignal sensors and the patients' history record, and the immediate notification of a doctor or a medical center in the case of an emergency. The other is toward dealing with the everyday practice in hospital settings, involving the monitoring of the inpatient's status and the mobile access to medical data.

However, the use of such pervasive healthcare raises several challenges. Personal data security and location privacy are considered to be the most important. Furthermore, pervasive healthcare systems are very critical systems, as they deal with a person's health, and therefore they raise high standards regarding reliability, scalability, privacy-enhancing, interoperability, and configurability, among other things. On the other hand, since healthcare systems are intended to be used by low or medium computer literate users, usability issues come to the foreground.

Regardless of the remaining challenges, pervasive healthcare systems are anticipated to be expanded in the near future by using the most recent technological advances in a more active and direct way for offering more comprehensive and higher quality health services to citizens.

REFERENCES

[1] P. Bahl and V. N. Padmanabhan, "RADAR: An in-building RF-based user location and tracking system," in *Proc. IEEE INFOCOM*, Tel Aviv, Israel in March 2000, pp. 775–784.

[2] P. Castro, P. Chiu, T. Kremenek, and R. Muntz, "A probabilistic room location service for wireless networked environments," in *Proc. Ubicomp 2001, LNCS 2201*, G. D. Abowd, B. Brumitt, and S. A. N. Shafer, Eds. Springer-Verlag, 2001.

[3] Electronic information available at http://www.ekahau.com

[4] J. Hightower and G. Borriello, "Location systems for ubiquitous computing," *IEEE Computer*, vol. 34, no. 8, pp. 57–66, Aug. 2001 (also in IEEE special report "Trends Technologies and Applications in Mobile Computing").

[5] G. Abowd, "Software engineering issues for ubiquitous computing," in *Proc. Int. Conf. Software Engineering*, 1999, pp. 5–84.

[6] Bluetooth Consortium, http://www.bluetooth.com

[7] http://standards.ieee.org/

[8] N. Maglaveras *et al.*, "Home care delivery through the mobile telecommunications platform: The citizen health system (CHS) perspective," *Int. J. Med. Inf.*, vol. 68, pp. 99–111, 2002.

[9] J. Birnbaum, "Pervasive information systems," *Commun. ACM*, vol. 40, no. 2, pp. 40–41, 1997.

[10] S. Khoor, K. Nieberl, K. Fugedi, E. Kail, "Telemedicine ECG-telemetry with Bluetooth technology," *Comput. Cardiol.*, pp. 585–588, 2001.

[11] E. S. Hall *et al.*, "Enabling remote access to personal electronic medical records," *IEEE Eng. Med. Biol. Mag.*, vol. 22, no. 3, pp. 133–139, May/Jun. 2003.

[12] J. Andreasson *et al.*, "Remote system for patient monitoring using Bluetooth/spl trade," Sensors, 2002, *Proc. IEEE*, vol. 1, pp. 304–307, Jun. 2002.

[13] C. Finch, "Mobile computing in healthcare," *Health Manage. Technol.*, vol. 20, no. 3, pp. 63–64, Apr. 1999.

[14] CHILI (tele-) radiology workstation (Steinbeis Transferzentrum Medizinische Informatik, Heidelberg, Germany). A joint project with the German Cancer Research Center in Heidelberg, Germany.

[15] U. Engelmann, A. Schroter, E. Boralv, T. Schweitzer, and H.-P. Meinzer, "Mobile teleradiology: All images everywhere," *Int. Congr. Series 1230*, pp. 844–850, 2001.

[16] R. Andrade, A. von Wangenheim, and M. K. Bortoluzzi, "Wireless and PDA: A novel strategy to access DICOM-compliant medical data on mobile devices," *Int. J. Med. Inf.*, vol. 71, pp. 157–163, 2003.

[17] I. Maglogiannis, N. Apostolopoulos, and P. Tsoukias, "Designing and implementing an electronic health record for personal digital assistants (PDA's)," *Int. J. Qual. Life Res.*, vol. 2, no. 1, pp. 63–67, 2004.

[18] V. Stanford, "Pervasive health care applications face tough security challenges," *Pervasive Comput.*, Apr./Jun. 2002.

[19] F. A. Mora, G. Passariello, G. Carrault and J.-P. Le Pichon "Intelligent patient monitoring and management systems: A review," *IEEE Eng. Med. Biol. Mag.*, vol. 12, no. 4, pp. 23–33, Dec. 1993.

[20] S. Barro, J. Presedo, D. Castro, M. Fernandez-Delgado, S. Fraga, M. Lama, and J. Vila, "Intelligent telemonitoring of critical-care patients," *IEEE Eng. Med. Biol. Mag.*, vol. 18, no. 4, pp. 80–88, Jul./Aug. 1999.

[21] A. Kara, "Protecting privacy in remote-patient monitoring," *Comput. Practices*, May 2001.

[22] J. Dan and J. Luprano, "Homecare: A telemedical application," *Med. Device Technol.*, Dec. 2003.

[23] J. Luprano, "Telemonitoring of physiological parameters: The engineer point-of-view (sensors and data transmission)," *2003 SATW Congr.*, Bern, Sep. 2003.

[24] K. Hung and Y. Ting, "Implementation of a WAP-based telemedicine system for patient monitoring," *IEEE Trans. Inf. Technol. Biomed.*, vol. 7, no. 2, Jun. 2003.

[25] R. Ashok, D. Agrawal, "Next-generation wearable networks," *IEEE Computer*, pp. 31–39, Nov. 2003.

CHAPTER 10

Telemedicine and Virtual Reality

10.1 INTRODUCTION

When one thinks of research fields related to life sciences, virtual reality (VR) may not be the first one that comes to mind, since it is usually related with the entertainment industry. However, the advent of low-cost specialized motion capture devices and powerful graphics rendering hardware have provided computers with advanced visualization capabilities, able to illustrate medical information with a multitude of techniques. Formally, VR can be thought of as a collection of capturing devices and interaction of hardware and software that caters for immersing and navigating a user within a synthetic virtual environment, possibly with novel, naturalistic means of interfacing, while also providing interaction capabilities with objects contained therein; in the case of augmented reality (AR), this notion can be extended with images, video footage, or models of real objects and environments [7, 9].

To fully immerse one into such a synthetic environment, usually takes more than a computer screen and a mouse; instead, users typically wear helmets and gloves, used to provide location-specific renderings of the virtual world in the form of a heads-up display (HUD), as well as provide tactile feedback in case the user interacts with a particular object. These devices also serve as input devices to the computer that renders the virtual environment, since they collect information concerning the users' gaze direction and capture their hand movements and gestures; as a result users, who are represented via an "avatar" within the virtual environment, can communicate their position, direction, and movement to the host computer and express interest or manipulate a model of an object using the same interaction means as they would in the physical world. The result of this naturalistic process is the complete immersion of the user into the virtual environment, where complex or ancient buildings or remote locations are easy to reproduce and offer as a setting. For example, models of ancient cities such as Pompeii or Olympia, museums like the Louvre, or distant planets such as Mars are typical "destinations" for the user of a virtual world. In science-related applications medical practitioners can plan and test difficult operations, robots can be tele-operated in space, and science teachers can demonstrate different forms of gravity in schools. As a result, VR is quickly catching up in related fields, especially where visualization, navigation, and interaction with multidimensional data are required.

10.2 OVERVIEW OF VIRTUAL REALITY TECHNOLOGY

In general, the minimum set of features for VR applications are visualization and navigation of data collected and rendered from a host computer. This information can be completely synthetic, as in the case of architecture, or remotely captured and reconstructed, as is the case with medical applications such as magnetic resonance imaging (MRI) or positron emission tomography (PET) data [1].

The computer-generated environment within which the use agent exists consists of a three-dimensional (3D) reconstruction, or rendering, of a containing environment, filled with objects and possibly other user agents; to create such an environment, one typically needs dedicated modeling software, even if models of real objects and environments are used. In order to achieve greater realism and thus enhance the immersion experience, texture maps can be used to reproduce the visual surface detail of a real object in the models, while real-time simulations of dynamics allow objects to be moved within the virtual environment and interact with each other following natural laws such as gravity and friction.

As mentioned before, the conventional method of working with computers, that is providing information via a mouse or keyboard and viewing results on a monitor, can hamper immersion into a virtual world. Therefore, VR developers tend to replace this interface with one that is more intuitive and allows interaction with everyday, human–human communication concepts; this naturalistic interface typically consists of related hardware containing capture sensors (cameras, IR, magnetic, etc.) for controlling user interaction within the virtual world and feedback effectors (such as headphones, goggles, or force-feedback gloves) for the user [2].

10.3 MEDICAL APPLICATIONS OF VIRTUAL REALITY TECHNOLOGY

In the field of medical applications, VR is usually put to use in surgical and nonsurgical training and planning, medical imaging and visualization, and rehabilitation and behavioural evaluation [9]. In the field of telemedicine, telesurgery and data visualization are the prevalent applications of VR, since its real power is shown in situations where multidimensional, heterogeneous information may be visualized collectively and transparently to the end user. In this framework, it is the medical practitioners who are immersed into the virtual environment and either navigate into 3D data sets such as MRI scans or transfer their knowledge and expertise into remote areas. For example, situations where a patient cannot be moved to a hospital for examinations or surgery or cases where specialists are located very far from the point of interest, e.g., an accident scene can benefit from telesurgery or remote data visualization [7, 8, 10].

Putting VR technologies to work in the field of telesurgery brings in a number of major advantages related to resources, safety, and interoperability. For example, there is a limited need to use corpses or experimental subjects that are typically costly to maintain. In addition to this, new surgical techniques can be tested without putting both patients and practitioners to danger; in this case, surgery progress can be monitored from a variety of viewpoints, even from within internal organs. Finally, all transactions can be recorded and then recreated, so that annotation and discussion in a later stage are possible.

In order to design and deploy usable VR applications in the field of telemedicine, one must take into account the following requirements:

1. *Display fidelity*: High-resolution graphics are essential in order to illustrate reconstructed organs in a truthful manner.
2. *Reconstruction of organs*: Modeled organs should have the same density and texture as the real ones.
3. *Organ reactions and dynamics*: Modeled organs should be in the same positions and have the same functionality as the real ones in simulated circumstances, for example, bleeding.
4. *Organ/object interaction*: In cases of simulated probing or surgery, a dynamics simulation should exist so that the modeled organs respond to pressure or cuts by the virtual instruments.
5. *Force or haptic feedback*: Touching a virtual organ with a virtual instrument should be communicated to the user via a glove or other feedback device [4, 5].

Another important field where VR applications can prove useful is exposure treatment and rehabilitation. In these cases, novel and untested or dangerous techniques can be designed and investigated before large-scale deployment. In exposure treatment, the patient is gradually immersed in situations or subjected to stimuli that may be the cause of phobias or extreme reactions (for example, acrophobia, the fear of height, or disorders in war veterans [11]), usually via mounting a HUD. Monitoring devices that capture a variety of possible factors, e.g., respiration, heart beat, gaze direction, etc. are connected to a host computer to help locate pathological patterns or identify the source of the problem. After this happens, patients may be gradually transferred to *in vivo* situations and exposed to real stimuli [3].

In a rehabilitation scenario, patients may be immersed in entertaining or collaborative VR environments, thus improving their performance in related situation, while still catering for easy monitoring and recording of their performance. In one of the best known practices [6], patients with Parkinson's in San Anselmo, California, wear head-mounted devices and are immersed in an environment where virtual objects are places near their feet; trying to avoid these virtual objects help patients improve their mobility skills, a factor which is severely debilitated by the particular disease. VR treatment, besides being safer and less costly, can assist in reducing related medication; for example, if mobility disorders are reversed via the aforementioned technique, drugs with questionable side effects, like Ldopa, cease to be required for treatment.

10.4 CONCLUSIONS

VR tools and techniques are being developed rapidly in the scientific, engineering, and medical areas. This technology will directly affect medical practice. Computer simulation will allow physicians to practice surgical procedures in a virtual environment in which there is no risk to patients, and where mistakes can be recognized and rectified immediately by the computer. Procedures can be reviewed from new, insightful perspectives that are not possible in the real world.

The innovators in medical VR will be called upon to refine technical efficiency and increase physical and psychological comfort and capability, while keeping an eye on reducing costs for healthcare. The mandate is complex, but like VR technology itself, the possibilities are very exciting. While the possibilities—and the need—for medical VR are immense, approaches and solutions using new VR-based applications require diligent, cooperative efforts among technology developers, medical practitioners, and medical consumers to establish where future requirements and demand will lie.

REFERENCES

[1] P. Aslan, B. Lee, R. Kuo, R. K. Babayan, L. R. Kavoussi, K. A. Pavlin, and G. M. Preminger, "Secured medical imaging over the Internet," in *Medicine Meets Virtual Reality: Art, Science, Technology: Healthcare Evolution*. San Diego, CA: IOS Press, 1998, pp. 74–78.

[2] E. A. Attree, B. M. Brooks, F. D. Rose, T. K. Andrews, A. G. Leadbetter, and B. R. Clifford, "Memory processes and virtual environments: I can't remember what was there, but I can remember how I got there. Implications for persons with disabilities," in *Proc. European Conf. Disability, Virtual Reality, and Associated Technology*, 1996, pp. 117–121.

[3] R. W. Bloom, "Psychiatric therapeutic applications of virtual reality technology (VRT): research prospectus and phenomenological critique," in *Medicine Meets Virtual Reality: Global Healthcare Grid*. San Diego, CA: IOS Press, 1997, pp. 11–16.

[4] G. Burdea, R. Goratowski, and N. Langrana, "Tactile sensing glove for computerized hand diagnosis," *J. Med. Virtual Real.*, vol. 1, pp. 40–44, 1995.

[5] M. Burrow, "A telemedicine testbed for developing and evaluating telerobotic tools for rural health care," in *Medicine Meets Virtual Reality II: Interactive Technology and Healthcare: Visionary Applications for Simulation Visualization Robotics*. San Diego, CA: Aligned Management Associates, 1994, pp. 15–18.

[6] A. Camurri, E. Cervetto, B. Mazzarino, P. Morasso, G. Ornato, F. Priano, C. Re, L. Tabbone, A. Tanzini, R. Trocca, and G. Volpe, "Application of multimedia techniques in the physical rehabilitation of Parkinson's patients," in *Proc. 1st Int. Workshop Virtual Reality Rehabilitation*, Lausanne, Switzerland, 2002, pp. 65–75.

[7] A. Rovetta, M. Canina, F. Lorini, R. Pegoraro *et al.*, "Demonstration of surgical telerobotics and virtual telepresence by Internet + ISDN from Monterey (USA) to Milan (Italy)," in *Proc. Medicine Meets Virtual Reality: Art, Science, Technology: Healthcare Evolution.* J. D. Westwood, H. M. Hoffman, D. Stredney, S. Weghorst, Eds. San Diego, CA: IOS Press, 1998, pp. 79–83.

[8] C. V. Edmond, D. Heskamp, D. Sluis, D. Stredney, D. Sessanna, G. Weit, R. Yagel, S. Weghorst, P. Openheimer, J. Miller, M. Levin, and L. Rosenberg, "ENT endoscopic surgical training simulator," in *Medicine Meets Virtual Reality: Global Healthcare Grid.* San Diego, CA: IOS Press, 1997, pp. 518–528.

[9] W. Greenleaf and T. Piantanida, "Medical applications of virtual reality technology," in *The Biomedical Engineering Handbook*, J. D. Bronzino, Ed., 2nd ed. Boca Raton, FL: CRC Press, 2000, pp. 69.1–69.23.

[10] J. W. Hill, J. F Jensen, P. S. Green, and A. S. Shah, "Two-handed tele-presence surgery demonstration system," in *Proc. ANS 6th Ann. Topical Meeting Robotics and Remote Systems*, vol. 2, Monterey, CA, 1995, pp. 713–720.

[11] S. Mrdeza and I. Pandzic, "Analysis of virtual reality contribution to treatment of patients with post-traumatic stress disorder," in *Proc. 7th Int. Conf. Telecommunications CONTel 2003*, Zagreb, Croatia.

[12] G. Riva, "The emergence of e-health: Using virtual reality and the Internet for providing advanced healthcare services," *Int. J. Healthc. Technol. Manage.*, vol. 4, no. 1–2, pp. 15–40, 2002.

The Authors

ILIAS G. MAGLOGIANNIS

Ilias G. Maglogiannis was born in Athens, Greece in 1973. He received a Diploma in Electrical and Computer Engineering and a Ph.D. degree in Biomedical Engineering and Medical Informatics from the National Technical University of Athens (NTUA), Greece in 1996 and 2000, respectively, with scholarship from the Greek Government. From 1996 until 2000 he worked as a researcher in the Biomedical Engineering Laboratory in NTUA and he has been engaged in several European and National Projects. From 1998 until 2000 he was also the head of the computer laboratory, in the Department of Civil Engineering in the NTUA. From February 2001 until March 2004 he served as a Visiting Lecturer in the Department of Information and Communication Systems Engineering in University of the Aegean. In March 2004 he was elected Lecturer in the same Department.

His published scientific work includes 5 lecture notes (in Greek) on Biomedical Engineering and Multimedia topics, 21 journal papers, and more than 40 international conference papers. He has served on program and organizing committees of national and international conferences and he is a reviewer for several scientific journals. His scientific activities include biomedical engineering, telemedicine and medical informatics, image processing, and multimedia. Dr. Maglogiannis is a member of IEEE—Societies: Engineering in Medicine and Biology, Computer, Communications, SPIE—International Society for Optical Engineering, ACM, the Technical Chamber of Greece, the Greek Computer Society, and the Hellenic Organization of Biomedical Engineering. Dr. Maglogiannis is also a national representative for Greece in the IFIP Working Group 12.5 (Artificial Intelligence—Knowledge-Oriented Development of Applications).

KOSTAS KARPOUZIS

Kostas Karpouzis is an associate researcher at the Institute of Communication and Computer Systems (ICCS) and holds an adjunct lecturer position at the University of Piraeus, teaching data warehousing and data mining. He graduated from the School of Electrical and Computer Engineering of the National Technical University of Athens in 1998 and received his Ph.D. degree in 2001 from the same University. His current research interests lie in the areas of human computer interaction, image and video processing, image interchange infrastructures using the MPEG-4 and MPEG-21 standards, sign language synthesis, and virtual reality. Dr. Karpouzis has published more than 70 papers in international journals and proceedings of international conferences. He is a member of the technical committee of the International Conference on Image Processing (ICIP), the IFIP Conference on Artificial Intelligence Applications and Innovations (AIAI) and a reviewer in many international journals. Since 1995 he has participated in more than 12 R&D projects at Greek and European level. He is also a national representative in IFIP Working Groups 12.5 "Artificial Intelligence Applications" and 3.2 "Informatics and ICT in Higher Education."

MANOLIS WALLACE

Manolis Wallace was born in Athens in 1977. He obtained his degree in Electrical and Computer Engineering from the National Technical University of Athens (NTUA) and his Ph.D. from the Computer Science Division of NTUA. He is with the University of Indianapolis, Athens Campus (UIA) since 2001 where he serves now as an Assistant Professor. Since 2004 he is also the Chair of the Department of Computer Science of UIA.

His main research interests include handling of uncertainty, information systems, data mining, personalization, and applications of technology in education. He has published more than 40 papers in the above-mentioned fields, 10 of which are in international journals. He is the guest editor of 2 journal special issues and a reviewer for more than 10 journals and various books. His academic volunteering work also includes participation in various conferences as organizing committee chair, session organizer, session chair, or program committee member.